Food Biopreservation: Analytical Techniques

Food Biopreservation: Analytical Techniques

Contributors

Mahdi Ghanbari and Mansooreh Jami et al.

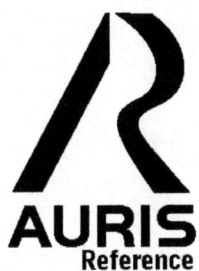

AURIS
Reference

www.aurisreference.com

Food Biopreservation: Analytical Techniques

Contributors: Mahdi Ghanbari and Mansooreh Jami et al.

Published by Auris Reference Limited
www.aurisreference.com

United Kingdom

Food Biopreservation: Analytical Techniques

ISBN: 978-1-78154-880-6

British Library Cataloguing in Publication Data
A CIP record for this book is available from the British Library

Printed in the United Kingdom

Exclusively distributed by CBS Publishers & Distributors Pvt. Ltd.

Sales & Distribution Rights only for India, Pakistan, Bangladesh, Sri Lanka, Nepal and Bhutan.This book is not to be sold outside these territories.

Contents

List of Abbreviations

AOF	Animal origin foods
ALB	Antilisterial Bacteriocin
AEFC	antimicrobial edible films and coatings
BLS	Bacteriocin-like substances
BHI	Brain heart infusion
CFS	Cell-free supernatant
CDC	Centers for Disease Control
CFU	colony-forming units
DGGE	denaturing gel electrophoresis
DAP	Diaminopimelic acid
EFC	Edible films and coatings
EPS	Extracellular polymeric substance
FDA	Food and Drug Administration's
GRAS	generally recognized as safe
GRAS	generally regarded as safe
HHP	high hydrostatic pressure
HPH	high pressure homogenization
HTST	high temperature short time
HIV	human immunodeficiency virus
LAB	Lactic acid bacteria
LAE	Lauric arginate
LDPE	low density polyethylene film
LTLT	Low temperature long time
MRS	Man, Rogosa and Sharp
MRSA	Methicillin-resistant Staphylococcus aureus
MAP	modified atmosphere packing
NADH	nicotinamide adenine dinucleotide
ODM	optical density method
PKE	palm kernel expeller
PCA	Principal Components Analysis
WDA	well diffusion assay
YPD	yeast extract-peptone-dextrose

List of Contributors

Mahdi Ghanbari
University of Zabol, Faculty of Natural Resources, Department of Fishery; Zabol, Iran
BOKU — University of Natural Resources and Life Sciences, Department of Food Sciences and Technology, Institute of Food Sciences, Vienna, Austria

Mansooreh Jami
University of Zabol, Faculty of Natural Resources, Department of Fishery; Zabol, Iran
BOKU — University of Natural Resources and Life Sciences, Department of Food Sciences and Technology, Institute of Food Sciences, Vienna, Austria

Lorena Rodrı´guez-Rubio
DairySafe Group, Department of Technology and Biotechnology of Dairy Products, Instituto de Productos La´cteos de Asturias (IPLA), Consejo Superior de Investigaciones Cientı´ficas (CSIC), Villaviciosa, Asturias, Spain

Beatriz Martı´nez
DairySafe Group, Department of Technology and Biotechnology of Dairy Products, Instituto de Productos La´cteos de Asturias (IPLA), Consejo Superior de Investigaciones Cientı´ficas (CSIC), Villaviciosa, Asturias, Spain

David M. Donovan
Animal Biosciences and Biotechnology Laboratory, Animal and Natural Resources Institute, Beltsville Agricultural Research Center (BARC), Agricultural Research Service (ARS), USDA, Beltsville, Maryland, United States of America

Pilar Garcı´a
DairySafe Group, Department of Technology and Biotechnology of Dairy Products, Instituto de Productos La´cteos de Asturias (IPLA), Consejo Superior de Investigaciones Cientı´ficas (CSIC), Villaviciosa, Asturias, Spain

Ana Rodrı´guez
DairySafe Group, Department of Technology and Biotechnology of Dairy Products, Instituto de Productos La´cteos de Asturias (IPLA), Consejo Superior de Investigaciones Cientı´ficas (CSIC), Villaviciosa, Asturias, Spain

Lipsy Chopra
Biochemical Engineering Research and Process Development Centre, CSIR-Institute of Microbial Technology, Sector-39A Chandigarh; 160036, India

Gurdeep Singh
Biochemical Engineering Research and Process Development Centre, CSIR-Institute of Microbial Technology, Sector-39A Chandigarh; 160036, India

Kautilya Kumar Jena
Biochemical Engineering Research and Process Development Centre, CSIR-Institute of Microbial Technology, Sector-39A Chandigarh; 160036, India

Debendra K. Sahoo
Biochemical Engineering Research and Process Development Centre, CSIR-Institute of Microbial Technology, Sector-39A Chandigarh; 160036, India

Djadouni Fatima
Laboratory of Applied Microbiology, Department of Biology, Faculty of Sciences, Es-Senia University, Oran, Algeria

Kihal Mebrouk
Laboratory of Applied Microbiology, Department of Biology, Faculty of Sciences, Es-Senia University, Oran, Algeria

Heddadji Miloud
Laboratory of Applied Microbiology, Department of Biology, Faculty of Sciences, Es-Senia University, Oran, Algeria

M. Ciani
Dipartimento di Biotecnologie Agrarie e Ambientali, Universita` di Ancona, Ancona 60131, Italy

f. Fatichenti
Dipartimento di Biotecnologie Agrarie e Ambientali, Universita` di Ancona, Ancona 60131, Italy

Antonio Gálvez
Área de Microbiología, Departamento de Ciencias de la Salud, Facultad de Ciencias Experimentales, Universidad de Jaén, Campus Las Lagunillas s/n. 23071-Jaén Spain

Hikmate Abriouel
Área de Microbiología, Departamento de Ciencias de la Salud, Facultad de Ciencias Experimentales, Universidad de Jaén, Campus Las Lagunillas s/n. 23071-Jaén Spain

Antonio Cobo
Área de Microbiología, Departamento de Ciencias de la Salud, Facultad de Ciencias Experimentales, Universidad de Jaén, Campus Las Lagunillas s/n. 23071-Jaén Spain

Rubén Pérez Pulido
Área de Microbiología, Departamento de Ciencias de la Salud, Facultad de Ciencias Experimentales, Universidad de Jaén, Campus Las Lagunillas s/n. 23071-Jaén Spain

Kamal Rai Aneja
Vaidyanath Research, Training and Diagnostic Centre, Kurukshetra 136118, India

Romika Dhiman
Department of Microbiology, Kurukshetra University, Kurukshetra 136119, India

Neeraj Kumar Aggarwal
Department of Microbiology, Kurukshetra University, Kurukshetra 136119, India

Ashish Aneja
University Health Centre, Kurukshetra University, Kurukshetra 136119, India

Swarnadyuti Nath
Faculty of Fishery Sciences, West Bengal University of Animal and Fishery Sciences, Kolkata, India

S. Chowdhury
Faculty of Fishery Sciences, West Bengal University of Animal and Fishery Sciences, Kolkata, India

Prof. K.C. Dora
Faculty of Fishery Sciences, West Bengal University of Animal and Fishery Sciences, Kolkata, India

S. Sarkar
Faculty of Fishery Sciences, West Bengal University of Animal and Fishery Sciences, Kolkata, India

Backialakshmi S
Department of Biotechnology, Dr. G R Damodaran College of Science, Coimbatore-14, Tamil Nadu, India

Meenakshi RN
Department of Biotechnology, Dr. G R Damodaran College of Science, Coimbatore-14, Tamil Nadu, India

Saranya A
Department of Biotechnology, Dr. G R Damodaran College of Science, Coimbatore-14, Tamil Nadu, India

Jebil MS
Department of Biotechnology, Dr. G R Damodaran College of Science, Coimbatore-14, Tamil Nadu, India

Krishna AR
Department of Biotechnology, Dr. G R Damodaran College of Science, Coimbatore-14, Tamil Nadu, India

Krishna JS
Department of Biotechnology, Dr. G R Damodaran College of Science, Coimbatore-14, Tamil Nadu, India

Suganthi Ramasamy
Department of Biotechnology, Dr. G R Damodaran College of Science, Coimbatore-14, Tamil Nadu, India

Tana Hintz
Department of Food Science, Rutgers, The State University of New Jersey, New Brunswick, NJ 08901, USA

Karl K. Matthews
Department of Food Science, Rutgers, The State University of New Jersey, New Brunswick, NJ 08901, USA

Rong Di
Department of Plant Biology, Rutgers, The State University of New Jersey, New Brunswick, NJ 08901, USA

Irais Sánchez-Ortega
DIPA, PROPACFacultad de Química, Universidad Autónoma de Querétaro, 76010 Querétaro, QRO, Mexico
Área Académica de Química, Instituto de Ciencias Básicas e Ingeniería, Universidad Autónoma del Estado de Hidalgo, Ciudad del Conocimiento, Carr. Pachuca-Tulancingo Km 4.5 Col Carboneras, 42184 Mineral de la Reforma, HGO, Mexico

Blanca E. García-Almendárez
DIPA, PROPACFacultad de Química, Universidad Autónoma de Querétaro, 76010 Querétaro, QRO, Mexico

Eva María Santos-López
Área Académica de Química, Instituto de Ciencias Básicas e Ingeniería, Universidad Autónoma del Estado de Hidalgo, Ciudad del Conocimiento, Carr. Pachuca-Tulancingo Km 4.5 Col Carboneras, 42184 Mineral de la Reforma, HGO, Mexico

Aldo Amaro-Reyes
DIPA, PROPACFacultad de Química, Universidad Autónoma de Querétaro, 76010 Querétaro, QRO, Mexico

J. Eleazar Barboza-Corona
División Ciencias de la Vida, Universidad de Guanajuato, Campus Irapuato-Salamanca, 36500 Irapuato, GTO, Mexico

Carlos Regalado
DIPA, PROPACFacultad de Química, Universidad Autónoma de Querétaro, 76010 Querétaro, QRO, Mexico

En Yang
South China Botanical Garden, Chinese Academy of Sciences, Guangzhou 510650, China
Kunming University of Science and Technology, Yunnan, China.

Lihua Fan
Agriculture and Agri-Food Canada, Atlantic Food and Horticulture Research Centre, 32 Main Street, Kentville, NS B4N 1J5, Canada

Yueming Jiang
South China Botanical Garden, Chinese Academy of Sciences, Guangzhou 510650, China

Craig Doucette
Agriculture and Agri-Food Canada, Atlantic Food and Horticulture Research Centre, 32 Main Street, Kentville, NS B4N 1J5, Canada

Sherry Fillmore
Agriculture and Agri-Food Canada, Atlantic Food and Horticulture Research Centre, 32 Main Street, Kentville, NS B4N 1J5, Canada

Mahdi Ghanbari
University of Zabol, Faculty of Natural Resources, Department of Fishery; Zabol, Iran

Mansoureh Jami
BOKU -University of Natural Resources and Life Sciences, Department of Food Sciences and Technology, Institute of Food Sciences; Vienna,, Austria

Masoud Rezaei
Tarbiat Modares University, Faculty of Marine Science, Department of Seafood Science and Technology, Noor, Mazandaran,, Iran

Belal J. Muhialdin
Universiti Putra Malaysia (UPM), Malaysia

Zaiton Hassan
Universiti Sains Islam Malaysia (USIM), Malaysia

Nazamid Saari
Universiti Putra Malaysia (UPM), Malaysia

Preface

Fermentation is the oldest traditional method in order to protect against spoilage and pathogenic microorganisms. Thermal treatment, pH and water activity lowering and preservative addition other food preservation techniques that are commonly used. The book Food Biopreservation Analytical Techniques explores possible solutions that are applicable not only to food science but to microbiology, food. First chapter focuses on lactic acid bacteria and their bacteriocins. In second chapter, antimicrobial activity in milk of HydH5 [a virion-associated peptidoglycan hydrolase (VAPGH) encoded by the Staphylococcus aureus bacteriophage vB_SauS-phiIPLA88], and three different fusion proteins created between HydH5 and lysostaphin has been assessed. In third chapter, sonorensin, predicted to belong to the heterocycloanthracin subfamily of bacteriocins, was found to be effectively killing active and non-multiplying cells of both Gram-positive and Gram-negative bacteria. Sonorensin showed marked inhibition activity against biofilm of Staphylococcus aureus. The aim of fourth chapter is to evaluate 40 lactic acid bacteria (LAB) and 20 Bacillus strains isolated from the fermented tomato (Solanum lycopersicum) for their capacity to produce antimicrobial activities against several bacteria and fungi. Fifth chapter focuses on killer toxin of kluyveromyces phaffii DBVPG 6076 as a biopreservative agent to control apiculate wine yeasts. In sixth chapter, we discuss how different biocontrol strategies can be applied either alone or in combination with other intervention strategies to improve the control of pathogenic bacteria most frequently found in sprouts. In seventh chapter, emerging preservation techniques for controlling spoilage and pathogenic microorganisms in fruit juices are presented. Eighth chapter presents the role of biopreservation in improving food safety and storage. In ninth chapter, antilisterial bacteriocin isolated from lactic acid bacteria has been targeted for its biopreservative potential. Tenth chapter focuses on the plant-derived products as antimicrobial agents for use in food preservation and to control foodborne pathogens in foods. Eleventh chapter emphasizes on antimicrobial edible films and coatings for meat and meat products preservation. In twelfth chapter, isolated and identified bacteriocinogenic LAB from cheeses and yogurts, then further evaluated their antimicrobial effects in vitro and on fresh-cut produce inoculated with L. innocua, a surrogate bacteria for L. monocytogenes. Last chapter focuses on the characterization of antimicrobial compounds produced by the lactobacilli isolates, in addition, their ability to inhibit the growth of relevant food borne pathogens as well as of spoilage bacteria and last but not least, of contaminants in aquaculture.

Chapter 1

LACTIC ACID BACTERIA AND THEIR BACTERIOCINS: A PROMISING APPROACH TO SEAFOOD BIOPRESERVATION

Mahdi Ghanbari[1, 2] and Mansooreh Jami[1, 2]

[1]University of Zabol, Faculty of Natural Resources, Department of Fishery; Zabol, Iran

[2]BOKU — University of Natural Resources and Life Sciences, Department of Food Sciences and Technology, Institute of Food Sciences, Vienna, Austria

INTRODUCTION

The growing interest in a correct life style, including alimentation, and the parallel attention on food quality have contributed to orientate consumers towards fishery products which are considered safe, of high nutritional value and capable of influencing human health in a positive way [1]. The diverse nutrient composition of seafood makes it an ideal environment for the growth and propagation of spoilage micro-organisms and common food-borne pathogens [2]. It has been estimated that as much as 25% of all food produced is lost post-harvest owing to microbial activity [1,2]. It has been mentioned that as many as 30% of people in industrialized countries suffer from a food borne disease each year and in 2000 at least two million people died from diarrhoeal disease worldwide. It is clear that indigenous bacteria present in marine environment as well as the result of post contamination during process are responsible for many cases of illnesses [3,4]. In the last years, the traditional processes applied to seafood like salting, smoking and canning have decreased in favor of mild technologies involving lower salt content, lower cooking temperature and vacuum (VP) or modified atmosphere packing (MAP). The treatments are usually not sufficient to destroy microorganisms and in some cases psychrotolerant pathogenic and spoiling bacteria can develop during the extended shelf-life of these products [2,5]. As several of these products are eaten raw, it is therefore essential that adequate preservation technologies are applied to maintain its safety and quality. Among alternative food preservation technologies, particular attention has been paid to biopreservation to extent the shelf-life and to enhance the hygienic quality, minimizing the impact on the nutritional and organoleptic properties of perishable food products such as seafood [1,6]. Biological preservation refers to the use of a natural or controlled microflora and/or its antimicrobial metabolites to extend the shelf life and improve the safety of food. Lactic acid bacteria (LAB) are particularly interesting candidates for this technique [1,2,6,7]. Indeed, they are frequently naturally present in food products and are often strong competitors, by producing a wide range of antimicrobial metabolites such as organic acids, diacetyl, acetoin, hydrogen peroxide, reuterin, reutericyclin, antifungal peptides, and bacteriocins [8-10). Hence, the last two decades have seen intensive investigation on LAB and their metabolites to discover new LAB strains that can be used in food preservation [1,7,11-13].

BACTERIAL HAZARDS ASSOCIATED WITH FISH AND FISH PRODUCTS

From the viewpoint of microbiology, fish and related products are a risky foodstuff group. Pathogenic bacteria associated with seafood can be categorized into three general groups [14]: 1) Bacteria (indigenous bacteria) that belong to the natural microflora of fish (*Clostridium botulinum*, pathogenic*Vibrio* spp., *Aeromonas hydrophila*); 2) Enteric bacteria (non-indigenous bacteria) that are present due to faecal contamination (*Salmonella* spp., *Shigella* spp., pathogenic *Escherichia coli*, *Staphylococcus aureus*); and 3) bacterial contamination during processing, storage, or preparation for consumption (*Bacillus cereus*, *Listeria monocytogenes*, *Staphylococcus aureus*, *Clostridium perfringens*, *Clostridium botulinum*, *Salmonella* spp.).

Vibrio parahaemolyticus has been isolated from sea and estuary waters on all continents with elevated sea water temperatures. *V. parahaemolyticus* is frequently isolated from fish, molluscs, and crustaceans throughout the year in tropical climates and during the summer months in cold or temperate climates [15]. Fish food associated with illnesses due to consumption of *V. parahaemolyticus* includes fish-balls, fried mackerel (*Scomber scombrus*), tuna (*Thunnus thynnus*), and sardines (*Sardina pilchardus*). These products include both raw and undercooked fish products and cooked products that have been substantially recontaminated [9,15]. The most affected by the pathogens are Japan, Taiwan, and other Asian coastal regions, though cases of disease have been described in many countries and on many continents [9,16]. Cases of diseases caused by *V. parahaemolyticus* are occasional in Europe. During 20 years, only two cases of gastroenteritis were recorded in Denmark. The interest in this organism has been widened by the finding that similar organisms, *V.*

alginolyticus and group of F *Vibrio* sp. also cause serious disease in humans [17]. *V. cholerae* is often transmitted by water but fish or fish products that have been in contact with contaminated water or faeces from infected persons also frequently serve as a source of infection [1,9,19]. The organism would be killed by cooking and recent cases of cholera in South America have been associated with the uncooked fish marinade seviche (*Cilus gilberti*) [18].

E. *coli* is a classic example of enteric bacteria causing gastroenteritis. E. *coli* including other coliforms and bacteria as *Staphylococcus* spp. and sometimes enterococci are commonly used as indices of hazardous conditions during processing of fish. Such organisms should not be present on fresh-caught fish [9,20,21]. The contamination fish derived food with pathogenic E. *coli* probably occurs during handling of fish and during the production process [20,22]. An outbreak of diarrhoeal illness caused by ingestion of food contaminated with enterotoxigenic E. *coli* was described in Japan [23]. The illness was strongly associated with eating tuna paste. Brazilian authors [24] isolated 18 enterotoxigenic strains of E. *coli* (ETEC) from 3 of 24 samples of fresh fish originating from Brazilian markets; 13 of them produced a thermolabile enterotoxin. Infection with verocytotoxin _ producing strains of E. *coli*(VTEC) after ingestion of fish was recorded in Belgium [25]. An outbreak caused by salted salmon roe contaminated, probably during the production process, with enterohaemorrhagic E. *coli* (EHEC) O157 occurred in Japan in 1998 [22]. The roe was stored frozen for 9 months but it appears that O157 could survive freezing and a high concentration of NaCl and retained its pathogenicity for humans [26].

Aeromonas spp. has been recognized as potential foodborne pathogens for more than 20 years. Aeromonads are ubiquitous in fresh water, fish and shellfish and also in meats and fresh vegetables [27]. The epidemiological results so far are, however, very questionable. The organism is very frequently present in many food products, including raw vegetables, and very rarely has a case been reported. Up to 8.1% of cases of acute enteric diseases in 458 patients in Russia were caused by *Aeromonas* spp. [28]. In this study, *Aeromonas* spp. isolates with the same pathogenicity factors were isolated from river water in the Volga Delta, from fish, raw meat, and from patients with diarrhoea. Most *Aeromonas* spp. isolates are psychrotrophic and can grow at refrigerator temperatures [29]. This could increase the hazard of food contamination, particularly where there is a possibility of cross-contamination with ready-to-eat food products.

Salmonella has been isolated from fish and fishery product, though it is not psychrotrophic or indigenous to the aquatic environment [30]. The relationship between fish and *Salmonella* has been described by several scientists; some believe that fish are possible carriers of *Salmonella* which are harbored in their intestines for relatively short periods of time and some believe that fish get actively infected by *Salmonella* [31]. Most outbreaks of food poisoning associated with fish derive from the consumption of raw or insufficiently heat treated fish and cross-contamination during processing and the U.S. Food and Drug Administration's (FDA) data showed that *Salmonella* was the most common contaminant of fish and fishery products [31]. The highest *Salmonella* incidence in fishery products was determined in Central Pacific and African countries while it was lower in Europe and including Russia, and North America [32]. The most common serovar found in the world was *S.* sub Weltvreden [30, 31]. In seafood the commonest serotype encountered was *S.* sub Worthington followed by *S.* sub Weltevreden.

Enterotoxins produced by *Staphylococcus aureus* are another serious cause of gastroenteritis after consumption of fish and related products. In 3 of 10 samples of fresh fish, higher counts of *Staph. aureus* were detected than permitted by Brazilian legislation [20, 33]. In the southern area of Brazil,*Staph. aureus* was isolated from 20% of 175 examined samples of fresh fish and fish fillets (*Cynoscion leiarchus*). *Staph. aureus* has also been detected during the process of drying and subsequent smoking of eels in Alaska in 1993 [34]. During the process, *S. aureus* populations increased to more than 10^5CFU g^{-1} of the analyzed sample, after 2 to 3 days of processing. Subsequent laboratory studies showed that a pellicle (a dried skin-like surface) formed rapidly on the strips when there was rapid air circulation in the smokehouse and that bacteria embedded in/under the pellicle were able to grow even when heavy smoke deposition occurred.

In ready-to-eat products, cooking, preservation ingredients, and storage atmosphere inhibit the Gram-negative organisms, resulting in a longer shelf life. Such conditions favor the growth of psychotropic pathogens such as *Listeria monocytogenes*, allowing them to grow to dangerous levels [9,35,36]. *L. monocytogenes* is a serious threat to consumer health and safety and has been implicated in several deadly outbreaks around the world [1,2]. This organism is halotolerant (up to 28% w/v for short periods), resistant to freezing temperatures, can grow and multiply during refrigeration, where other competing organisms cannot, and is able to survive at low water activity (aw) [9,14,37]. *L. monocytogenes* is widely distributed in the general environment including fresh water, coastal water and live fish from these areas. Contamination or recontamination of seafood may also take place during processing [37-39]. Moreover, *L. monocytogenes* is a psychrotrophic pathogen with the ability to grow from under 0 to 45°C [40]. This ability to grow at storage temperatures means that this bacterium is the main hazard in this kind of product. The pathogenic bacteria *L. monocytogenes* may grow on fresh seafood. *Listeria* has been found in farmed rainbow trout [41]. The outbreak of listeriosis related to vacuum packed gravad and cold-smoked fish was described in at least eight human cases for 11 months in Sweden [42]. Cold-smoked and gravad rainbow trout (*Oncorhynchus mykiss*) and salmon (*Salmo salar*) have been focused on during recent years as potential sources of infection with *L. monocytogenes*and there are several report on isolation of this food borne pathogen from fish-processing plants environments [14,37,39,43-45]. Seafood treatment is necessary to prevent food-borne illness. However, the pervasive nature of *L. monocytogenes* makes it difficult for processors to fully eliminate the organism from the environment.

Development of new-generation foods, which are mildly processed, contain few or no preservatives, are packaged in vacuum or modified atmospheres to ensure long shelf life and rely primarily on refrigeration for preservation, has raised concerns of potential increases in botulism risk caused by psychrotrophic nonproteolytic group II *Clostridium botulinum* [46]. An average of 450 outbreaks of foodborne botulism with 930 cases have been reported annually worldwide [47]. The main habitat of clostridia is the soil but they are also found in sewage, rivers, lakes, sea water, fresh meat, and fish [48,49]. Most critical are the hygienic conditions for handling the product after smoking. There is a risk of botulism due to the growth of *C. botulinum* type E in smoked fish. The bacterium becomes a hazard when processing practices are insufficient to eliminate botulinal spores from raw fish, particularly improper thermal processing [21]. The growth of *C. botulinum* and toxin production then depends on appropriate conditions in food before eating: the temperature, oxygen level, water activity, pH, the presence of preservatives, and competing microflora [21]. A problem with *C. botulinum* has been encountered with some traditional fermented fish products. These rely on a combination of salt and reduced pH for their safety. If the product has insufficient salt, or fails to achieve a rapid pH drop to below 4.5, *C. botulinum* can grow. There was no evidence that the fish had been mishandled, but a low salt environment in the viscera allowed the bacterium to multiply and to produce toxin. *C. perfringens*, an important cause of both food poisoning and non-food-borne diarrhoeas in humans, was found in a number of fish owing to contamination with sewage, which is the main source of this organism [21].

BIOPRESERVATION

Seafood products are known to be especially susceptible to both microbiological and biochemical spoilage pathways. The development of effective processing treatments to extend the shelf life of fresh fish products is a must [2]. Additionally, the consumers' demand for high-quality and minimally processed seafood has recently captivated great attention [5, 9]. However, an increase in foodborne illness outbreaks is concomitant with the increase in consumer demand for less processed foods [1]. These trends highlight the importance of studying new microbial stress factors to extend the shelf-life of foods. Until now, approaches to reduce the risk of outbreaks of food poisoning have relied on the search for addition of more efficient chemical preservatives or on the application of more drastic physical treatments such as heating, refrigeration, high hydrostatic pressure (HHP), ionising radiation, pulsed-light, ozone, ultrasound, etc [1,5,50]. In spite of some possible advantage, these types of treatments have many drawbacks and limitation in seafood products: the proven toxicity of many of the commonest chemical preservatives (e.g. nitrites) (3), the alteration of the organoleptic and nutritional properties of seafood by physical treatments due to their delicate nature (e.g. freezing damage, discolouration in case of HHP and ionising radiation) [50,51] and especially recent consumer trends in purchasing and consumption, with demands for healthy seafood products that have been subjected to less extreme treatments (less heat and chill damage), with lower levels of salts, fats, acids, and sugars and/or the complete or the partial removal of chemically synthesized additives [1,2,7]. To harmonize consumer demands with the necessary safety standards, traditional means of controlling microbial spoilage and safety hazards in seafood are being replaced by an alternative solution that is gaining more and more attention: "biopreservation technology" [2,9,13,52,53]. It consists in inoculating food with microorganisms, or their metabolites, selected for their antibacterial properties and may be an efficient way of extending shelf life and food safety through the inhibition of spoilage and pathogenic bacteria without altering the nutritional quality of raw materials and food products [54, 55].

Lactic acid bacteria (LAB) possess a major potential for use in biopreservation because they are safe to consume, and during storage they naturally dominate the microbiota of many foods. Certain LAB species and strains isolated from seafood have been shown to exert strong antagonistic activity against spoilage and pathogenic microorganisms such as *Listeria, Clostridium, Staphylococcus,* and *Bacillus*spp [56-58]. The antagonistic and inhibitory properties of LAB are due to the competition for nutrients and the production of one or more antimicrobially active metabolites such as organic acids (lactic and acetic acid), hydrogen peroxide, and antimicrobial peptides (bacteriocins) [10]. Certain LAB are able to grow at refrigeration temperatures and are tolerant to modified-atmosphere packaging, low pH, high salt concentrations, and the presence of certain additives such as lactic acid, acetic acid, and ethanol. Because of these benefits, LAB can be used as protective cultures to restrict the growth of undesired organisms such as certain spoilage and pathogenic bacteria, with the subsequent benefits in terms of food safety [9,10,58]. Moreover, these microorganisms may have additional functional properties and, in some circumstances, they can be beneficial for the consumers [6]. LAB represent the microbial group most commonly used as protective cultures, as they are present in all fermented foods and have a long history of safe use [8]. Safety for the consumers is an aspect of great importance, in particular for some seafood products which are not cooked before consumptions, but also for other types of foods.

THE ROLE OF LACTIC ACID BACTERIA IN BIOPRESERVATION TECHNOLOGY

Characterization and Classification

Lactic acid bacteria (LAB) encompass a heterogeneous group of microorganisms having as a common metabolic property the production of lactic acid as the majority end - product from the fermentation of carbohydrates [59]. LAB are Gram (+), usually nonmotile, non - sporulating, catalase - negative, acid - tolerant, facultative anaerobic organisms and have less than 55 mol% G+C content in their DNA [60-62]. Except for a few species, LAB members are nonpathogenic organisms with a reputed generally recognized as safe status (GRAS). Taxonomic revisions of these genera and the description of new genera mean that LAB could, in their broad physiological definition, comprise around 20 genera [10]. However, from a practical, food-technology point of view, the following genera are considered the principal LAB: *Aerococcus, Carnobacterium, Enterococcus, Lactobacillus, Lactococcus, Leuconostoc,Oenococcus, Pediococcus, Streptococcus, Tetragenococcus, Vagococcus,* and *Weissella* [61]. The classification of lactic acid bacteria into different genera is largely based on morphology, mode of glucose fermentation, growth at different temperatures, configuration of the lactic acid produced, ability to grow at high salt concentrations, and acid or alkaline tolerance [62, 63]. An important characteristic used in the differentiation of the LAB genera is the mode of glucose fermentation under standard conditions. In this regard, the accepted definition is that given by Hommes and Vogel [64]: obligately homofermentative LAB are able to ferment hexoses almost exclusively to lactic acid by the Embden–Meyerhof–Parnas (EMP) pathway while pentoses and gluconate are not fermented as they lack phosphoketolase; facultatively heterofermentative LAB degrade hexoses to lactic acid by the EMP pathway and are also able to degrade pentoses and often gluconate as they possess both aldolase and phosphoketolase; finally, obligately heterofermentative degrade hexoses by the phosphogluconate pathway producing lactate, ethanol or acetic acid and carbon dioxide; moreover, pentoses are fermented by this pathway [62]. Several strains of groups 1 and 2 and some of the hetero fermentative group 3 are either used in fermented foods, but group 3 are also commonly associated with food spoilage. (For a more detailed discussion concerning the metabolic pathways, see [59].

Antimicrobial Components from Lab

Bacteriocins

Bacteriocins are ribosomally synthesized peptides, that exert their antimicrobial activity against either strains of the same species as the bacteriocin producer (narrow range), or to more distantly related species (broad range) [1,2,7]. It has been estimated that between 30% and 99% of all bacteria and archaea produce bacteriocins; their production by LAB is very significant from the point of view of their potential applications in food systems and thus, unsurprisingly, these have been most extensively investigated [6,10,12,60,65,66]. It has been noted that the activity of bacteriocins is frequently directed against bacteria that are related to the bacteriocin - producing strain or against bacteria found in similar environments [67]. It has also been noted that some bacteriocins can also play a role in cell signaling. Microorganisms that produce bacteriocins also possess immunity mechanisms to confer self - protection, that is, to protect bacteriocin producers from committing "suicide" [10,68,69]. Besides concern about antibiotic resistance, increasing consumer awareness of potential health risks associated with chemical preservatives has increased interest in bacteriocins. Bacteriocins are naturally produced so they are more easily accepted by consumers [54]. Bacteriocins are usually classified combining various criteria. The main ones being the producer bacterial family, their molecular weight and finally their amino acid sequence homologies and/ or gene cluster organization [59,70]. Based on a relatively recent approach [69,71,72] bacteriocins produced by LAB have been categorized into two major classes: the lanthionine - containing bacteriocins or lantibiotics (class I) and the largely unmodified linear peptide antimicrobials (class II).

Organic Acid Production

An important role of meat LAB starter cultures is the rapid production of organic acids; this inhibits the growth of unwanted flora and enhances product safety and shelf life. The types and levels of organic acids produced during the fermentation process depend on the LAB strains present, the culture composition, and the growth conditions [74]. Fermentation of the carbohydrates, glucose, glycogen, glucose-6-phosphate and small amounts of ribose, in meat and meat products, produces organic acids by glycolysis (Embden-Meyerhof Parnas pathway, EMP pathway) or the Hexose Monophosphate, HMP pathway. L (+) lactic acid is more inhibitory than its D (-) counterpart [68]. The antimicrobial effect of organic acids lies in the reduction of pH, and in the action of undissociated acid molecules [75]. It has been proposed that low external pH causes acidification of the cytoplasm. The lipophilic nature of the undissociated acid allows it to diffuse across the cell membrane collapsing the electrochemical proton gradient. Alternatively, cell membrane permeability may be affected, disrupting substrate transport systems [72]. The LAB in particular are able to reduce the pH to levels where putrefactive (e.g. clostridia and pseudomonads), pathogenic (e.g. *Salmonella* s and *Listeria* spp.) and toxinogenic bacteria (*Staphylococcus aureus. Bacillus cereus, Clostridium botulinum*) will be either inhibited or killed [7]. Also, the undissociated acid, on account of its fat solubility, will diffuse into the bacterial cell, thereby reducing the intracellular pH and slowing down metabolic activities, and in the case of Enterobacteriaceae such as *E. coli* inhibiting growth at around pH 5.1.

Other Antimicrobials of Lab

Hydrogen peroxide is produced from lactate by LAB in the presence of oxygen as a result of the action of flavoprotein oxidases or nicotinamide adenine dinucleotide (NADH) peroxidise [76]. The antimicrobial effect of H_2O_2 may result from the oxidation of sulfhydryl groups causing denaturing of a number of enzymes, and from the peroxidation of membrane lipids thus increasing membrane permeability [8]. Most undesirable bacteria such as *Pseudomonas* spp. and *S. aureus* are many times sensitive to H_2O_2. Carbon dioxide (CO_2) is mainly produced by heterofermentative LAB. CO_2 plays a role in creating an anaerobic environment which inhibits enzymatic decarboxylations, and the accumulation of CO_2 in the membrane lipid bilayer may cause a dysfunction in permeability [8]. CO_2 can effectively inhibit the growth of many food spoilage microorganisms, especially Gram-negative psychrotrophic bacteria [77]. Diacetyl, an aroma component, is produced by strains within all genera of LAB by citrate fermentation. It is produced by heterofermentative lactic acid bacteria as a by-product along with lactate as the main product [8]. Diacetyl is a high value product and is extensively used in the dairy industry as a preferred flavour compound. Diacetyl also has antimicrobial properties. Diacetyl was found to be more active against gram-negative bacteria, yeasts, and molds than against gram-positive bacteria. Diacetyl is thought to react with the arginine-binding protein of gram-negative bacteria and thereby interfering with the utilization of this amino acid [78].

LAB IN FISH AND FISH PRODUCTS

LAB are not considered as genuine microflora of the aquatic environment, but certain genera, including*Carnobacterium, Lactobacillus, Enterococcus,* and *Lactococcus,* have been found in fresh and sea water fresh fish [61,63,79-83]. The number of lactobacilli in the gastro-intestinal tract of Arctic char was smaller in those reared in sea water than in fresh water, while the number of *Leuconostoc* and enterococci remained the same [84]. It is well documented that lactobacilli are part, not dominant, of the native intestinal microbiota of Arctic charr (*Salvelinus alpinus* L.), Atlantic cod, Atlantic salmon (*Salmo salar* L.), and brown trout (*Salmo trutta*) [82,85]. Several studies have shown the presence of other lactic acid bacteria, specially carnobacteria such as *Carnobacterium maltaromaticum* and*Carnobacterium divergence* within the intestinal content of salmonid species like Arctic charr (*Salvelinus alpinus*), Atlantic salmon (*Salmo salar*), rainbow trout (*Oncorhyncus mykiss*) [63,86-89], Atlantic cod [89], common wolffish (*Anarhichas lupus* L.) [85], brown trout [82] and also wild pike [63,82]. Bacteria of the genus *Enterococcus* have been isolated from the intestine of common carp (*Cyprinus carpio*) and brown trout [80,82].

LAB dominating in spoiled vacuum-packaged cold-smoked fish products include the genera of*Lactobacillus, Leuconostoc, Lactococcus* and *Carnobacterium* [9]. Magnússon & Traustadóttir [91] reported the complete dominance of homofermentative lactobacilli in vacuum-packaged cold-smoked herring. In vacuum packaged cold-smoked salmon and herring, *Lactobacillus curvatus* has been found in majority together with lower numbers of *Lactobacillus sakei, Lactobacillus plantarum, Lactococcus*spp. and *Leuconostoc mesenteroides* [58,]. Paludan-Müller, Huss, & Gram [92] identified*Carnobacterium piscicola* as the dominant microorganism isolated from spoiled vacuum-packaged cold-smoked salmon. Leroi et al. [93] also isolated carnobacteria during the first stage of storage of vacuum-packaged cold-smoked salmon, whereas *Lactobacillus farciminis, Lactobacillus sakei,* and*Lactobacillus alimentarius* were isolated at advanced storage times. Other studies have also confirmed that most bacteria in vacuum-packaged "gravad" fish products stored at refrigeration temperatures are carnobacteria [94] and *L. sakei,* and to a lesser extent *Leuconostoc* spp., *L. curvatus,* and *Weissella viridescens* [95]. Gancel et al [90] have isolated 78 strains belonging to the genus *Lactobacillus* from fillets of vacuum packed smoked and salted herring *(Clupea harengus).* LAB has been found to occur in marinated herring, herring fillets and cured stockfish [58]. In marinated or dried fish, the lactic acid bacteria flora maybe quite diverse since the presence of *Lactobacilli* and *Pediococci* has been reported [90]. Thai fermented fishery products were screened for the presence of LAB by Ostergaard et al. [96]. LAB was found to occur in the low salted fermented products in the range of 10^7-10^9 cfu/g. The high salt product "hoi dorng" had a lower LAB count of 10^3-10^5 cfu/g. Olympia et al [97] have isolated 10^8LAB/g from a Philippine low salt rice-fish product burong bangus. Several studies have been mentioned that some species of *Carnobacteriuim* such as *C. divergens* and *C. maltaromaticum* are present in seafood and are able to grow to high concentrations in different fresh and lightly preserved products such as modified atmosphere-packed (MAP) [98-100], chilled MAP [101,102], high-pressure processing treated seafood products [103] and vacuum-packed cold smoked or sugar-salted ('gravad') seafood [53,93,95]. These studies clearly highlight the ability of LAB fish isolates to grow on different harsh condition rather than other organisms. Obviously many investigation have been shown that carnobacteria are common in chilled fresh and lightly preserved seafood, but at higher storage temperatures (15–25°C) other species could be dominate the spoilage microbial community of seafood.

APPLICATION OF LAB IN SEAFOOD

Treating catfish fillets with of 0.50% sodium acetate, 0.25% potassium sorbate with 2.50% lactic acid culture completely inhibited growth of Gram negative bacteria, improved catfish odor and appearance during 13 days storage [110]. Einarsson & Lauzon [111] treated shrimps with various bacteriocins from lactic acid bacteria and reported shelf life extension except carnocin UI49. Total mesophilic and psychotropic bacteria and MRS counts of the samples treated with carnocin UI49 were not different than those of controls at 4.5°C. In a study with five strains of lactic acid bacteria (four *Lactobacillus* and one *Carnobacterium*) on fermented salmon fillets, *L. sake* LAD and *L. alimentarius* BJ33 was regarded as suitable starters for fermentation of salmon fillets [112] based on starter growth (increase of more than 1log in 3 days) and acidification of muscle (e.g. pH reduction of approximately 0.7 units in 5 days) as well as sensory evaluation. Kisla & Ünlütürk [113] studied the microbial shelf life of rainbow trout treated with nisin-containing aqueous solution of *Lactococcus lactis* subsp. lactis NCFB 497and lactic acid. They reported the dipping of rainbow trout fillets into a lactic culture did not prolonged the shelf life due to the low inoculum level and type of lactic culture used. Elotmani & Assobhei [114] evaluated the inhibition of the microbial flora of sardine by using nisin and a lactoperoxidase system (LP), observing the efficiency of the nisin–LP combination in inhibiting fish spoilage flora. In another study growth of *L. monocytogenes* was significantly inhibited (P < 0.05) by *L. sakei* Lb706 in rainbow trout fillets stored under vacuum at 4°C during 10 days of storage while bacteriocin negative Lb706-B did not affect the growth of *L. monocytogenes*. In the presence of the sakacin A-producing strain of *L. sakei*(Lb706), the growth of *L. monocytogenes* was significantly inhibited (P < 0.05) in the first 3 days of storage at 10°C, after which its count increased to 10^7 CFU g^{-1} [115]. Altieri et al. [106] succeeded in inhibiting *Pseudomonas* spp. and *P. phosphoreum* in VP fresh plaice fillets at low temperatures by using a *Bifidobacterium bifidum* starter, and extending the shelf-life, especially under MAP. Bifidobacteria combined with sodium acetate (SA) extended refrigerated shelf-life of catfish fillets at 4°C [116]. The application of two *Lactobacillus sakei* CECT 4808 and *L. curvatus* CECT 904T protective cultures on refrigerated vacuum-packed rainbow trout (*Oncorhynchus mykiss*) fillets resulted in extension of shelf-life by 5 days by significantly improved in the counts of all microbiological spoilage indicator organisms (Enterobacteriaceae, *Pseudomonas* spp., H2S-producing bacteria, yeasts and moulds) and also significantly improved in all examined chemical parameters and off-odour [117].

Under biopreservation, combined coating of *Lactobacillus casei* DSM 120011 and *Lactobacillusacidophilus* 1M in *Streptomces* sp. NIOF metabolites, played effective role in lowering the biochemical and microbiological changes, extended shelf-life and safety of stored fish under low temperature as reported by Daboor & Ibrahim [118]. Tahiri et al. [119] suggest that selection of protective strains to improve the sensory quality of seafood products should focus on specific spoilage microorganism's inhibition. This approach was chosen by Matamoros et al., [120] who have isolated seven strains from various marine products on the basis of their activity against many spoiling and pathogenic, Gram positive and Gram negative marine bacteria. Among strains, two *Le. gelidum,* and two *Lc. piscium*demonstrated promising effect in delaying the spoilage of tropical shrimp and of VP CSS. However, no correlation with the classical quality indices measured was evidenced. A recent study demonstrated that this protective effect could be due to the inhibition of *B. thermosphacta* identified as one of the major spoiler organisms in cooked shrimp stored under MAP [121]. The inoculation of Tilapia (*Oreochromis niloticus*) fillets with *Lactobacillus casei* DSM 120011 and *Lactobacillus acidophilus* 1M at 2% concentration decreased both total volatile basic nitrogen (TVB-N), trimethylamine nitrogen (TMA-N) and thiobarbituric acid (TBA) values and improved the biochemical quality criteria, microbial aspects and safety of frozen fish fillets during 45 and 90 days storage. [122].

For Shirazinejad et al. [123] 2.0% lactic acid combined with nisin indicated the highest reduction in population of *Pseudomonas* spp. and H_2S producing bacteria during storage time of Chilled Shrimp. Fall et al. [121] evidenced the in situ inhibition of *B. thermosphacta*, a major spoiling bacterium, by *L.piscium* that could explain the protective effect observed in shrimp. Additionally, those strains also showed an inhibitory effect on *L. monocytogenes* [124] and *Staph. aureus*. Recently, Sudalayandi & Manja [109] succeeded to preserve fresh fish through controlling spoilage bacteria and amines of Indian mackerel fish chank for two days at 37°C by inoculating them with different strains of LAB such as *Pediococcus acidilactici, Pediococcus pentosaceous, Streptococcus thermophilus, Lactococcus lactis, Lactobacillus plantarum, Lactobacillus acidophilus* and *Lactobacillus helveticus*. Using bacteriocin-like metabolite producer and non-producer strains of *Pediococcus* spp. [125] only slightly improved sensory quality of Horse Mackerel during cold storage. It was concluded that *Pediococcus* strains used in this study were not proper for preserving horse mackerel fillets especially at low storage temperatures. EntP-producing enterococci isolated from farmed turbot, under a spray-dried format exhibited antilisterial, antistaphylococcal, and antibacilli activities in turbot fillets either vacuum-packaged or subjected to modified-atmosphere packaging [2].

LAB Protective cultures have not been applied in many other seafood products except for cold smoked salmon (CSS), as they are normally flora of such products at the end of storage, and *L. monocytogenes* control. The effectiveness of bacteriocins to control growth of *L. monocytogenes* in vacuum packed cold smoked salmon has also been demonstrated by several researchers. Among them, Sakacin P has been found to be very potent against *L. monocytogenes* and is one of the most extensively studied bacteriocins [126-131]. Leroi et al. [132] succeeded in increasing the sensory use-by-date of CSS slices by inoculating them with strains of *Carnobacterium* sp. However the results varied depending on the batch treated. Addition of nisin to CO_2 packed cold smoked salmon resulted in a 1 to 2 log_{10} reduction of *L. monocytogenes* [11]. Using a strain of *C. maltaromaticum*, Paludan-Müller et al. [92] only slightly extended the shelf-life of smoked salmon. Budu-Amoako et al. [133] tested nisin combined with heat as anti Listerial treatment in cold- packed lobster meat, finding decimal reductions of inoculated *L. monocytogenes* of 3 to 5 logs, whereas heat or nisin alone resulted in decimal reductions of 1 to 3 logs.

Duffes et al. [65] isolated *C. divergens* and *C. maltaromaticum* strains that exhibited listericidal activity in a model experiment with cold-smoked fish. They found that *C. piscicola* V1 inhibited *L. monocytogenes* by the in situ production of bacteriocins in vacuum-packed cold-smoked salmon stored at 4°C and 8°C. In contrast, another related species, namely, *C. divergens* V41 and its divercin V41, only exhibited a bacteriostatic effect on the target microorganism. Two strains of *C. maltaromaticum* isolated from CSS demonstrated their efficiency to limit the growth of *L. monocytogenes* in VP

CSS during 31 days of storage at 5°C [134]. In a study using vacuum-packed cold smoked rainbow trout, the combination of nisin and sodium lactate injected into smoked fish decreased the count of *L. monocytogenes* from 3.3 to 1.8 $\log_{10}$ over 16 days of storage at 8°C [135]. Sakacin P was added to vacuum-packed cold smoked salmon, a lightly processed high-fat (15–20%) product, together with a sakacin P-producing *L. sakei* culture in order to study the effect on the growth of *L. monocytogenes*. In this product, the combination of purified sakacin P and a live culture was found to be bactericidal against *L. monocytogenes*. The addition of sakacin P alone inhibited the growth of *L. monocytogenes* on this product for about 1 week [126]. Silva et al. [136] used a bacteriocin-producing *Carnobacterium*strain under a spray-dried format. This strain survived the process and retained antilisterial ability, although it lost activity against other Gram-positive targets such as *Staph. aureus*. Some authors have evaluated the antimicrobial activity of nisin combined with other bacteriocins. Bouttefroy & Milliere [137] tested combinations of nisin and curvaticin 13 produced by *L. curvatus* SB13 for preventing the regrowth of bacteriocin-resistant cells of *L. monocytogenes*, finding that this combination induced a greater inhibitory effect than the use of a single bacteriocin. Aasen et al. [131] studied the interactions of the bacteriocins sakacin P and nisin with food constituents in cold-smoked salmon, chicken cold cuts, and raw chicken. They stated that owing to the amphiphilic nature of these peptides, they can be adsorbed to food macromolecules and undergo proteolytic degradation, which may limit their use as preservation agents. More than 80% of the added sakacin P and nisin were rapidly adsorbed by proteins in the food matrix that had not been heat-treated, less than 1% of the total activity remaining after 1 week in cold- smoked salmon. In heat-treated foods, they found that, bacteriocin activity was stable for more than 4 weeks. No important differences were observed between sakacin P and nisin, but less nisin was adsorbed by muscle proteins at low pH. The growth of *L. monocytogenes* was completely inhibited for at least 3 weeks in both chicken cold cuts and cold-smoked salmon by the addition of sakacin P (3.5 μ/g), despite proteolytic degradation in the salmon.

In the presence of the bacteriocinogenic strain *C. maltaromaticum* CS526 isolated from surimi, the population of *L. monocytogenes* in CSS decreased from 10^3 to 50 CFU g^{-1} after 7 days at 4°C [138]. This activity could be linked to the production of the bacteriocin piscicocin CS526, since a non-bacteriocin producing strain had a lower effect on the growth of the pathogenic bacteria [138, 139]. The growth of the protective *Carnobacterium* strains did not modify the sensory characteristic of the product. One of these strains showing the strongest inhibition activity produces a bacteriocin, named Carnobacteriocin B2 that was involved in the antilisterial activity [105]. Three strains of bacteriocin producing *Carnobacterium* have been tested with the agar diffusion test method against a wide collection of *L. monocytogenes* (51 strains) isolated from seafood. All of the *Listeria* strains were sensitive. The inhibition was confirmed in co-culture with a mix of *L. monocytogenes* strains in sterile CSS [140]. One of these strains, *C. divergens* V41 showed its ability to maintain *L. monocytogenes* at the initial inoculating level of 20 CFU g^{-1} during 28 days of storage at 4°C and 8°C. The effect of this strain on sensory characteristics and physico-chemical parameters revealed that it did not spoiled the product [56].

A bacteriocinogenic strain of *L. sakei* isolated from CSS allowed a 4 log reduction of *Listeria innocua*after 14 days of storage at 4°C. A reduction of 2 log units after 24 h at 5°C was also demonstrated with that strain in CSS juice towards *L. monocytogenes* [141]. Mix of bacteriocin-producing LAB like *L. casei, L. plantarum* and *C. maltaromaticum* were successfully used to limit the growth of *L. innocua* in CSS [142]. *C. maltaromaticum* had no effect on the inhibition of the Gram positive spoilage bacteria *B. thermosphacta* in cooked shrimps [143]. The anti-listerial activity of 3 LAB strains used individually or as co-cultures was assayed on cold-smoked salmon artificially contaminated with *L. innocua* and stored under vacuum at 4°C [142]. The association of *L. casei* T3 and *L. plantarum* PE2 was the most effective, probably due to a competition mechanism against the pathogen. In their study Tomé et al. [144] have also selected a strain of *Enterococcus faecium* among five bacteriocinogenic LAB strains for its ability to induce a decrease of the population of *L. innocua* inoculated in CSS. However in these studies the inhibition activities were not confirmed on *L. monocytogenes*. For Matamoros et al. [145] two LAB strains, *Lactococcus piscium* EU2241 and *Leuconostoc gelidum* EU2247 were efficient to limit the growth of both pathogenic bacteria *L. monocytogenes* and *S. aureus* in a challenge test in cooked shrimp stored under VP from 2 to 3 log CFU g^{-1} units after 4 weeks at 8°C followed by 1 week at 20°C. The strain of *Leuconostoc* produced a bacteriocin-like compound but its activity was slight lower than the *Lactococcus* strain that was non-bacteriocinogenic. In another study, the application of *C. divergens* M35 towards *L. monocytogenes* in CSS resulted in a maximal decrease of 3.1 log CFU g^{-1} of the pathogenic bacteria after 21 days of storage at 4°C whereas a non bacteriocinogenic strain had no effect [119].

CONCLUSION AND FUTURE PROSPECTIVE

The presence of LAB in many processed seafood product is now well documented and the bio-protective potential of many strains and/or their bacteriocin has been highlighted in the last years. In situ production is readily cost-effective provided that the bacteriocin producers are technologically suitable. To date, only nisin and pediocin PA - 1 have been applied commercially in food applications where they are used to protect against spoilage and pathogenic organisms. However, other bacteriocins could be at least as effective for food processors as it is possible to apply them with hurdle approaches, particularly in light of consumer demands for minimally processed, safe, preservative - free foods. Control of pathogenic bacteria has widely focused on *L. monocytogenes* considered as the main risk in ready-to-eat seafood. However, in these minimally processed products, the new combination of hurdles can give selective advantages to enhance food safety and quality, particularly effective against other pathogenic bacteria like clostridia, vibrio or staphylococci. These goals can be facilitated through the incorporation of live bacteriocin - producing strain(s) or through the use of bacteriocins as concentrated preparations, either through direct addition to the seafood or in an immobilized form on packaging as well as in conjunction with other factors such as high pressure or pulse electric fields, to achieve more effective preservation of foods. The great results obtained with protective culture, bacteriocins for improving safety and quality of seafood products clearly indicate that the application of LAB protective culture and/or their bacteriocins in seafood product can suggest several important benefits; 1) extended shelf life of seafood during storage time, 2) decrease the risk for transmission of foodborne pathogens in lightly preserved seafood products, 3) ameliorate the economic losses due to seafood spoilage, 4) reduce the application of chemical preservatives and drastic physical treatments such as heating, refrigeration, etc. causing better preservation nutritional quality of food, 5) good option for industry due to cost effective way and finally 6) a good response to consumer demands for minimally processed, safe, preservative - free foods. At present the new techniques and disciplines emerging in the post – genomic era, such as genomics, proteomics, metabolomics, and system biology, open new avenues for interpretation of biological data. In combination with classical and molecular techniques, these new methods will be invaluable in the rational optimization of LAB function in order to obtain safer traditional and new seafood products.

REFERENCES

1. M. L Cortesi, A Panebianco, A Giuffrida, A Anastasio, 2009Innovations in seafood preservation and storage.Veterinary Research CommunicationsSupplement 1: S15S23

2. A Campos, P Castro, S. P Aubourg, J. B Velázquez, 2012Use of Natural Preservatives in Seafood. In McElhatton A, Sobral. Novel Technologies in Food Science, Integrating Food Science and Engineering Knowledge

Into the Food Chain.: © Springer Science+Business Media;. 325360

3. F Feldhusen, 2000The role of seafood in bacterial foodborne diseases. Method. Microbiol. 216511660

4. ICMSF (2011Fish and Seafood ProductsIn International Commission on Microbiological Specifications for Foods (ICMSF). Microorganisms in Foods.: Springer Science+Business Media, LLC. 107133

5. S Alzamora, J Welti-chanes, S Guerrero, 2012Rational Use of Novel Technologies:A Comparative Analysis of the Performance of Several New Food Preservation Technologies for Microbial Inactivation. In McElhatton A, Sobral PJA(). Novel Technologies in Food Science, Integrating Food Science and Engineering Knowledge Into the Food Chain.: © Springer Science+Business Media, LLC.

6. A. H Soomro, T Masud, K Anwaar, 2002Role of lactic acid bacteria (LAB) in food preservation and human health-A review.Pak. J. Nut. 12024

7. A Gálvez, H Abriouel, R López, N Omar, 2007Bacteriocin-based strategies for food biopreservationInt. J. Food Microbiol. 1205170

8. W. H Holzapfel, R Geisen, U Schillinger, 1995A review paper: biological preservation of foods with reference to protective cultures, bacteriocins and food-grade enzymes. Int. J. Food Microbiol. 24: 343362

9. P Calo-mata, S Arlindo, K Boehme, T Miguel, A Pascola, J Barros-velazquez, 2008Current Applications and Future Trends of Lactic Acid Bacteria and their Bacteriocins for the Biopreservation of Aquatic Food ProductsFood Biopro. Tech.14363

10. B Collins, P Cotter, Hill, Paul Ross R (2010Applications of Lactic Acid Bacteria- Produced Bacteriocins. In Mozzi F, Raya R, GM V. Biotechnology of Lactic Acid Bacteria Novel Applications.: Blackwell Publishing. 89109

11. L Nilsson, 1997Control of Listeria monocytogenes in cold-smoked salmon by biopreservation: Danish Institute for Fisheries Research and The Royal Veterinary and Agricultural University of Copenhagen, Denmark, Ph. D Dissertation.

12. J Cleveland, T Montville, I Nes, M Chikindas, 2001Bacteriocins: Safe, natural antimicrobials for food preservationInt. J. Food Microbiol. 71: 120

13. C Dortu, P Thonart, 2009Bacteriocins from lactic acid bacteria: interest for food products biopreservationBiotech. Agr. Society Environ. 13: 143154

14. A Beaufort, S Rudelle, N Gnanou-besse, M. T Toquin, A Kerouanton, H Bergis, 2007Prevalence and growth of Listeria monocytogenes in naturally contaminated cold-smoked salmon. Lett. Appl. Microbiol. 44406411

15. W Baffone, A Pianei, F Bruscolini, E Barbieri, B Cierio, 2000Occurrence and expression of virulence-related properties of Vibrio species isolated from widely consumed seafood products.Int. J. Food Microbiol. 54918

16. InternationalDiseaseSurveillanceCenter   (IDSC)1999Vibrio parahaemolyticus, Japan 1996-1998, Infectious Agents Surveillance Report (IASR), 2012

17. S. W Joseph, R. R Colwell, J. B Kaper, 1982Vibrio parahaemolyticus and related halophilic Vibrios.Crit. Rev. Microbiol. 10177124

18. K. M Kam, T. H Leung, Y. Y Ho, N. K Ho, T. A Saw, 1995Outbreak of Vibrio cholerae 01 in Hong Kong related to contaminated fish tank waterPublic Health3389395

19. R. R Colwell, 1996Global climate and infectious diseases: the cholera paradigm. Sci. 2742025203 1

20. A. M Ayulo, R Machado, V Scussel, 1994Entero toxigenic Escherichia coli and Staphylococcus aureus in fish and seafood from the southern region of Brazil. Int. J. Food Microbiol. 24171178

21. P Chattopadhyay, 2000Fish- catching and handling. In Robinson RK. Encyclopedia of Food Microbiol. London: Academic Press; 153 p.

22. Y Asai, T Murase, R Osawa, T Okitsu, R Suzuki, S Sata, J Terajima, H Izumiya, H Watanabe, 1999Isolaton of Shiga toxin-producing Escherichia coli O157:H7 from processed salmon roe associated with the outbreakes in Japan, 1988, and a molecular typing of the isolates by pulsed-field gel electrophoresis. Kansenshogaku Zasshi. 732024

23. MitsudaT; Muto, T; Yamada, M; Kobayashi, N; Toba, M; Aihara, Y; Ito, A; Yokota, S (1998Epidemiological study of a food-borne outbreak of enterotoxigenic Escherichia coli O25:NM by pulsed-field gel electrophoresis and randomly amplified polymorphic DNA analysis.J. Clin. Microbiol. 36652656

24. Vieira RHSFRodrigues DP, Gocalves FA, Menezes FGR, Aragao JS, Sousa OV (2001Microbicidal effect of medicinal plant extracts (Psidium guajava Linn. and Carica papaya Linn.)upon bacteria isolated from fish muscle and known to induce diarrhea in childrednRev. Inst. Med. trop. S. Paulo 433145148

25. D Pierard, N Crowcroft, S De Bock, D Potters, G Crabbe, VLF, Lauwers S

(1999A case-control study of sporadic infection with O157 and non-O157 verocytotoxin-producing Escherichia coli.Epid. Infec. 122359365

26. J. J Semanchek, D. A Golden, 1998Influence of growth temperature on inactivation and injury of Escherichia coli O157:H7 by heat, acid, and freeezingJ. Food Prot. 61395401

27. J. H Isonhood, M Drake, 2002Aeromonas species in foods.J. Food Prot. 65575582

28. N. P Pogorelova, L. A Zhuravleva, Ibragimov FKH, Iushchenko GV (1995Bacteria of the genus Aeromonas as the causative agents of saprophytic infection. Zh Mikrobiol Epidemiol Immunobiol. 4912

29. C. F Fernandes, G. J Flick, T. B Thomas, 1998Growth of inoculated psychrotrophic pathogens on refrigerated fillets of aquacultured rainbow trout and channel catfish.J. Food Prot. 613313317

30. L Novotny, R Halouzka, L Matlova, O Vavra, L Dvorska, M Bartos, 2010Morphology and distribution of granulomatous inflammation in freshwter oramental fish infected with mycobacteria. J. Fish Dis. 33947955

31. I. A Olgunoglu, 2012Salmonella in Fish and Fishery Products. In Mahmoud BSM. Salmonella- A Dangerous Foodborne Pathogen.: InTech; 2012.

32. M. L Heinitz, R. D Ruble, D. E Wagner, S. R Tatini, 2000Incidence of Salmonella in fish and seafood.J. Food Prot. 635579592

33. Vieira RHSFRodrigues DP, Gocalves FA, Menezes FGR, Aragao JS, Sousa OV (2001Microbicidal effect of medicinal plant extracts (Psidium guajava Linn. and Carica papaya Linn.) upon bacteria isolated from fish muscle and known to induce diarrhea in children. Revista do Instituto de Medicina Tropical da Sao Paulo. 43145148

34. M. W Eklund, M. E Peterson, F. T Poysky, R. N Paranjpye, G. A Pelroy, 2004Control of bacterial pathogens during processing of cold-smoked and dried salmon strips.J. Food Prot. 67347351

35. G. A Francis, O Beirne, D (1998Effects of the indigenous microflora of minimallu processed lettuce on the survival and growth of L. monocytogenes. Int. J. Food. Sci, Tech. 33477488

36. V. F Alves, De Martinis ECP, Destro MT, Vogel BF, Gram L (2005Antilisteral activity of a Carnobacterium piscicola isolated from brazilian smoked fish (Surubim (Pseudoplatystoma sp.)) and its activity against a persistent strain of Listeria monocytogenes isolated from surubim. J. Food Prot.1120682077

37. M Zunabovic, K Domig, W Kneifel, 2011Practical relevance of methodologies for detecting and tracing of Listeria monocytogenes in ready-to-eat foods and manufacture environments- A review. LWT- Food Sci. Tech. 442351362

38. H. H Huss, L. V Jorgensen, B. F Vogel, 2000Control options for Listeria monocytogenes in seafoods. Int. J. Food Microbiol. 62326774

39. S Gudmundsdóttir, B Gudbjörnsdottir, H Lauzon, H Einarsson, K. G Kristinsson, M Kristjansson, 2005Tracing Listeria monocytogenes isolates from cold smoked salmon and its processing environment in Iceland using pulsed-field gel electrophoresisIntJ. Food Microbiol. 1014151

40. D. O Bayles, B. A Annous, B. J Wilkinson, 1996Cold stress proteins induced in Listeria monocytogenes in response to temperature downshock and growth at low temperatures.Appl. Environ. Microbiol. 6211161119

41. H Miettinen, G Wirtanen, Prevalence and location of Listeria monocytogenes in farmed rainbow trout (2005Int. J. Food Microbiol. 104135143

42. W Tham, H Ericsson, S Loncarevic, H Unnerstad, M. L Danielsson-tham, 2000Lessons from an outbreak of listeriosis related to vacuum-packed gravad and cold-smoked fishInt. J. Food Microbiol. 623173175

43. Fonnesbech Vogel BHuss HH, Ojeniyi B, Ahrens P, Gram L (2001Elucidation of Listeria monocytogenes contamination routes in cold-smoked salmon processing plants detected by DNA-based typing methods.Appl. Environ. Microbiol. 67(6): 25862595

44. A. D Hoffman, K. L Gall, D. M Norton, M Wiedmann, 2003Listeria monocytogenes contamination patterns for the smoked fish processing environment and for raw fish.J. Food Prot. 66: 652670

45. J Thimothe, Kerr Nightingale K, Gall K, Scott VN, Wiedmann M (2004Tracking of Listeria monocytogenes in smoked fish processing plants.J. Food Prot. 67328341

46. M. W Peck, 1997Clostridium botulinum and the safety of refrigerated processed foods of extended durabilityTrend.Food Sci. Tech. 8186192

47. C. L Hatheway, 1995Hath Botulism: the present status of the disease. Curr. Top. Microbiol. Imm. 195.

48. J Haagsma, 1991The distribution of Pathogenic anaerobic bacteria and the environment. Sci. Technic. Rev. Office Int. des. 1049764

49. H Sramova, C Benes, 1998Occurrence of botulism in the Czech Republic (in Czech). Zpravy CEM (SZU Praha). 7395397

50. G. H Zhou, X. L Xu, Y Liu, 2010Preservation technologies for fresh meat. Meat Sci.86119128

51. F Devlieghere, L Vermeiren, J Debevere, 2004New preservation technologies: Possibilities and limitationsInt. Dairy J.14273285

52. S Rodgers, 2001Preserving non-fermented refrigerated foods with microbial cultures- a review. Trends. Food Sci. Tech. 12276284

53. M. F Pilet, F Leroi, 2011Applications of protective cultures, bacteriocins and bacteriophages in fresh seafood and seafood product. In Lacroix C. Protective cultures, antimicrobial metabolites and bacteriophages for food and beverage biopreservation.: © 2011 Woodhead Publishing Limited.

54. A Galvez, H Abriouel, N Benomar, R Lucas, 2010Microbial antagonists to food-borne pathogens and biocontrol.Cur. Opin. Biotech. 21142148

55. P Garcia, L Rodriguez, A Rodriguez, B Martinez, 2010Food biopreservation:Promising strategies using bacteriocins, bacteriophage and endolysins.Trends. Food Sci. Tech. 373382

56. A Brillet, M. F Pilet, H Prévost, M Cardinal, F Leroi, 2005Effect of inoculation of inoculation of Carnobacterium divergens 41a biopreservative strain against Listeria monocytogenes risk, on the microbiological, and sensory quality of cold-smokedsalmon. Int. J. Food Microbiol. 104: 309-324.

57. A. L Pinto, M Fernandes, C Pinto, H Albano, F Castilho, P Teixeira, P. S Gibbs, 2009Characterization od anti- Listeria bacteriocins isolated from shellfish. Int. J. Microbiol. 1295058

58. F Leroi, 2010Occurrence and role of lactic acid bacteria in seafood productsFood Microbiol. 27698709

59. F Mozzi, R. R Raya, G. M Vignolo, editors (2010Biotechnology of Lactic Acid Bacteria: Novel Applications: Blackwell Publishing.

60. E Stiles, 1996Biopreservation by lactic acid bacteria.Antonie van Leeuwenhoek70331345

61. M Ghanbari, M Rezaei, M Jami, M Nazari, 2010Isolation and characterization of Lactobacillus species from intestinal content of Beluga(Huso huso) and persian sturgeon (Acipenser persicus).Iran. J. Vet. Res. 102152157

62. B Mayo, T Aleksandrzak- Piekarczyk, M Fernández, M Kowalczyk, P Álvarez- Martín, J Bardowski, 2010Updates in the Metabolism of Lactic Acid Bacteria. In Mozzi F, Raya RR, Vignolo GM, editors. Biotechnology of Lactic Acid Bacteria: Novel Applications.: Blackwell Publishing.

63. E Ringo, F Gatesoupe, 1998Lactic acid bacteria in fish: a reviewAquacult. 160177203

64. W. P Hammes, R. F Vogel, 1995The genus Lactobacillus Glasgow: Blackie Academic & Professional.

65. F Duffes, C Corre, F Leroi, X Dousset, P Boyaval, 1999Inhibition of Listeria monocytogenes by in situ produced and semipurifi ed bacteriocins of Carnobacterium spp. on vacuum-packed, refrigerated. J. Food Prot. 623941403

66. C Campos, O Rodríguez, P Calo-mata, M Prado, J Barros-velazquez, 2006Preliminary characterizationof bacteriocins from Lactococcus lactis, Enterococcus faecium and Enterococcus mundtii strains isolated from turbot (Psetta maxima). Food Res. Int. 3935664

67. D Drider, G Fimland, Y Hechard, L Mcmullen, H Prevost, 2006The continuing story of class IIa bacteriocinsMicrobiology and Molcular Biology Reviews. 70564582

68. A Ouwehand, S Vesterlund, 2004Antimicrobial Components from Lactic Acid BacteriaIn Salminen S, Wright v, Ouwehand A. Lactic Acid Bacteria Microbiological and Functional Aspects.: Marcel Dekker, Inc.

69. P. D Cotter, C Hill, R. P Ross, 2005Bacteriocins: Developing innate immunity for food. Nat. Rev. Microbiol 3777788

70. I Nes, S Yoon, D Diep, 2007Ribosomally Synthesiszed Antimicrobial Peptides (Bacteriocins) in Lactic Acid Bacteria: A ReviewFood Sci. Biotech.. 165675690

71. P. D Cotter, L. A Draper, E. M Lawton, O Mcauliffe, C Hill, R. P Ross, 2006Overproduction of wild-type and bioengineered derivatives of the lantibiotic lacticin 3147. Appl. Environ. Microbiol. 7244924496

72. O Gillor, A Etzion, M Riley, 2008The dual role of bacteriocins as anti- and probioticsAppl. Microbiol. Biotech.. 81591606

73. I Nes, 2011History, Current Knowledge, and Future Directions on Bacteriocin Research in Lactic Acid Bacteria. In Drider D, RS, (eds.). Prokaryotic Antimicrobial Peptides: From Genes to Applications.: Springer Science+Business Media, LLC 312

74. S. E Lindgren, W. J Dobrogosz, 1990Antagonistic activities of lactic acid bacteria in food and feed fermentations.FEMS Microbiol. Lett. 87(1-2): 149 EOF63 EOF

75. P. K Podolak, J. F Zayas, C. L Kastner, Fung DYC (1996Inhibition of Listeria monocytogenes and Escherichia coli O157:H7 on beef by application of organic acidsJ. Food Prot. 59370373

76. M. S Ammor, B Mayo, 2007Selection criteria for lactic acid bacteria to be used as functional cultures in dry sausage production: An update. Meat Sci. 76: 138–146.

77. F Devlieghere, J Debevre, 2000Influence of dissolved carbon dioxide on the growth of spoilage bacteriaLebensmittel- und Wissenschaft-Technologie. 33531537

78. E Lanciotti, C Santini, E Lupi, D Burrini, 2003Actinomycetes, cyanobacteria and algae causing tastes and odours in water of the River Arno used for the water supply of Florence. J. Water Sup. Res. Tech. 527489500

79. E. I Kvasnikov, N. K Kovalenko, L. G Materinskaya, 1997Lactic acid bacteria of freshwater fish. Microbiol. 46619624

80. Y Cai, P Suyanandana, P Saman, 1999Classification and characterization of lactic acid bacteria isolated from the intestines of common carp and freshwater prawnsThe J. Gen. Appl. Microbiol. 45177184

81. H. H Huss, V. F Jeppesen, C Johansen, L Gram, 1995Biopreservation of fish products a review of recent approaches and resultsJ. Aquat. Food. Prod. Tech. 4526

82. C. J González, J. P Encinas, M. L García-lópez, A Otero, 2000Characterization and identification of lactic acid bacteria from freshwater fishesFood Microbiol. 17383391

83. A Bucio, R Hartemink, J. W Schrama, J Verreth, F. M Rombouts, 2006Presence of lactobacilli in the intestinal content of freshwater fish from a river and from a farm with a recirculation systemFood Microbiol. 235476482

84. E Ringo, E Strom, 1994Microflora of Arctic char, Salvelinus alpinus (L.); gastrointestinal microflora of free-living fish, and effect of diet and salinity on the intestinal microflora. Aquacult. Fish. Manag. 25623629

85. E Ringo, 2004Lactic acid bacteria in fish and fish farmingIn Salminen S, Wright A, Ouwehand A, editors. Lactic acid bacteria : Microbiological and Functional Aspects. 3rd ed. New-York: CRC Press; 581610

86. E Ringo, E Strom, J. A Tabachek, 1995Intestinal microflora of salmonids: a review. Aqua. Res. 26773789

87. E Ringo, R. E Olsen, 1999The effect of diet on aerobic bacterial flora associated with intestine of Arctic charr (Salvelinus alpinus L.). J. Appl. Microbiol. 861228

88. B Spanggaard, I Huber, J Nielsen, T Nielsen, K. F Appel, L Gram, 2000The microflora of rainbow trout intestine. A comparison of traditional and molecular identificationAquacult. 182115

89. M Seppola, R. E Olsen, E Sandaker, P Kanapathippillai, W Holzapfel, E Ringo, 2006Random amplification of polymorphic DNA (RAPD) typing of carnobacteria isolated from hindgut chamber and large intestine of Atlantic cod (Gadus morhua L.). Sys. Appl. Microbiol. 29: 131137

90. F Gancel, F Dzierszinski, R Tailliez, 1997Identification and characterization of Lactobacillus species isolated from fillets of vacuum-packed smocked and salted herring (Clupea harengus). J.appl. Microbiol. 82722728

91. H Magnússon, K Traustadóttir, 1982The microbial flora of vacuum-packed smoked herring fillets. J. Food Tech. 17695702

92. C Paludan-müller, P Dalgaard, H Huss, L Gram, 1998Evaluation of the role of Carnobacterium piscicola in spoilage of vacuum and modified atmosphere-packed-smoked salmon stored at 5°C. Int.J. Food Microbiol. 39155166

93. F Leroi, J. J Joffraud, F Chevalier, M Cardinal, 1998Study of the microbial ecology of cold smoked salmon during storage at 8°CInt. J. Food Microbiol. 39111121

94. J Leisner, B Laursen, H Prevost, D Drider, P Dalgaard, 2007Carnobacterium:positive and negative effects in the environment and in foods. FEMS Microbiol. Rev. 13592613

95. J. J Leisner, J. C Millan, H. H Huss, L. M Larsen, 1994Production of histamine and tyramine by lactic acid bacteria isolated from vacuum-packed sugar-salted fish.J. Appl. Bacter. 76417423

96. A Ostergaard, Ben Embarek PK, Wedel-Neergaard C, Huss HH, Gram L (1998Characterization of anti-listerial lactic acid bacteria isolated from Thai fermented fish productsFood Microbiol. 15223233

97. M Olympia, H Ono, A Shinmyo, M Takano, 1992Lactic acid bacteria in fermented fishery, burong bangus. J. Fer.Bioeng. 733193197

98. S Mauguin, G Novel, 1994Characterization of lactic acid bacteria isolated from seafoodJ. Appl. Bacter. 76616625

99. J Emborg, B. G Laursen, T Rathjen, P Dalgaard, 2002Microbial spoilage and formation of biogenic amines in fresh and thawed modified atmosphere-packed salmon (Salmo salar) at 2 degrees C.J. Appl. Microbiol. 92(4). 790799

100. L Franzetti, M Scarpellini, D Mora, A Galli, 2003Carnobacterium spp. in seafood packaged in modified atmosphereAnnal. Microbiol. 53189193

101. J Emborg, B. G Laursen, T Rathjen, P Dalgaard, 2002Microbial spoilage and formation of biogenic amines in fresh and thawed modified

atmosphere-packed salmon (Salmo salar) at 2°C. J. Appl. Microbiol. 92790799

102. P Dalgaard, H. L Madsen, N Samieian, J Emborg, 2006Biogenic amine formation and microbial spoilage in chilled garfish (Belone belone) effect of modified atmosphere packaging and previous frozen storage. J. Appl. Microbiol. 1018095

103. R Lakshmanan, P Dalgaard, 2004Effect of high-pressure processing on Listeria monocytogenes, spoilage microflora and multiple compound quality indices in chilled cold-smoked salmon. J. Appl. Microbiol. 96398408

104. S Wessels, H. H Huss, 1996Suitability of Lactococcus lactis ATCC 11454 as a protective culture for lightly preserved fish products. Food Microbiol. 13323332

105. L Nilsson, Y. Y Ng, J. N Christiansen, B. L Jorgensen, D Grotinum, L Gram, 2004The contribution of bacteriocin to inhibition of Listeria monocytogenes by Carnobacterium piscicola strains in cold-smoked salmon systemsJ. Appl. Microbiol. 96133143

106. C Altieri, B Speranza, Del Nobile MA, Sinigaglia M (2005Suitability of bifidobacteria and thymol as biopreservatives in extending the shelf life of fresh packed plaice filletsJ. Appl. Microbiol. 9912941302

107. L. J Yin, C. W Wu, S. T Jiang, 2007Biopreservative effect of pediocin ACCEL on refrigerated seafoodFish. Sci. 73907912

108. E Ringo, (2008) The ability of carnobacteria isolated from fish intestine to inhibit growth of fish pathogenic bacteria. Aqua. Res. 39 171180 .

109. K. M Sudalayandi, 2011Efficacy of lactic acid bacteria in the reduction of trimethylamine-nitrogen and related spoilage derivatives of fresh Indian mackerel fish chunksAfr. J. Biotech. 104247

110. C. R Kim, J. O Hearnsberger, 1994Gram negative bacteria inhibition by lactic acid culture and food preservatives on catfish fillets during refrigerated storageJ. Food Sci. 59513516

111. H Einarsson, H. L Lauzon, 1995Biopreservtaion of brined shrimp (Pandalus borealis) by bacteriocins from lactic acid bacteria. Appl. Environ. Microbiol. 61669675

112. M Morzel, N. G Fransen, E. K Arendt, 1997Defined starter cultures for fermentation of salmon fillets. J.Food Sci. 62612141217

113. D Kişla, A Ünlütürk, 2004Microbial shelf life of rainbow trout fillets treated with lactic culture and lactic acidAdv. Food Sci. 261720

114. F Elotmani, O Assobhei, 2004In vitro inhibition of microbial flora of fish by nisin and lactoperoxidase systemLett. Appl. Microbiol. 386065

115. Aras Husar SKaban G, Husar O, Yanik T, Kaya M (2005Effect of Lactobacillus sakei Lb706 on Behavior of Listeria monocytogenes in Vacuum-Packed Rainbow Trout Fillets. Tur. J. Vet. Anim. Sci. 2910391044

116. Y Kim, T Ohta, T Takahashi, A Kushiro, K Nomoto, T Yokokura, N Okada, H Danbara, 2006Probiotic Lactobacillus casei activates innate immunity via NF-κB and 38MAP kinase signaling pathways.Microb. Infec.. 8: 994-1005.

117. P Katikou, Ambrosiadis IGD, Koidis P, Georgakis SA 2(2007Effect of Lactobacillus cultures on microbiological, chemical and odour changes during storage of rainbow trout filletsJ. Sci. Food Agri. 87477484

118. S. M Daboor, S. M Ibrahim, 2008Biochemical and microbial aspects of tilapia (Oreochromis niloticus L.) biopreserved by Streptomces sp. metabolites. In 4th International Conference of Veterinary Research Division, National Research Center (NRC); Cairo, Egypt. 3949

119. M Tahiri, E Desbiens, C Kheadr, I. F Lacroix, 2009Comparison of different application strategies of divergicin M35 for inactivation of Listeria monocytogenes in cold-smoked wild salmon.Food Microbiol. 26: 783793

120. S Matamoros, M. F Pilet, F Gigout, H Prévost, F Leroi, 2009evaluation of seafood-borne psychrotrophic lactic acid bacteria as inhibitors of pathogenic and spoilage bacteria. Food Microbiol. 26638644

121. P. A Fall, F Leroi, M Cardinal, F Chevalier, M. F Pilet, 2010Inhibition of Brochothrix thermosphacta and sensory improvement of tropical peeled cooked shrimp by Lactococcus piscium CNCM I-4031.Lett. Appl. Microbiol. 50357361

122. S. M Ibrahim, G. D Salha, 2009Effect of antimicrobial metabolites produced by lactic acid bacteria on quality aspects of frozen Tilapia (Oreochromis niloticus) fillets. World Journal of Fish and Marine Sciences. 14045

123. A. R Shirazinejad, I Noryati, A Rosma, I Darah, 2010Inhibitory Effect of Lactic Acid and Nisin on Bacterial Spoilage of Chilled ShrimpWorld Acad. Sci. Eng. Tech.

124. P. A Fall, F Leroi, F Chevalier, C G Pilet, MF (2010Protective effect of a non-bacteriocinogenic Lactococcus piscium CNCM I-4031 strain against Listeria monocytogenes in sterilised tropical cooked peeled shrimp. J. Aquat. Food Prod. Tech. 198492

125. S Cosansu, S Mol, Ucok Alakavuk D, Tosun ŞY (2011Effects of Pediococcus spp. on the quality of vacuum-packed Horse Mackerel during Cold Storage. J. Agri. Sci. 175966

126. T Katla, T Moretro, I. M Aasen, A Holck, L Axelsson, K Naterstad, 2001Inhibition of Listeria monocytogenes in cold smoked salmon by addition of sakacin P and/or live Lactobacillus sakei culturesFood Microbiol. 18431439

127. H Blom, T Katla, B. F Hagen, L Axelsson, 1997A model assay to demonstrate how intrinsic factors affect diffusion of bacteriocins.Int. J. Food Microbiol. 38103109

128. M. B Brurberg, I. F Nes, Eijsink VGH (1997Pheromone-induced production of antimicrobial peptides in Lactobacillus.Mol. Microbiol. 26347360

129. Eijsink VGHSkeie M, Middelhoven H, Brurberg MB, Nes IF (1998Comparative studies of pediocin-like bacteriocins.. Appl. Environ. Microbiol. 6432753281

130. M. G Ganzle, S Weber, W. P Hammes, 1999Effect of ecological factors on the inhibitory spectrum and activity of bacteriocinsInt. J. Food Microbiol. 46207217

131. I. M Aasen, T Moretro, T Katla, L Axelsson, I Storro, 2000Infuence of complex nutrients, temperature and pH on bacteriocin production by Lactobacillus sakei CCUG 42687. Appl. Microbiol. Biotech.. 53159166

132. F Leroi, N Arbey, J Joffraud, F Chevalier, 1996Effect of inoculation with lactic acid bacteria on extending the shelf-life of vacuum-packed cold-smoked salmonInt. J. Food Sci. Tech. 1996; 31: 497504

133. B Budu-amoako, R. F Albert, J Harris, J Delves-broughton, 1999Combined effect of nisin and moderate heat on destruction of Listeria monocytogenes in cold-pack lobster meat.J. Food Prot. 624650

134. L Nilsson, L Gram, H Huss, 1999Growth control of Listeria monocytogenes on cold smoked salmon using a competitive lactic acid bacteria flora.J. Food Prot. 62336342

135. A Nykanen, K Weckman, A Lapvetelainen, 2000Synergistic inhibition of Listeria monocytogenes on cold-smoked rainbow trout by nisin and sodium lactateInt. J. Food Microbiol. 616372

136. J Silva, A. S Carvalho, P Teixeira, P. A Gibbs, 2002Bacteriocin production by spray-dried lactic acid bacteria.Lett. Appl. Microbiol. 3427781

137. A Bouttefroy, J. B Milliere, 2000Nisin-curvaticin 13 combinations for avoiding the regrowth of bacteriocin resistant cells of Listeria monocytogenes ATCC 15313. Int. J. Food Microbiol. 626575

138. K Yamazaki, M Suzuky, Y Kawai, N Inoue, T. J Montville, 2003Inhibition of Listeria monocytogenes in cold-smoked salmon by Carnobacterium piscicola CS526 isolated from frozen surimi.J. Food Prot. 6614201425

139. K Yamazaki, M Suzuki, Y Kawai, I. N Montville, T. J Purification, and characterization of a novel class IIa bacteriocin, piscicocin CS526, from surimi associated Carnobacterium piscicola CS526. Appl. Environ. Microbiol. 71: 554557

140. A Brillet, M. F Pilet, H Prevost, A Bouttefroy, F Leroi, 2004Biodiversity of Listeria monocytogenes sensitivity to bacteriocin-producing Carnobacterium strains and application in sterile cold-smoked salmonJ. Appl. Microbiol. 9710291037

141. A Weiss, W. P Hammes, 2006Lactic acid bacteria as protective cultures against Listeria spp. on cold-smoked salmonEur. Food Res. Tech. 222343346

142. M Vescovo, G Scolari, C Zacconi, 2006Inhibition of Listeria innocua growth by antimirobial-producing lactic acid cultures in vacuum-packed cold-smoked salmon. Food Microbiol. 23689693

143. B. G Laursen, L Bay, I Cleenwerck, M Vancanneyt, J Swings, P Dalgaard, 2005Carnobacterium divergens and Carnobacterium maltaromicum as spoilers or protective cultures in meat and seafood: phenotypic and genotypic characterisation. Sys. Appl. Microbiol. 28151164

144. E Tomé, V. L Pereira, C. I Lopes, P. A Gibbs, P. C Teixeira, 2008In vitro tests of suitability of bacteriocin-producing lactic acid bacteria, as potential biopreservation cultures in vacuum-packaged cold-smoked salmonFood Control19535543

145. S Matamoros, F Leroi, M Cardinal, F Gigout, Kasbi Chadli F, Cornet J, Prevost F, Pilet M.F (2009Psychrotrophic lactic acid bacteria used to improve the safety and quality of vacuum-packaged cooked and peeled tropical shrimp and cold smoked salmon.J. Food Prot. 72365374

Chapter 2

POTENTIAL OF THE VIRION-ASSOCIATED PEPTIDOGLYCAN HYDROLASE HYDH5 AND ITS DERIVATIVE FUSION PROTEINS IN MILK BIOPRESERVATION

Lorena Rodrı´guez-Rubio[1] , Beatriz Martı´nez[1] , David M. Donovan[2] , Pilar Garcı´a[1] , Ana Rodrı´guez[1]

[1] DairySafe Group, Department of Technology and Biotechnology of Dairy Products, Instituto de Productos La´cteos de Asturias (IPLA), Consejo Superior de Investigaciones Cientı´ficas (CSIC), Villaviciosa, Asturias, Spain

[2] Animal Biosciences and Biotechnology Laboratory, Animal and Natural Resources Institute, Beltsville Agricultural Research Center (BARC), Agricultural Research Service (ARS), USDA, Beltsville, Maryland, United States of America

ABSTRACT

Bacteriophage lytic enzymes have recently attracted considerable interest as novel antimicrobials against Gram-positive bacteria. In this work, antimicrobial activity in milk of HydH5 [a virion-associated peptidoglycan hydrolase (VAPGH) encoded by the *Staphylococcus aureus* bacteriophage vB_SauSphiIPLA88], and three different fusion proteins created between HydH5 and lysostaphin has been assessed. The lytic activity of the five proteins (HydH5, HydH5Lyso, HydH5SH3b, CHAPSH3b and lysostaphin) was confirmed using commercial whole extended shelf-life milk (ESL) in challenge assays with 10^4 CFU/mL of the strain *S. aureus* Sa9. HydH5, HydH5Lyso and HydH5SH3b (3.5 μM) kept the staphylococcal viable counts below the control cultures for 6 h at 37°C. The effect is apparent just 15 minutes after the addition of the lytic enzyme. Of note, lysostaphin and CHAPSH3b showed the highest staphylolytic protection as they were able to eradicate the initial staphylococcal challenge immediately or 15 min after addition, respectively, at lower concentration (1 μM) at 37°C. CHAPSH3b showed the same antistaphyloccal effect at room temperature (1.65 μM). No re-growth was observed for the remainder of the experiment (up to 6 h). CHAPSH3b activity (1.65 μM) was also assayed in raw (whole and skim) and pasteurized (whole and skim) milk. Pasteurization of milk clearly enhanced CHAPSH3b staphylolytic activity in both whole and skim milk at both temperatures. This effect was most dramatic at room temperature as this protein was able to reduce *S. aureus* viable counts to undetectable levels immediately after addition with no re-growth detected for the duration of the experiment (360 min). Furthermore, CHAPSH3b protein is known to be heat tolerant and retained some lytic activity after pasteurization treatment and after storage at 4°C for 3 days. These results might facilitate the use of the peptidoglycan hydrolase HydH5 and its derivative fusions, particularly CHAPSH3b, as biocontrol agents for controlling undesirable bacteria in dairy products.

INTRODUCTION

Staphylococcus aureus is a bacterial pathogen responsible for a wide range of human and animal infections, including food poisoning caused by the ingestion of enterotoxins produced in food by enterotoxigenic strains [1], [2]. Staphylococcal enterotoxins are notoriously thermostable and maintain their stability even after the thermal treatments customarily utilized in the food industry. This represents a threat to consumers and makes necessary the control of staphylococcal contaminants to avoid the production of high risk levels of enterotoxins [3].

Humans and domestic animals are the primary reservoirs of *S. aureus*, as this microorganism colonizes mucous membranes and skin. Thus, food handlers and animals are usually the primary source of *S. aureus* contamination of food products of animal origin [4]. *S. aureus* is also an important etiological agent of mastitis in cattle, goats and sheep [5], with the mastitic udder being a source of contaminated milk and milk-derived dairy products, along with the dairy farm environment and processing facilities [6]. Although an important food safety concern, *S. aureus* mastitis is difficult to eradicate and constitutes a serious economic problem for dairy herd management [7]. Several antimicrobial treatments are available for clinical mastitis differing in the antimicrobial agent, route of application, duration, probability of cure or recurrence, and cost [8]. However, this problem remains unsolved in part due to the ability of *S. aureus* to invade and reside intracellularly [9], within mammary cells, thereby evading most antibiotics, but also because of the high frequency of antibiotic resistance among *S. aureus*strains [10], [11].

Bacteriophage endolysins have been proposed as antimicrobials to control Gram positive bacteria due to their ability to degrade the bacterial cell wall resulting in lysis of the pathogen[12]. This bactericidal activity has been successfully used to control antibiotic-resistant pathogenic bacteria in animal models [13]. For instance, the pneumococcal lysin Cpl-1 protected a mouse model against pneumococcal bacteraemia and colonization by intravenous administration and topical nasal treatment, respectively [14]. More recently, staphylococcal lysins have also been used against staphylococcal infections in mouse models. This is the case for endolysins MV-L [15], LysGH14 [16] or the chimeric lysin ClyS [17] that protected mice against lethal doses of methicillin-resistant *S. aureus* (MRSA) by intraperitoneal injections. The effectiveness shown by the staphylococcal phage K endolysin, LysK, CHAP domain construct and the CHAP domain construct from the phage K tail-associated muralytic enzyme to eliminate *S. aureus* from the nares of challenged mice and rats, respectively, supports the potential use of phage lytic proteins' catalytic domains as antimicrobials [18], [19].

Lysostaphin is a well characterized peptidoglycan hydrolase produced by *Staphylococcus simulans* biovar. *staphylolyticus*. Its lytic action against *S. aureus* relies mainly on its N- terminal domain with glycylglycine endopeptidase activity that cleaves the pentaglycine cross bridges present in staphylococcal peptidoglycan, while its C- terminal domain promotes its specific binding to staphylococcal peptidoglycan [20]. It was shown to protect mammary glands against*S. aureus* challenge in both mice [21] and cattle [22]. It has also been shown that antimicrobial synergy exists *in vitro* between some phage endolysins and antibiotics or antimicrobial peptides of bacterial origin against *S. aureus* [18], [23], [24]. In this regard, the *in vitro* synergy observed between phage lytic proteins and lysostaphin was recently expanded to include the *in vivo*protection of murine mammary glands from an *S. aureus* challenge [25].

In addition to mammary gland protection, phage endolysins might also serve to inhibit undesirable bacterial growth for food biocontrol purposes [24]. The staphylococcal phage vB_SauS-phiIPLA88 endolysin, LysH5, has been demonstrated to control *S. aureus* growth in milk. The purified protein was able to rapidly kill *S. aureus* growing in pasteurized milk with a 10^6 CFU/ ml inoculum undetectable after 4 h of co-incubation with 1.6 µM LysH5 at 37°C [26].

In addition to endolysins, there is a largely untapped group of phage lytic proteins the virion-associated peptidoglycan hydrolases (VAPGHs) that are involved in local cell-wall degradation to facilitate the injection of phage DNA into the cell cytoplasm [27]. These PGHs have been reported to be encoded by phages infecting *S. aureus* [28], [29] and other bacterial species[30]–[33]. Their antimicrobial activity was first postulated in 1940, with 'lysis from without' that takes place when a very high number of phages are adsorbed onto the host cell [34].

In a previous work, we have described the VAPGH HydH5 encoded by the *S. aureus* phage vB_SauS-phiIPLA88. Bioinformatic analysis of the protein sequence revealed two putative domains, a cysteine, histidine-dependent amidohydrolase/peptidase (CHAP) domain [35], [36]; and a LYZ2 (lysozyme subfamily 2) domain [28], which conferred antimicrobial activity against *S. aureus* [37]. Three different fusion proteins obtained between lysostaphin and HydH5 (CHAPSH3b, HydH5SH3b and HydH5Lyso) showed significantly greater activity than the parental protein HydH5, and lysed both bovine and human *S. aureus*, including MRSA N315, and human *Staphylococcus epidermidis* strains [38].

In this work, we have assessed the antimicrobial ability of HydH5 and its derivative fusion proteins in milk, in order to explore new biopreservation strategies to effectively inhibit *S. aureus* growth in dairy products.

MATERIALS AND METHODS

Bacterial Strains and Culture Conditions

S. aureus Sa9, isolated from a mastitic milk sample, was used as the indicator strain for lytic activity [39]. This bacterium was grown in TSB broth (Tryptic Soy Broth, Difco, Franklin Lakes, NJ, US) at 37°C for up to 18 h with vigorous shaking. For selective counting Baird–Parker agar supplemented with egg yolk tellurite (Scharlau Chemie, S.A. Barcelona, Spain) was used in commercial whole extended shelf life (ESL) (125°C, 4 sec) milk samples, and ChromoID *S. aureus* plates (Biomérieux, Marcy l'Etoile, France) in raw and pasteurized (72°C, 15 sec) (whole and skim) milk samples. *E. coli* BL21(DE3)/pLysS containing the pET21a-HydH5 and derivative plasmids were used to overproduce the lytic proteins HydH5, HydH5SH3b, HydH5Lyso and CHAPSH3b [38].

Microbiological and Physicochemical Analyses of Milk

Microbiological and physicochemical analyses were performed in commercial cow's whole ESL milk whole and skim raw milk (the latter was centrifuged at 6,000×g for 20 min to remove fat) and whole and skim pasteurized milk supplied by a collaborating farm. Samples of milk (500 ml) were aseptically sampled. Serial dilutions of milk were made in quarter-strength Ringer solution (Oxoid, Basingstoke, Hampshire, UK) and plated in duplicate on the appropriate agar medium. Total bacterial counts were performed in the different types of milk by deep-plating appropriate dilutions on Plate Count Agar (32°C, 72 h). *S. aureus* counting was performed as indicated above.

Total solids, fat and protein content were determined according to the International Dairy Federation [40]–[42].

Protein Purification

Protein purification was performed as previously described [38]. Purity of each preparation was determined in 15% (w/v) SDS-PAGE gels. Electrophoresis was conducted in Tris–Glycine buffer at 20 mA for 1 h in the BioRad Mini-Protean gel apparatus. Protein was quantified by the Quick Start Bradford Protein Assay (BioRad, Hercules, CA). Quantification of lytic activity was performed by turbidity reduction assays against live *S. aureus* Sa9 cells prepared as previously described [43], [44].

Challenge Tests in Milk

HydH5, HydH5SH3b or HydH5Lyso proteins (3.5 µM), CHAPSH3b protein (1 µM) and lysostaphin (1 µM) were individually added to 2 ml of whole ESL milk inoculated with 10^4CFU/ml of *S. aureus* Sa9 and incubated at 37°C for 6 h. CHAPSH3b (1.65 µM) was also assayed at room temperature (RT) for the same period. The anti-staphylococcal activity of CHAPSH3b (1.65 µM) was further assayed in whole and skim raw milk and in whole and skim pasteurized milk inoculated with 10^3 CFU/ml at 37°C and RT for 2 h. Challenged milk without lytic protein additions were used as controls.

Samples were taken at different times throughout the incubation period and survival of *S. aureus* Sa9 was determined by serial dilution plating onto Baird-Parker plates for ESL milk samples (37°C, 48 h) and ChromoID *S. aureus* plates (37°C, 24 h) for raw and pasteurized milk samples, respectively. ChromoID *S. aureus* is a selective and differential culture medium for *Staphylococcus* sp, in which different staphylococcal species are distinguished by the colour of colonies (green for *S. aureus*; pink in *S. saprophyticus*; purple in *S. xylosus*; white in *S. epidermidis*). This chromogenic medium inhibits other Gram positive bacteria, Gram negative bacteria and yeasts.

CHAPSH3b Fusion Protein Stability in Milk

To test the stability of CHAPSH3b in milk, 1.65 µM protein was added to 2 ml whole raw milk and kept at 4°C for 3 days. Samples (250 µL) were taken every day, challenged with 10^3CFU/ml of *S. aureus* Sa9 and incubated at RT for 15 min. Staphylococcal viable counts in the presence and in the absence of the antimicrobial protein were determined by serial dilution plating onto ChromoID *S. aureus* plates. Results were expressed as the percentage of viable counts reduction compared to the untreated control.

CHAPSH3b Pasteurization Treatment

Commercial whole ESL milk and whole raw milk containing CHAPSH3b (1.97 µM) were pasteurized (72°C, 15 s) in a thermo cycler (BioRad Laboratories, Hercules, CA, USA). Samples were cooled at RT for 15 min, further inoculated with 10^3 CFU/ml of *S. aureus* Sa9 and incubated for 0, 15, 30, 60 and 120 minutes at RT. Staphylococcal viable counts were determined as indicated above and results also expressed as the percentage reduction of viable counts.

Statistical Analysis

Statistical analysis was performed using the SPSS-PC +11.0 software (SPSS, Chicago, IL, USA). Staphylococcal CFU data were subjected to one-way ANOVA within each sampling time. Types of anti-staphylococcal protein (HydH5, HydSH3b, HydH5lyso and CHAPSH3b) were compared against the untreated control. Data of cold storage stability of protein CHAPSH3b were compared with one-way ANOVA and the LSD test was used for a comparison of means at a level of significance $P<0.05$.

RESULTS

Microbiological and Physicochemical Characteristics of Milk

Total viable bacterial counts were below the detection limit (<10 CFU/ml) in ESL milk, whereas about 7.08×10^4 CFU/ml and 3.89×10^1 CFU/ml were detected in raw and pasteurized milk, respectively. Viable counts were lower in both raw skim (1.90×10^3 CFU/ml); and pasteurized skim milk (1.20×10^1 CFU/ml). *S. aureus* counts were only detected in whole raw milk ($2.84–4.0\times10^1$ CFU/ml) and they kept below 10^2 CFU/ml throughout 2 h of incubation.

Results of gross composition are shown in Table 1. Mean values of total solids, fat and protein contents of whole commercial ESL milk, and raw (whole and skim) and pasteurized (whole and skim) milk were within the standards of commercial and farmhouse milk.

Table 1: Gross composition of milk used in the HydH5 and its derivative fusion proteins antistaphyloccal assays[a]

ESL[b]	Whole		Skim	
	Raw	Pasteurized	Raw	Pasteurized
Total solids[c] 12.28±0.08	11.14±0.01	11.17±0.03	11.04±0.12	11.01±0,09
Fat[d] 3.54±0.02	3.18±0.03	3.20±0.05	0.1±0.02	0.15±0.04
Protein[d] 3.06±0.09	3.09±0.12	3.09±0.14	3.07±0.12	3.08±0.16

[a]Data reported are means ±standard deviations of two independent milk samples.
[b]ESL: extended shelf life milk.
[c]Total solids: data expressed as g/100 g milk.
[d]Fat/Protein: expressed as g/100 g milk.
doi:10.1371/journal.pone.0054828.t001

HydH5 and its Derivative Fusion Proteins have Antimicrobial Activity against *S. aureus* Sa9 in Commercial ESL Milk

The antimicrobial activity of HydH5, its derivative fusion proteins and lysostaphin was assessed in commercial whole ESL milk inoculated with 10^4 CFU/ml of *S. aureus* Sa9. The effect of the different proteins on *S. aureus* growth was first tested at 37°C. In the absence of antimicrobial proteins (control cultures) the staphylococcal strain grew from 10^4 to 6.5×10^4 CFU/ml during the first hour of incubation with a more robust increase in CFU subsequently, achieving 8.9×10^7 CFU/ml at the end of six hours (Fig. 1A). The addition of HydH5, HydH5SH3b and HydH5Lyso (3.5 μM) to the *S. aureus* inoculated milk resulted in an immediate effect on *S. aureus* viability, with the viable counts maintained below the time zero control counts (immediately after addition of the antimicrobials). At time 0, only the viable counts in HydH5Lyso treated cultures were significantly different ($P<0.05$) compared to the control cultures. From 15 min onwards, the inhibitory effect of each of the proteins on *S. aureus* viability was significant ($P<0.01$ at 15 min and $P<0.001$ thereafter). The greatest reduction in viable counts (about 2.34±0.01 log CFU/ml) was detected at the end of the 6 h incubation period (Fig. 1A). These activities are, however, far from lysostaphin antistaphylococcal activity since 1 μM of this bacterial peptide resulted in an immediately kill of the *S. aureus* population and no viable counts were detected even at time 0. In addition, no-re-growth was observed afterwards (data not shown). Only the fusion protein CHAPSH3b showed an inhibitory effect on *S. aureus* similar to lysostaphin since 1 µM resulted in a complete clearance of the pathogen 15 min after addition without further re-growth throughout the assay period (6 h) (Fig. 1A). Likewise, viable counts became undetectable immediately after the addition of 1.65 µM CHAPSH3b (data not shown). At RT, the inhibitory effect of CHAPSH3b decreased slightly with a higher protein concentration (1.65 µM) required to fully eliminate *S. aureus* in 15 min after addition (Fig. 1B), whereas a continuous proliferation of the staphylococcal population occurred in the control cultures.

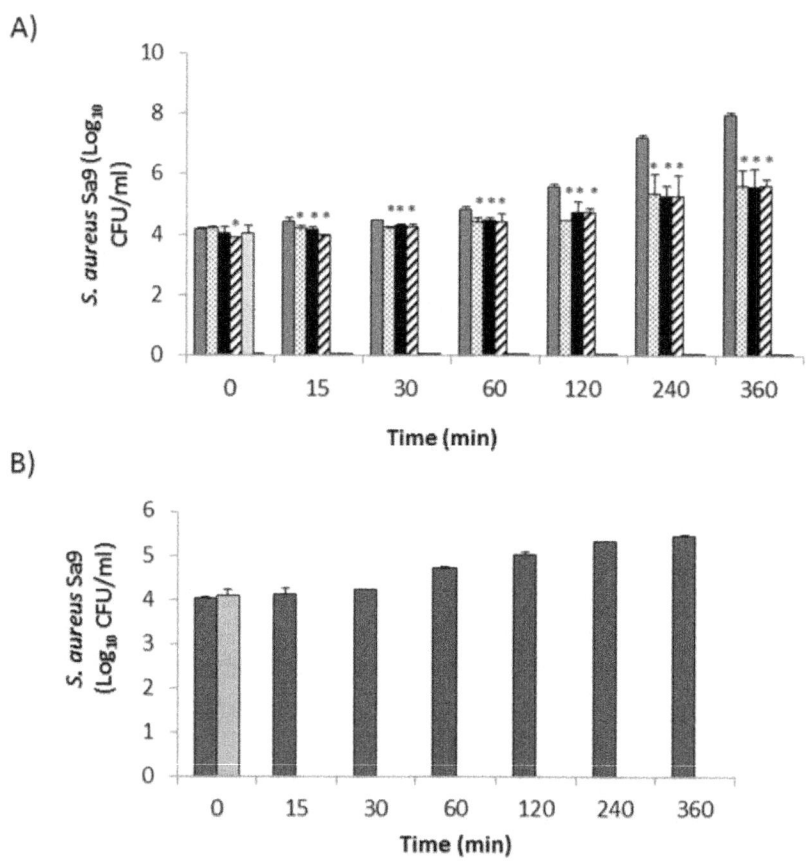

Figure 1: Antimicrobial activity of HydH5, lysostaphin, and its derivative fusion proteins in commercial whole ELS milk.

Milk was inoculated with 10^4 CFU/ml of *S. aureus* Sa9 and incubated for 0, 15, 30 min, 1, 2, 4 and 6 h at 37°C either without lytic protein (control; dark grey bars) or in the presence of : A) 1 µM CHAP-SH3b (light grey bars), and 3.5 µM HydH5 (stippled bars), 3.5 µM HydH5SH3b (black bars) and 3.5 µM HydH5Lyso (diagonal stripes bars); 1 µM lysostaphin (gross line on X axis); (B) 1.65 µM CHAPSH3b (light grey bars), control without protein (dark grey bars), at room temperature. Values expressed as $\log_{10}$ CFU/ml are the means ± standard deviations of two independent experiments. Bars having an asterisk are significantly different from the control (*$P \leq 0.05$). *S. aureus* detection threshold (<10 CFU/ml).

CHAPSH3b Fusion Protein is Effective in Raw Milk and Highly Effective in Pasteurized Milk

The effect of the CHAPSH3b protein was tested against *S. aureus* Sa9 strain (10^3 CFU/ml) in whole and skim raw milk at both 37°C and RT. As shown in Figure 2A, the staphylococcal growth observed in the whole raw milk control cultures was immediately inhibited by addition of 1.65 µM of CHAPSH3b as no viable counts were detected at 37°C or RT and re-growth was prevented for ~30 min of incubation. Thereafter, *S. aureus* growth was observed at both temperatures but CHAPSH3b treatment kept viable counts below the control counts throughout the 2 h experiment. CHAPSH3b showed higher growth inhibition at RT than at 37°C (Fig. 2A). Significant differences between control and treated cultures were observed throughout the remaining 2 h incubation period at both RT ($P<0.001$) and 37°C ($P<0.001$ at 30 min and $P<0.01$ at 60 and 120 min of sampling time). At the end of the incubation period, the presence of the antimicrobial protein resulted in a reduction of 1.09±0.12 and 0.7±0.17 log CFU/ml at RT and 37°C, respectively, compared to the control cultures. The level of indigenous *S. aureus* in raw milk was also monitored through the incubation period as an additional control. This population remained below 10^2 CFU/ml and was also sensitive to CHAPSH3b (data not shown). Similar staphylococcal growth kinetics was observed in skim raw milk in the presence of CHAPSH3b (Fig. 3A). As in whole milk (Fig. 2), re-growth also occurred after 30 min and the antimicrobial protein exhibited higher inhibitory activity at RT. Differences in staphylococcal viable counts between VAPGH-treated and control samples were significant ($P<0.001$) at both 37°C and RT. The final reduction in staphylococcal CFU was similar at 37°C (0.75±0.23 log CFU/ml) and lower (0.41±0.09 log CFU/ml) at RT than in whole raw milk (Fig. 3A).

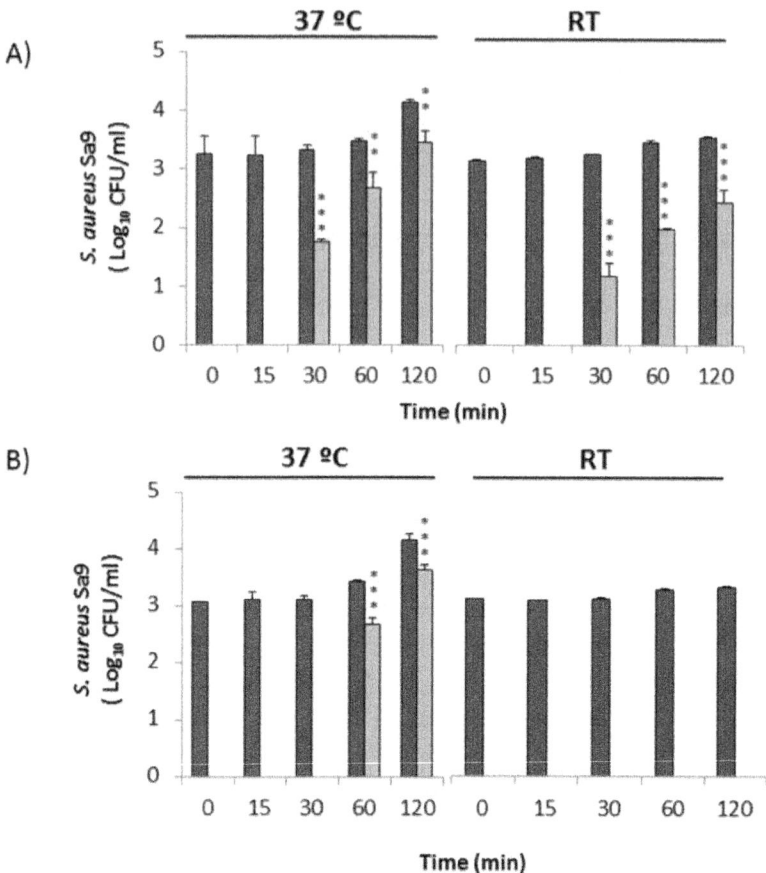

Figure 2: Antimicrobial activity of CHAPSH3b fusion protein in whole milk.

Milk was inoculated with 10^3 CFU/ml of *S. aureus* Sa9 and incubated in the presence of 1.65 µM CHAPSH3b for 0, 15, 30, 60 and 120 min either at RT or 37°C in: A) raw milk and B) pasteurized milk. Dark grey bars indicate *S. aureus* Sa9 control culture and light grey bars *S. aureus* Sa9+ CHAPSH3b. Values, expressed as log CFU/ml, are the means ± standard deviations of two independent experiments. Bars having asterisks are significantly different from the control (**$P<0.01$; ***$P<0.001$). *S. aureus* detection threshold (<10 CFU/ml).

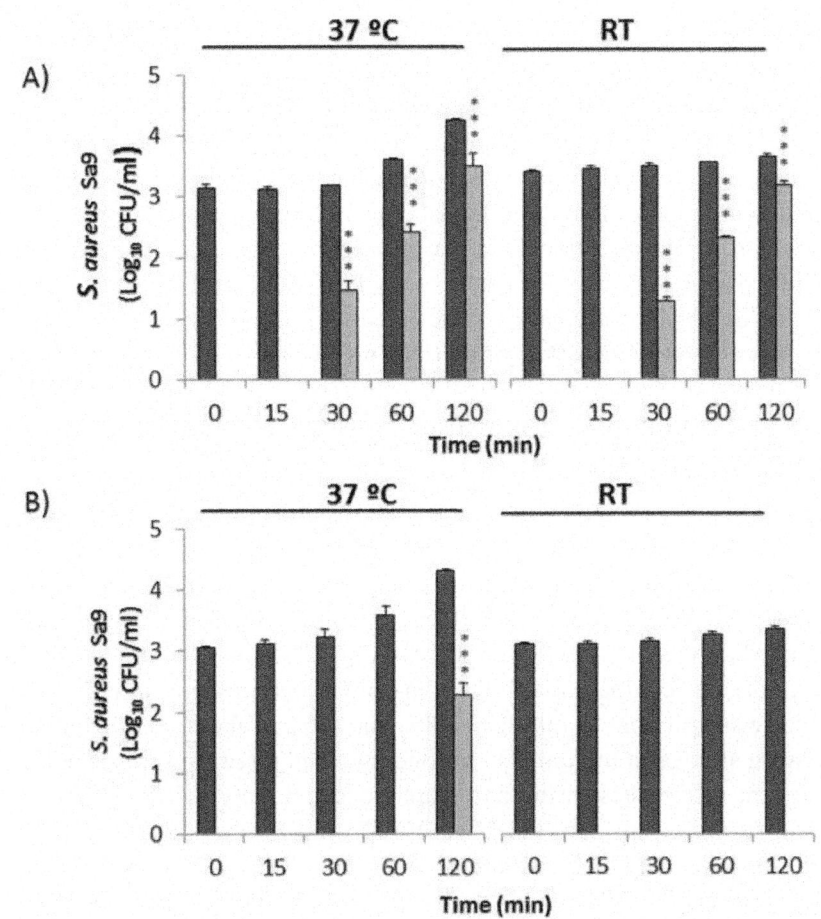

Figure 3: Antimicrobial activity of CHAPSH3b protein in skim milk.

Milk was inoculated with 10^3 CFU/ml of *S. aureus* Sa9 and incubated in the presence of 1.65 µM CHAPSH3b for 0, 15, 30, 60 and 120 min either at RT or 37°C in: A) raw milk and B) pasteurized milk. Dark grey bars indicate *S. aureus* Sa9 control culture and light grey bars *S. aureus* Sa9+ CHAPSH3b. Values, expressed as log CFU/ml, are the means ± standard deviations of two independent experiments. Bars having asterisks are significantly different from the control (***P<0.001). *S. aureus* detection threshold (<10 CFU/ml).

CHAPSH3b was more effective in reducing *S. aureus* Sa9 in pasteurized milk (whole and skim) as is shown in Figure 2B and 3B. At RT, CHAPSH3b (1.65 µM) was able to reduce *S. aureus* viable counts to undetectable levels in whole and skim milk immediately after addition, and no re-growth was detected for 2 h thereafter. At 37°C, the presence of CHAPSH3b prevented staphylococcal re-growth for over 30 min in whole milk and for more than 1 h in skim milk. At the end of the incubation period, the final staphylococcal population in whole and skim pasteurized milk was 0.75±0.23 ($P<0.001$) and 2.02±0.21 log CFU/ml ($P<0.001$) lower than the control, respectively, in the presence of CHAPSH3b. In order to rule out the possibility of CHAPSH3b-resistant colonies confounding the data, ten *S. aureus* colonies were randomly selected from the selective agar plates used for viable count determination. Turbidity reduction assays of each colony performed in the presence of 1 μM of CHAPSH3b indicate that all were sensitive to lysis by CHAPSH3b to the same extent as the inoculated strain (data not shown).

CHAPSH3b Remains Active after Storage at 4°C in Milk and after Pasteurization Treatment

To assess the CHAPSH3b stability in milk, challenge assays were performed after storage of the protein (1.65 μM) in refrigerated raw milk for 3 days. As shown in Figure 4A, the non cold-stored protein reduced the initial staphylococcal population (10^3 CFU/ml) by 94% after just 15 min at RT. The inhibitory activity of CHAPSH3b decreased significantly ($P<0.05$) with the cold storage time as compared with the non-cold stored protein but the remaining activity was still able to kill 42%, 33% and 32% of the *S. aureus* population following storage in milk at 4°C for one, two and three days, respectively. No significant differences in the anti-staphylococcal activity after 2 and 3 days of cold storage were detected ($P>0.05$). To test the stability of CHAPSH3b under high temperature treatment, the protein (2 μM) was subjected at pasteurization (72°C, 15 s) in both commercial ESL whole milk and raw whole milk and further challenged with *S. aureus* at RT. Pasteurization in ELS milk did not affect the inhibitory activity of CHAPSH3b since no viable counts were detected in treated cultures throughout the incubation period (Fig. 4B). However, the heat treatment in raw milk clearly reduced CHAPSH3b activity as only partial inhibition of *S. aureus* CFUs was observed in the first 15 min of treatment (Fig. 4B) compared to the undetectable CFUs that was observed in untreated raw milk cultures spiked with a lower concentration (1.65 μM) of unpasteurized protein (Fig. 2A). Nevertheless, the reductions in viable counts between control and treated cultures were significant throughout the incubation period *(P<0.001* at time 0 and 60 min; $P<0.01$ at time 15, 30 and 120 min).

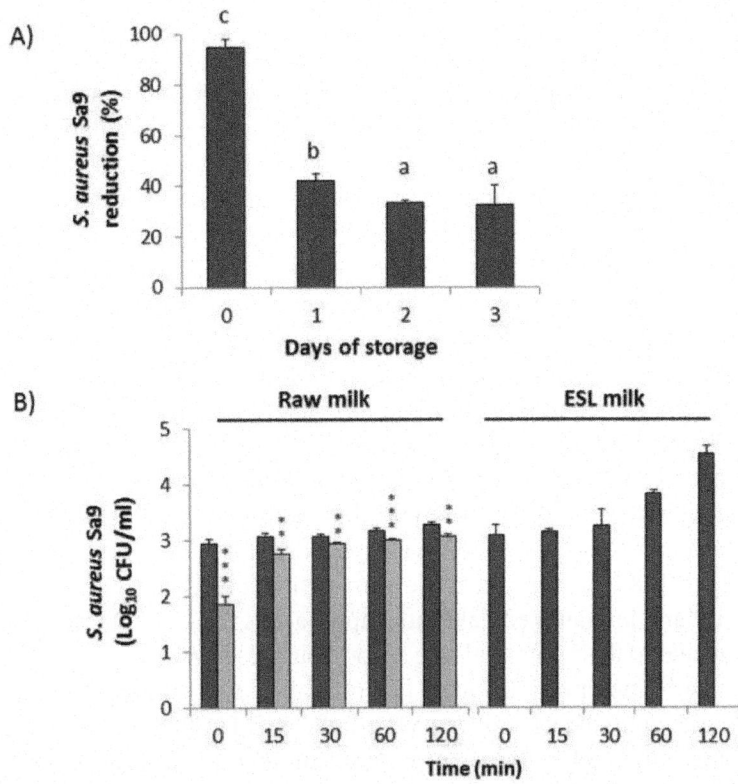

Figure 4: Cold storage stability and pasteurization resistance of CHAPSH3b in milk.

A) 1.65 µM CHAPSH3b was stored in raw milk at 4°C for 3 days. Samples were taken every day, inoculated with 10^3 CFU/ml of *S. aureus* Sa9 and incubated for 15 min at room temperature before plating. Non-cold storage protein was used as control. Cold storage stability was expressed as the percentage reduction of *S. aureus* Sa9 CFU/ml after CHAPSH3b addition. Values are the means ± standard deviations of two independent experiments. Bars having different letters are significantly different (*P*<0.05). B) 1.97 µM CHAPSH3b was pasteurized at 72°C for 15 s in raw milk (left) and commercial pasteurized milk (right). Samples were inoculated with 10^3 CFU/ml and incubated for 0, 15, 30, 60 and 120 min at room temperature before plating. *S. aureus* inoculated cultures without lytic protein addition were used as control (dark grey bars). Light grey bars indicate *S. aureus* Sa9+ CHAPSH3b. Data from pasteurized samples with CHAPSH3b activity was expressed as log CFU/ml. Values are the means ± standard deviations of two independent experiments. Bars having asterisks are significantly different from the control (**P*<0.01; ***P*<0.001).

Discussion

Current food safety depends on a combination of preventive hygiene-based approaches that are focused on minimizing the microbial contamination of raw material that mainly include physical and chemical decontamination treatments aimed to remove the microbial contamination in food products [45]. Due to the increasing consumer demand for natural, nutritious and fresh-tasting foods, the food industry is interested in replacing traditional preservation techniques (e.g. heat and chemical treatments) whenever possible, to avoid the risk of sensory quality changes or the presence of unwanted chemical residues in foods [46]. Food preservation treatments based on natural antimicrobials such as bacteriocins, bacteriophages or phage-derived lytic enzymes could help to fight against pathogenic and spoilage bacteria along the food chain and are not expected to alter the sensory change or other undesirable effects of traditional treatments [24]. The application of bacteriocins in food safety has been widely studied for the last two decades [47], but food biopreservation based on phages and phage-derived lytic enzymes is a more recent avenue of research [48]. So far, phage derived lysins have been mainly assayed in veterinary and human medical model approaches [12], [49], and less attention has been paid to their potential role as food biopreservatives [50]. Nevertheless, some phage lytic enzymes have shown antibacterial activity in milk. This is true for *S. aureus* bacteriophage vB_SauS-phiIPLA88 endolysin LysH5 that completely inhibited *S. aureus* growth in commercial pasteurized milk after 4 h of treatment[26], or the fusion proteins λSA2-E-Lyso-SH3b (streptococcal λSA2 endolysin endopeptidase domain fused to the lysostaphin SH3b domain) and λSA2-E-LysK-SH3b (streptococcal λSA2 endolysin endopeptidase domain fused to the staphylococcal phage K endolysin SH3b domain) that showed anti-staphylococcal activity in ultra-high temperature (UHT) milk by reducing the bacterial load by 3 and 1 log CFU/ml, respectively, within 3 h of incubation [51]. Recently, it has been also reported that *Listeria* bacteriophage endolysin LysZ5 was able to kill 4 log CFU/ml of *L. monocytogenes* within 3 h at 4°C in soya milk [52].

The peptidoglycan hydrolase HydH5 encoded by the *S. aureus* phage vB_SauS-phiIPLA88[37] and the fusion proteins between HydH5 and lysostaphin (CHAPSH3b, HydH5SH3b and HydH5Lyso) have all been shown to yield staphylolytic activity in zymogram, plate lysis and turbidity reduction assays [38]. In this work, these constructs have been assessed as antimicrobial additives for preventing the growth of *S. aureus* in milk. HydH5, HydH5SH3b and HydH5Lyso showed staphylolytic activity in commercial whole ESL milk but were clearly less effective than lysostaphin and CHAPSH3b activities. Lysostaphin and CHAPSH3b (1 μM) were able to reduce the *S. aureus* load by 4-log CFU/ml immediately or 15 min after addition at 37°C, respectively, while nearly four times as much (3.5 µM) of the other constructs were needed to obtain just a reduction of the staphylococcal counts throughout the incubation period compared to the untreated cultures.

These findings are consistent with our previous results. In fact, the CHAP domain of HydH5 when fused to the lysostaphin SH3b domain showed a 4.8-fold higher activity, compared to full length HydH5 [38]. The high activity shown by CHAPSH3b in ESL milk, prompted us to broaden the assays on milk with a broader range of treatments. Accordingly, CHAPSH3b activity was assessed in raw (whole and skim) and pasteurized (whole and skim) milk.

CHAPSH3b is active in whole and skim raw milk at 37°C and RT, as the protein was able to reduce 10^3 CFU/ml below the detection limit (<10 CFU/ml) for 30 min. The staphylolytic activity, however, was lower than in high-heat treated milk yielding less of a reduction in viable counts, despite a higher concentration of enzyme (1.65 µM *versus* 1 µM). Of note, the indigenous *S. aureus* contamination of raw milk do not seem to have interfered in the CHAPSH3b activity because it was shown to be sensitive and hardly accounts for the total *S. aureus* population once Sa9 was added. Pasteurization of milk clearly enhanced CHAPSH3b staphylolytic activity in both whole and skim milk at both temperatures. Apparently, something in the raw milk is hampering the CHAPSH3b activity. One possibility is heat-sensitive components such as immunoglobulin M and agglutinins in so far as they have been reported to promote the formation of cell clumps [53] that would likely make it more difficult for the antibacterial protein to reach the staphylococcal cells sequestered inside the clumps. These components of raw milk have also been previously reported to hamper phage adsorption [54]. In contrast, CHAPSH3b activity does not seem to be affected by fat globules in milk since similar kinetics of staphylococcal inhibition were observed in whole or skim milk despite the fact that bacterial clumps have also been associated with fat globules [55].

Although structural and chemical composition of food can negatively affect the ability of antimicrobials to reach the pathogen [56], the addition of CHAPSH3b to different types of milk yielded an immediate reduction in *S. aureus* viable counts with the only exception being heat-treated protein in raw milk. This suggests a quick reaction to CHAPSH3b which is reminiscent with previous data with *Listeria monocytogenes* phage endolysins Ply118 and Ply500. The cell binding domain of these proteins showed a rapid and saturation-dependent binding to *L. monocytogenes* cell surface within 15 s, with no further increase [57]. A high affinity for staphylococcal cells, especially MRSA, was also described for the endolysin LysGH15 [17]. The staphylococcal re-growth observed at latter sampling times could be attributed to those cells that were sequestered and did not see the CHAPSH3b at the beginning of the treatment. However, it should be noted that the number of bacteria attained by the end of the assay period (clearly below the critical threshold of 10^5 CFU/ml for production of hazardous enterotoxins levels to consumers) does not present a high-risk of enterotoxin contamination of milk [58]. In addition, no CHAPSH3b resistant bacteria were isolated from lysin-treated milk. Therefore, insensitivity to CHAPSH3b appears to be a rare event under the experimental conditions tested. Other researchers also failed to detect resistance against endolysins used to control the growth of Gram positive bacteria, such as *Bacillus anthracis* [59] and *Streptococcus pneumoniae* [14].

Of note, CHAPSH3b showed higher activity at RT than at 37°C in raw and pasteurized milk. The lower growth rate of *S. aureus* at RT could account for the higher effectiveness of CHAPSH3b. The demonstrated 3 day longevity of CHAPSH3b in cold milk supports the notion of using CHAPSH3b as a potential staphylolytic agent to prevent *S. aureus* development during an unexpected breakdown in cold storage and thus, enhance food safety.

The ability of CHAPSH3b (1.65 µM) to kill up to 10^3 CFU/ml in raw milk in the first 30 min of treatment at 37°C along with its proven activity against MRSA strains [38] points to CHAPSH3b as a potential candidate to control *S. aureus* infections in cows' mammary glands. Previous reports have shown the effectiveness of chimeric phage lysins to kill mastitis-causing *S. aureus* in murine mammary glands [25]. The safety of endolysins has also been determined since experimental mice to which endolysin was administered did not exhibit adverse physiological effects [60].

The stability of CHAPSH3b after exposure at high temperatures (72°C, 15 s) and cold storage in milk has a clear technical interest for dairy products protection since this staphylolytic protein could be added to raw milk before thermal processing to control any potential contamination by *S. aureus*. CHAPSH3b thermostability is consistent with our previous results which showed that HydH5 retained activity after heat treatment (5 min treatment at 100°C) [37]. Recently, *Listeria* bacteriophage peptidoglycan hydrolases also revealed a high thermostability retaining up to 35% activity after 30 min of incubation at 90°C [51]. By contrast, the lytic activity of some phage endolysins was destroyed by heat treatment [26], [61]. However, none of these assays were performed in milk, work that is sorely needed.

Regarding the lytic proteins' stability in cold storage, prior to this study, the existing data were obtained in aqueous solutions but not in milk. This is the case for $CHAP_k$ that retained up to 70% of its lytic activity after being stored at 4°C for one month [18]. By contrast, the remaining lytic activity of CHAPSH3b was ~33% after 3 days of storage a 4°C in raw milk. As indicated above, the reduction of lytic activity in raw milk could be due to the presence of heat-sensitive components in raw milk that hamper the access of the lytic protein to its target on the cell wall of the bacterial host [56]. Increasing the concentration of the lytic protein might enable us to overcome this limited activity in raw milk. Indeed, 3.30 µM (100 µg/ml) CHAPSH3b was able to kill 10^3 CFU/ml of *S. aureus* in raw milk and no re-growth was observed within 2 h (data not shown).

Our findings demonstrate the ability of HydH5-derived proteins to inhibit the development of *S. aureus* in milk, with CHAPSH3b being particularly effective. The high anti-staphylococcal activity of CHAPSH3b along with its thermostability might enable this protein to be applied directly to raw milk after milking. Overall, our results suggest that phage lytic proteins might be useful as a valuable hurdle to prevent *S. aureus* growth in milk and presumably in other dairy products.

ACKNOWLEDGMENTS

We would like to thank Pablo González (IPLA-CSIC) for performing microbiological and physicochemical analyses of milk. Mentioning of trade names or commercial products in this article is solely for the purpose of providing specific information and does not imply recommendation or endorsement by the United States Department of Agriculture. The USDA is an equal opportunity provider and employer.

AUTHOR CONTRIBUTIONS

Conceived and designed the experiments: PG BM DMD AR. Performed the experiments: LR-R PG AR. Analyzed the data: PG BM DMD AR. Wrote the paper: PG DMD AR.

REFERENCES

1. Lowy FD (1998) *Staphylococcus aureus* infections. N Engl J Med 339: 520–532. doi: 10.1056/nejm199808203390806

2. Le Loir Y, Baron F, Gautier M (2003) *Staphylococcus aureus* and food poisoning. Genet Mol Res 2: 63–76.

3. Belay N, Rasooly A (2002) *Staphylococcus aureus* growth and enterotoxin A production in an anaerobic environment. J Food Prot 65: 199–204.

4. Montville TJ, Matthews KR (2008) *Staphylococcus aureus*. In: Montville

TJ, Matthews KT, editors. Food Microbiology: An Introduction, 2nd ed. Washington, DC: ASM Press. 189–201.

5.   5.Mørk T, Tollersrud T, Kvitle B, Jørgensen HJ, Waage S (2005) Comparison of*Staphylococcus aureus* genotypes recovered from cases of bovine, ovine, and caprine mastitis. J Clin Microbiol 43: 3979–3984. doi: 10.1128/jcm.43.8.3979-3984.2005

6.   Rosengren A, Fabricius A, Guss B, Sylvén S, Lindqvist R (2010) Occurrence of foodborne pathogens and characterization of *Staphylococcus aureus* in cheese produced on farm-dairies. Int J Food Microbiol 144: 263–269. doi: 10.1016/j.ijfoodmicro.2010.10.004

7.   Hogeveen H, Huijps K, Lam TJ (2011) Economic aspects of mastitis: New developments. N Z Vet J 59: 16–23. doi: 10.1080/00480169.2011.547165

8.   Barkema HW, Schukken YH, Zadoks RN (2006) Invited review: The role of cow, pathogen, and treatment regimen in the therapeutic success of bovine *Staphylococcus aureus* mastitis. J Dairy Sci 89: 1877–1895.

9.   Hebert A, Sayasith K, Senechal S, Dubreuil P, Lagace J (2000) Demonstration of intracellular Staphylococcus aureus in bovine mastitis alveolar cells and macrophages isolated from naturally infected cow milk. FEMS Microbiol Lett 193: 57–62. doi: 10.1111/j.1574-6968.2000.tb09402.x

10.   Türkyilmaz S, Tekbiyik S, Oryasin E, Bozdogan B (2010) Molecular epidemiology and antimicrobial resistance mechanisms of methicillin-resistant *Staphylococcus aureus*isolated from bovine milk. Zoonoses Public Health 57: 197–203. doi: 10.1111/j.1863-2378.2009.01257.x

11.   Kluytmans JA (2010) Methicillin-resistant *Staphylococcus aureus* in food products: cause for concern or case for complacency? Clin Microbiol Infect 16: 11–15. doi: 10.1111/j.1469-0691.2009.03110.x

12.   Fischetti VA (2010) Bacteriophage endolysins: a novel anti-infective to control Gram-positive pathogens. Int J Med Microbiol 300: 357–362. doi: 10.1016/j.ijmm.2010.04.002

13.   Hermoso JA, García JL, García P (2007) Taking aim on bacterial pathogens: from phage therapy to enzybiotics. Curr Opin Microbiol 10: 461–472. doi: 10.1016/j.mib.2007.08.002

14.   Loeffler JM, Nelson D, Fischetti VA (2001) Rapid killing of *Streptococcus pneumoniae*with a bacteriophage cell wall hydrolase. Science 294: 2170–2172. doi: 10.1126/science.1066869

15.   Rashel M, Uchiyama J, Ujihara T, Uehara Y, Kuramoto S, et al. (2007) Efficient elimination of multidrug-resistant *Staphylococcus aureus* by

cloned lysin derived from bacteriophage phi MR11. J Infect Dis 196: 1237–1247. doi: 10.1086/521305

16. Gu J, Xu W, Lei L, Huang J, Feng X, et al. (2011) LysGH15, a novel bacteriophage lysin, protects a murine bacteremia model efficiently against lethal methicillin-resistant*Staphylococcus aureus* infection. J Clin Microbiol 49: 111–117. doi: 10.1128/jcm.01144-10

17. Daniel A, Euler C, Collin M, Chahales P, Gorelick KJ, et al. (2010) Synergism between a novel chimeric lysin and oxacillin protects against infection by methicillin-resistant*Staphylococcus aureus*. Antimicrob Agents Chemother 54: 1603–1612. doi: 10.1128/aac.01625-09

18. Fenton M, Casey PG, Hill C, Gahan CG, Ross RP, et al. (2010) The truncated phage lysin CHAP(k) eliminates *Staphylococcus aureus* in the nares of mice. Bioeng Bugs 1: 404–407. doi: 10.4161/bbug.1.6.13422

19. Paul VD, Rajagopalan SS, Sundarrajan S, George SE, Asrani JY, et al. (2011) A novel bacteriophage Tail-Associated Muralytic Enzyme (TAME) from Phage K and its development into a potent antistaphylococcal protein. BMC Microbiol 11: 226. doi: 10.1186/1471-2180-11-226

20. Kumar JK (2008) Lysostaphin: an antistaphylococcal agent. Appl Microbiol. Biotechnol 80: 555–561. doi: 10.1007/s00253-008-1579-y

21. Kerr DE, Plaut K, Bramley AJ, Williamson CM, Lax AJ, et al. (2001) Lysostaphin expression in mammary glands confers protection against staphylococcal infection in transgenic mice. Nat Biotechnol 19: 66–70.

22. Wall RJ, Powell AM, Paape MJ, Kerr DE, Bannerman DD, et al. (2005) Genetically enhanced cows resist intramammary *Staphylococcus aureus* infection. Nat Biotechnol 23: 445–451. doi: 10.1038/nbt1078

23. Becker SC, Foster-Frey J, Donovan DM (2008) The phage K lytic enzyme LysK and lysostaphin act synergistically to kill MRSA. FEMS Microbiol Lett 287: 185–191. doi: 10.1111/j.1574-6968.2008.01308.x

24. García P, Rodríguez L, Rodríguez A, Martínez B (2010) Food biopreservation: promising strategies using bacteriocins, bacteriophages and endolysins. Trends Food Sci Technol 21: 373–382. doi: 10.1016/j.tifs.2010.04.010

25. Schmelcher M, Powell AM, Becker SC, Camp MJ, Donovan DM (2012a) Chimeric phage lysins act synergistically with Lysostaphin to kill mastitis causing Staphylococcus aureus in murine mammary glands. Appl Environ Microbiol 78: 2297–2305. doi: 10.1128/aem.07050-11

26. Obeso JM, Martínez B, Rodríguez A, García P (2008) Lytic activity of the recombinant staphylococcal bacteriophage PhiH5 endolysin active

against *Staphylococcus aureus* in milk. Int J Food Microbiol 128: 212–218. doi: 10.1016/j.ijfoodmicro.2008.08.010

27. Rodríguez-Rubio L, Martínez B, Donovan DM, Rodríguez A, García P (2012) Bacteriophage virion-associated peptidoglycan hydrolases: potential new enzybiotics. Crit Rev Microbiol (DOI: 10.3109/1040841X.2012.723675).

28. Rashel MJ, Uchiyama I, Takemura H, Hoshiba H, Ujihara T, et al. (2008) Tail-associated structural protein gp61 of *Staphylococcus aureus* phage ΦMR11 has bifunctional lytic activity. FEMS Microbiol Lett 284: 9–16. doi: 10.1111/j.1574-6968.2008.01152.x

29. Takac M, Blasi U (2005) Phage P68 virion-associated protein 17 displays activity against clinical isolates of *Staphylococcus aureus*. Antimicrob Agents Chemother 49: 2934–2940. doi: 10.1128/aac.49.7.2934-2940.2005

30. Steinbacher S, Miller S, Baxa U, Budisa N, Weintraub A, et al. (1997) Phage P22 tailspike protein: crystal structure of the head-binding domain at 2.3 A, fully refined structure of the endorhamnosidase at 1.56 A resolution, and the molecular basis of O-antigen recognition and cleavage. J Mol Biol 267: 865–880. doi: 10.1006/jmbi.1997.0922

31. Molineux IJ (2001) No syringes please, ejection of phage T7 DNA from the virion is enzyme driven. Mol Microbiol 40: 1–8. doi: 10.1046/j.1365-2958.2001.02357.x

32. Kanamaru S, Leiman PG, Kostychenko VA, Chipman PR, Mesyanzhinov VV, et al. (2002) Structure of the cell-puncturing device of bacteriophage T4. Nature 415: 553–557. doi: 10.1038/415553a

33. Kenny JG, McGrath S, Fitzgerald GF, van Sinderen D (2004) Bacteriophage Tuc2009 encodes a tail-associated cell wall-degrading activity. J Bacteriol 186: 3480–3491. doi: 10.1128/jb.186.11.3480-3491.2004

34. Delbrück M (1940) The growth of bacteriophage and lysis of the host. J Gen Physiol 23: 643–660. doi: 10.1085/jgp.23.5.643

35. Bateman A, Rawlings ND (2003) The CHAP domain: a large family of amidases including GSP amidase and peptidoglycan hydrolases. Trends Biochem Sci 5: 234–237. doi: 10.1016/s0968-0004(03)00061-6

36. Rigden DJ, Jedrzejas MJ, Galperin MY (2003) Amidase domains from bacterial and phage autolysins define a family of γ-d,l-glutamate-specific amidohydrolases. Trends in Biochemical Sciences. 28: 230–234. doi: 10.1016/s0968-0004(03)00062-8

37. Rodríguez L, Martínez B, Zhou Y, Donovan DM, Rodríguez A, et al. (2011) Lytic activity of the virion-associated peptidoglycan hydrolase HydH5 of *Staphylococcus aureus*bacteriophage vB_SauS-phiIPLA88. BMC Microbiol 11: 138. doi: 10.1186/1471-2180-11-138

38. Rodríguez-Rubio L, Martínez B, Rodríguez A, Donovan DM, García P (2012) Enhanced staphylolytic activity of the *Staphylococcus aureus* bacteriophage vB_SauS-phiIPLA88 virion associated peptidoglycan hydrolase: fusions, deletions and synergy with LysH5. Appl Environ Microbiol 78: 2241–2248. doi: 10.1128/aem.07621-11

39. García P, Madera C, Martínez B, Rodríguez A, Suárez JE (2009) Prevalence of bacteriophages infecting *Staphylococcus aureus* in dairy samples and their potential as biocontrol agents. J Dairy Sci 92: 3019–3026.

40. IDF Standard 4A (1982) Cheese and processed cheese. Determination of the total solids content (reference method). International Dairy Federation, Brussels, Belgium.

41. IDF Standard 152 (1991) Milk and milk products. Determination of fat content. General guidance on the use of butyrometric methods. International Dairy Federation, Brussels, Belgium.

42. IDF Standard 20B (1993) Milk. Determination of nitrogen content. Part 1: Kjeldahl method. International Dairy Federation, Brussels, Belgium.

43. Donovan DM, Foster-Frey J (2008) LambdaSa2 prophage endolysin requires Cpl-7-binding domains and amidase-5 domain for antimicrobial lysis of streptococci. FEMS Microbiol Lett 287: 22–33. doi: 10.1111/j.1574-6968.2008.01287.x

44. Becker SC, Dong S, Baker JR, Foster-Frey J, Pritchard DG, et al. (2009) LysK CHAP endopeptidase domain is required for lysis of live staphylococcal cells. FEMS Microbiol Lett 294: 52–60. doi: 10.1111/j.1574-6968.2009.01541.x

45. Bernard D, Scott VN (2007) Hazard analysis and critical control point systems: use in controlling microbiological hazards. In: Doyle P, Beuchat LR editors. Food Microbiology: Fundamentals and Frontiers. ASM Press, Washington DC. 971–985.

46. Devlieghere F, Francois K, Vereecken KM, Geeraerd AH, Van Impe JF, et al. (2004) Effect of chemicals on the microbial evolution in foods. J Food Prot 67: 1977–1990.

47. Gálvez A, López RL, Abriouel H, Valdivia E, Omar NB (2008) Application of bacteriocins in the control of foodborne pathogenic and spoilage bacteria. Crit Rev Biotechnol 28: 125–152. doi:

10.1080/07388550802107202

48. García P, Martínez B, Rodríguez L, Rodríguez A (2011) Bacteriophages and phage-encoded proteins: prospects in food safety and food quality. In: Mahendra R, Chikindas M, editors. Natural antimicrobials in Food safety and Food quality. United Kingdom: CAB International. 10–26.

49. O'Flaherty S, Ross RP, Coffey A (2009) Bacteriophage and their lysins for elimination of infectious bacteria. FEMS Microbiol Rev 33: 801–819. doi: 10.1111/j.1574-6976.2009.00176.x

50. Coffey B, Mills S, Coffey A, McAuliffe O, Ross RP (2010) Phage and their lysins as biocontrol agents for food safety applications. Annu Rev Food Sci Technol 1: 449–468. doi: 10.1146/annurev.food.102308.124046

51. Schmelcher M, Waldherr F, Loessner MJ (2012b) Listeria bacteriophage peptidoglycan hydrolases feature high thermoresistance and reveal increased activity after divalent metal cation substitution. Appl Microbiol Biotechnol 93: 633–643. doi: 10.1007/s00253-011-3372-6

52. Zhang H, Bao H, Billington C, Hudson JA, Wang R (2012) Isolation and lytic activity of the Listeria bacteriophage endolysin LysZ5 against *Listeria monocytogenes* in soya milk. Food Microbiol 31: 133–136. doi: 10.1016/j.fm.2012.01.005

53. Korhonen H, Marnila P, Gill HS (2000) Milk immunoglobulins and complement factors. Br J Nutr 84 (Suppl. 1)S75–S80. doi: 10.1017/s0007114500002282

54. O'Flaherty S, Coffey A, Meaney WJ, Fitzgerald GF, Ross RP (2005) Inhibition of bacteriophage K proliferation on *Staphylococcus aureus* in raw bovine milk. Lett Appl Microbiol 41(3): 274–279. doi: 10.1111/j.1472-765x.2005.01762.x

55. Walstra P, Jenness R (1984) Dairy Chemistry and Physics. New York: John Wiley and Sons Publ. Inc. 467 p.

56. Guenther S, Huwyler D, Richard S, Loessner M (2009) Virulent bacteriophage for efficient biocontrol of *Listeria monocytogenes* in ready-to-eat foods. Appl Environ Microbiol 75: 93–100. doi: 10.1128/aem.01711-08

57. Loessner MJ, Kramer K, Ebel F, Scherer S (2002) C-terminal domains of *Listeria monocytogenes* bacteriophage murein hydrolases determine specific recognition and high-affinity binding to bacterial cell wall carbohydrates. Mol Microbiol 44: 335–349. doi: 10.1046/j.1365-2958.2002.02889.x

58. Anunciaçao LL, Linardi WR, do Carmo LS, Bergdoll MS (1995) Production of Staphylococcal enterotoxin A in cream-filled cake. Int J Food Microbiol 26: 259–263. doi: 10.1016/0168-1605(94)00122-m

59. Schuch R, Nelson D, Fischetti VA (2002) A bacteriolytic enzyme that detects and kills Bacillus anthracis. Nature 418: 884–889. doi: 10.1038/nature01026

60. Borysowski J, Weber-Dabrowska B, Gorski A (2006) Bacteriophage endolysins as a novel class of antibacterial agents. Exp Biol Med (Maywood) 231: 366–377.

61. Donovan DM, Dong S, Garrett W, Rousseau GM, Moineau S, et al. (2006) Peptidoglycan hydrolase fusions maintain their parental specificities. Appl Environ Microbiol 72: 2988–2996. doi: 10.1128/aem.72.4.2988-2996.2006

Chapter 3

SONORENSIN: A NEW BACTERIOCIN WITH POTENTIAL OF AN ANTI-BIOFILM AGENT AND A FOOD BIOPRESERVATIVE

Lipsy Chopra[1], Gurdeep Singh[1], Kautilya Kumar Jena[1] & Debendra K. Sahoo[1]

[1]Biochemical Engineering Research and Process Development Centre, CSIR-Institute of Microbial Technology, Sector-39A Chandigarh; 160036, India

ABSTRACT

The emergence of antibiotic resistant bacteria has led to exploration of alternative therapeutic agents such as ribosomally synthesized bacterial peptides known as bacteriocins. Biofilms, which are microbial communities that cause serious chronic infections, form environments that enhance antimicrobial resistance. Bacteria in biofilm can be upto thousand times more resistant to antibiotics than the same bacteria circulating in a planktonic state. In this study, sonorensin, predicted to belong to the heterocycloanthracin subfamily of bacteriocins, was found to be effectively killing active and non-multiplying cells of both Gram-positive and Gram-negative bacteria. Sonorensin showed marked inhibition activity against biofilm of Staphylococcus aureus. Fluorescence and electron microscopy suggested that growth inhibition occurred because of increased membrane permeability. Low density polyethylene film coated with sonorensin was found to effectively control the growth of food spoilage bacteria like Listeria monocytogenes and S. aureus. The biopreservative effect of sonorensin coated film showing growth inhibition of spoilage bacteria in chicken meat and tomato samples demonstrated the potential of sonorensin as an alternative to current antibiotics/ preservatives.

Bacteria in nature usually dwell in complex and dynamic surface-associated, sessile microbial communities called biofilms that are encaged in a self-produced extracellular polymeric substance (EPS), and create problems in clinical therapeutics[1]. Bacterial cells growing in a biofilm are physiologically diverse from planktonic cells of the same bacteria[2], and the presence of EPS escalates antibiotic resistance by up to thousand folds[3]. Biofilms have immense negative impact on the world's economy and pose severe problems to industry, public health and medicine[4] due to increased rates of genetic exchange, altered biodegradability[5], increased resistance to antibiotics and chemical biocides and increased production of secondary metabolites[4,5,6]. Many bacteria produce bioactive peptides or proteins called bacteriocins. Bacteriocins can be the next generation of antibiotics for combating multi-drug resistant and/or biofilm forming bacterial infections due to their different mechanisms of action, which include membrane-disrupting action, functional inhibition of proteins, binding with DNA, and detoxification of polysaccharides[7]. Some bacteriocins can be transferred in biofilm EPS through pores formed in the lipid component of the EPS, while others can disperse biofilms[7].

In an infection, multiplying and non-multiplying bacteria exist side by side[8]. Non-multipliers are characterized by their lack of multiplication, survival in the presence of antibiotics and low metabolic activity[8]. It is also known that antibiotics kill multiplying bacteria and are inefficient at killing non-multipliers[9], leading to slow or partial death of the target population in an infected tissue resulting in requirement of repeated doses of antibiotics. This extends the period of therapy and enhances the emergence of resistance. Targeting non-multiplying bacteria is a new approach to antibacterial therapy intended to swiftly destroy all of the non-multiplying and multiplying bacteria in an infection, thereby shortening antibiotic regimes that would slow the emergence of genetic resistance as mutation cannot occur if there are no live target bacteria[8].

Numerous food preservation methods such as addition of preservatives (antibiotics, organic compounds such as sorbate, propionate, benzoate, acetate, and lactate), reduction of pH and water activity (acidification, dehydration) and thermal treatment (pasteurization, sterilization, heating) have been used to prevent food poisoning and spoilage[10]. Although these methods have been proven to be successful, however, consumers have been consistently concerned about the possible adverse health effects due to the presence of chemical additives in processed foods. This has led to the exploration of new means of preservation involving minimally processed foods with extended shelf-life[10]. For more than 50 years nisin produced by *Lactococcus lactis* has been used as a food preservative as it has been proven to be highly effective against microbial agents causing food poisoning and spoilage[11].

Amalgamation of bacteriocins into packaging films to prevent food spoilage and to control pathogens has been an area of dynamic research for the last decade. Bioactive packaging film prevents microbial growth on food surface by direct contact of the package with the surface of foods, such as meats and cheese. It prolongs the lag phase and reduces the growth rate of microbes in order to extend shelf life and to maintain product quality[12]. There are two methods which have been commonly used to prepare packaging films with bacteriocins[13]. The first method incorporates bacteriocins directly into polymers (e.g., incorporation of nisin into biodegradable films[14]) and the other incorporates bacteriocins into packaging films by coating or adsorbing bacteriocins to polymer surfaces (e.g., nisin/methylcellulose coatings for polyethylene films, adsorption of nisin on polyamide, ethylene vinyl acetate, acrylics, polyester and polyvinyl chloride[13]). An et al. claimed that a polymer-based solution coating would be the most desirable method in terms of stability and adhesiveness of attaching a bacteriocin to a plastic film[15].

Earlier, we have reported isolation, purification and characterization of sonorensin, a bacteriocin predicted to belong to heterocycloanthracin subfamily of bacteriocins from marine isolate *Bacillus sonorensis*MT93, and optimization of its production[16,17]. We have also reported the efficacy of sonorensin as a biopreservative in fruit products and as shelf life extender of pasteurised milk[17]. In the present study, the effectiveness of this bacteriocin as an anti biofilm agent and a food biopreservative has been demonstrated. We have also evaluated the effect of sonorensin against non-multiplying bacteria and an insight into its probable mode of action. It is the first bacteriocin of this subfamily to be characterized.

RESULTS

Biofilm Inhibition by Sonorensin

When different concentrations of sonorensin were incubated with *S. aureus* for 4 h at 37 °C for adherence to the wells of microtiter plates, it inhibited biofilm attachment in a concentration dependent manner (Fig. 1a). About $1.8 \pm 0.05\%$ attachment of biofilm was observed in the presence of 1X MIC (~50 µg/ml) of sonorensin. Sonorensin showed significant inhibitory activity against *S. aureus* biofilm formation at 24 h, in relation to its concentration (Fig. 1b). When the sonorensin treated biofilms were subjected to 2, 3-Bis (2-methoxy-4-nitro-5-sulfophenyl)—2H-tetrazolium-5-carbox-anilide (XTT) assay, reduced XTT conversion was observed in wells with higher concentration of sonorensin while the negative control showed the maximum reduction of XTT indicating the effect of sonorensin on the viability of cells in biofilm (Fig. 1c).

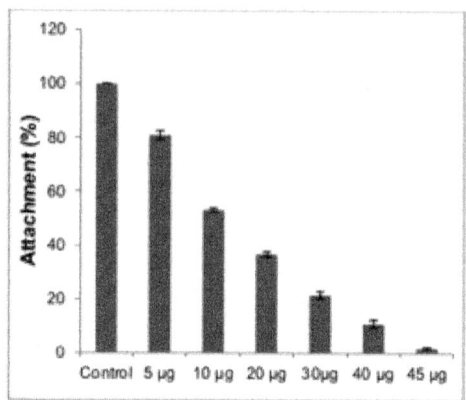

a

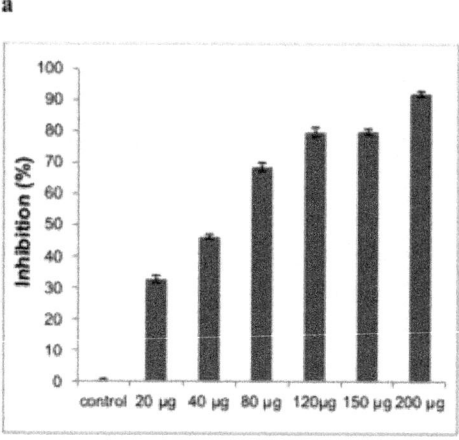

b

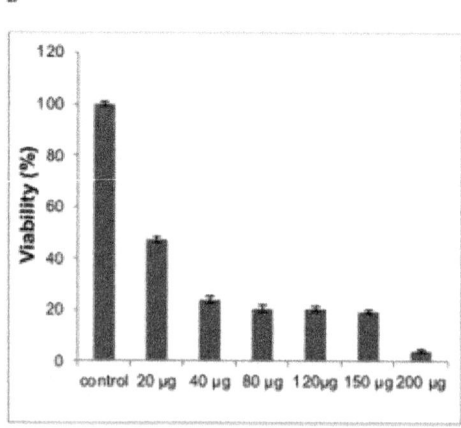

Figure 1: Effect of sonorensin on *S. aureus*biofilms.

Scanning electron micrographs of biofilms of *S. aureus* formed on cover slips and the effect of sonorensin on the preformed biofilm are shown in Fig. 2. For the non-treated controls, a biofilm formed consisted of nearly uniform, thick layer of cells (Fig. 2a), while the biofilm treated with sonorensin (50 µg/ml) was much less dense, and individually formed colonies could be seen (Fig. 2b).

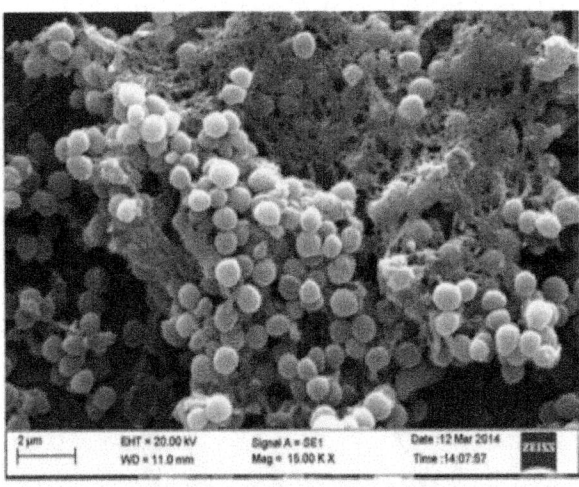

a

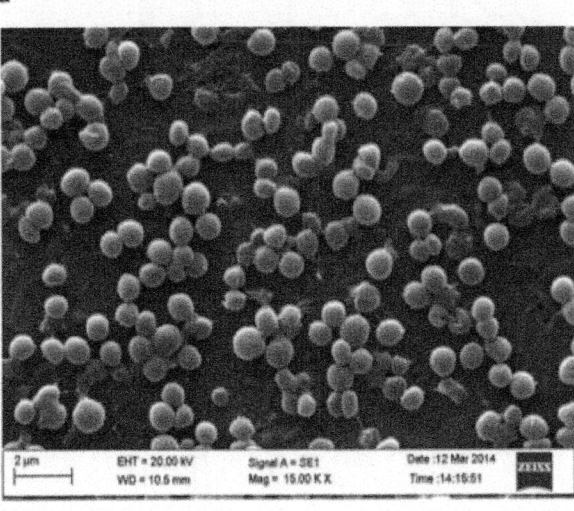

b

Figure 2: The scanning electron micrographs of mature (48 h old) biofilm of *S. aureus*cells (a) without sonorensin treatment and (b) after sonorensin treatment (50 µg/ml).

Sonorensin Is Effective Against Non-Multiplying Bacteria

Sonorensin was investigated for its efficacy against bacterial cells in dormant stage. *Escherichia coli* and *S. aureus* were used as indicator strains to produce long duration stationary phase cells of Gram-negative and Gram-positive bacteria, respectively. An extented lag phase compared to their vegetative counterparts in a regrowth experiment (Fig. 3a), sensitivity to nisin (Fig. 3b,c) and tolerance to ampicillin (Fig. 3b,c) confirmed their non-multiplying state. Susceptibility of vegetative cells of both *E. coli* and *S. aureus* to ampicillin is shown inSupplementary Fig. S1 online. The revived cultures regained sensitivity to antibiotics signifying that no further antibiotic resistance was attained through any genetic alteration during dormancy. Sonorensin was found to kill non-multiplying cells of both *E. coli* and *S. aureus* (Fig. 3b,c). On comparing its antimicrobial activity with nisin, it was observed that sonorensin was as effective as nisin in killing non-multiplying cells of *S. aureus* and *E. coli*.

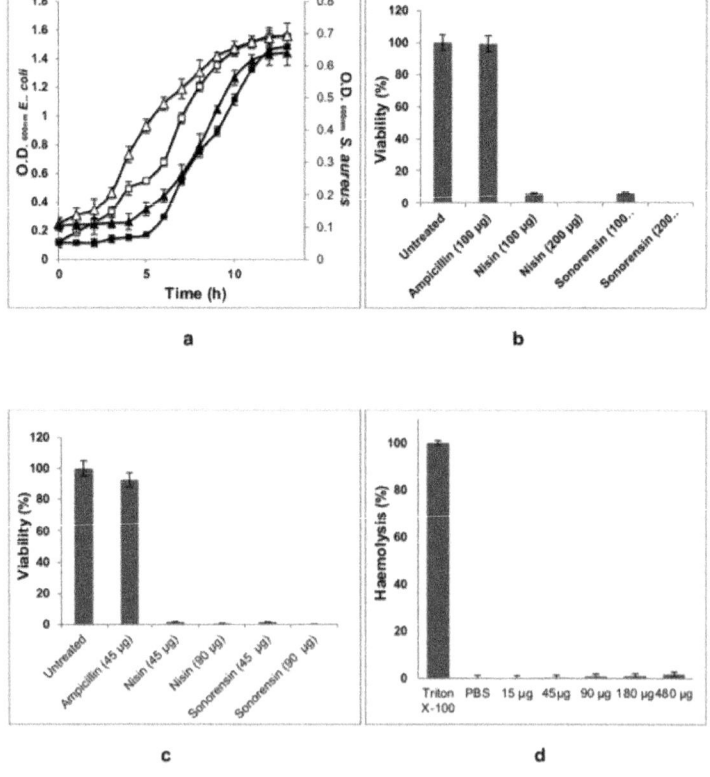

Figure 3: Effect of sonorensin on non-multiplying bacteria.

As the MIC of sonorensin against both the vegetative and non-multiplying cells of *E. coli* was higher as compared to that of *S. aureus*, the inhibitory activity of treatment combining sonorensin and the chelating agent EDTA on *E. coli* was investigated. It was found that sonorensin in combination with 20 mM EDTA showed more antimicrobial activity against both multiplying and non-multiplying cells of *E. coli* as compared to sonorensin alone (see Supplementary Fig. S2 online). The controls, EDTA and buffer individually, did not show any activity.

The haemolytic activity (if any) of sonorensin was determined by testing its toxicity against mammalian cells. It was found that sonorensin, at concentrations at which it destroys vegetative and dormant cells of *E. coli* and *S. aureus*, had virtually no effect on red blood cells (RBCs) and only $1.7 \pm 0.04\%$ haemolysis was observed at high concentrations of sonorensin (Fig. 3d).

Mode of Bactericidal Action: Increased Cytoplasmic Membrane Permeability

When the cytoplasmic membrane is permeable, ortho-Nitrophenyl-β-galactoside (ONPG), a non membrane—permeative chromogenic substrate, enters the cytoplasm and is degraded by β-galactosidase, producing O-nitrophenol that shows absorbance at 405 nm[18]. As shown in Fig. 4, sonorensin induced an increase in the permeability of *S. aureus* cytoplasmic membrane over time and in case of nisin (at same concentration), which has almost same MIC value against *S. aureus*, produced similar results of permeability. This suggested that sonorensin could permeabilize the cytoplasmic membrane of *S. aureus*.

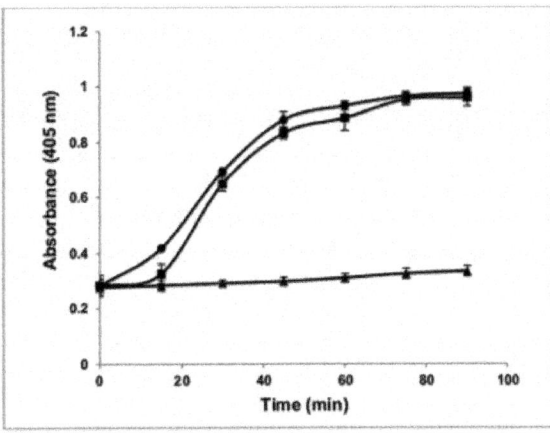

Figure 4: The cytoplasmic membrane permeabilization of *S. aureus* **cells treated with** sonorensin (squares) and nisin (circles).

PI is a viability-fluorescent marker that can penetrate impaired cells and intercalate into nucleic acid[18]. Sonorensin and nisin induced membrane damage of *S. aureus* cells was determined by staining the cells with PI after treatment with the sonorensin and nisin at $37\,°C$ for 1 h using Flow Cytometer. As shown in Fig. 5, in the absence of any antibacterial agent, 76.5% of untreated control *S. aureus* cells showed no PI fluorescence signal. However, a significant increase in PI fluorescence was observed for the cells treated with sonorensin (70.0% cells stained with PI) and nisin (68.1% cells stained with PI) as depicted in Fig. 5. These results indicated that the membrane integrity of *S. aureus* cells was destroyed by treatment with sonorensin and the effectiveness was comparable to that of nisin.

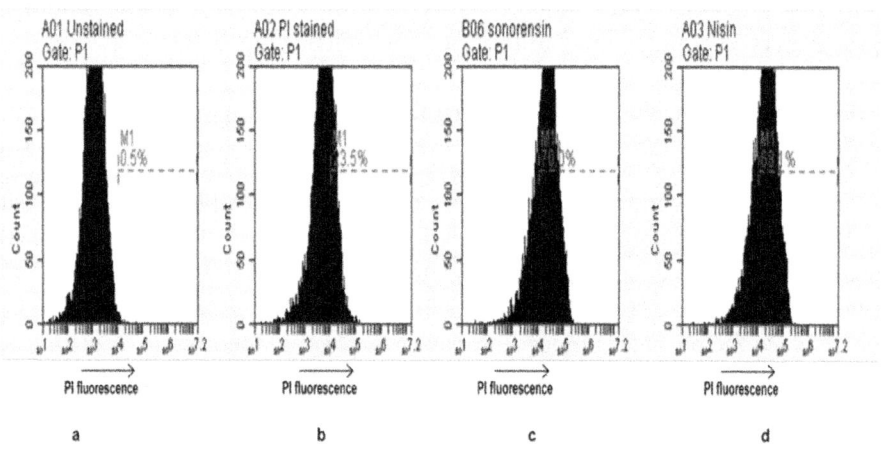

Figure 5: Flow cytometry analysis of effects of sonorensin and nisin on membrane integrity of *S. aureus* **cells.**

To further gain insight into the mode of bactericidal action of the sonorensin, SEM of *S. aureus* treated with lethal dose of sonorensin was performed. When compared to untreated cells, *S. aureus* cells pre-incubated with $50\,\mu g/ml$ of purified sonorensin for 4 h displayed major alterations like roughening of the cell surface with accumulation of cell debris and cell lysis (Fig. 6).

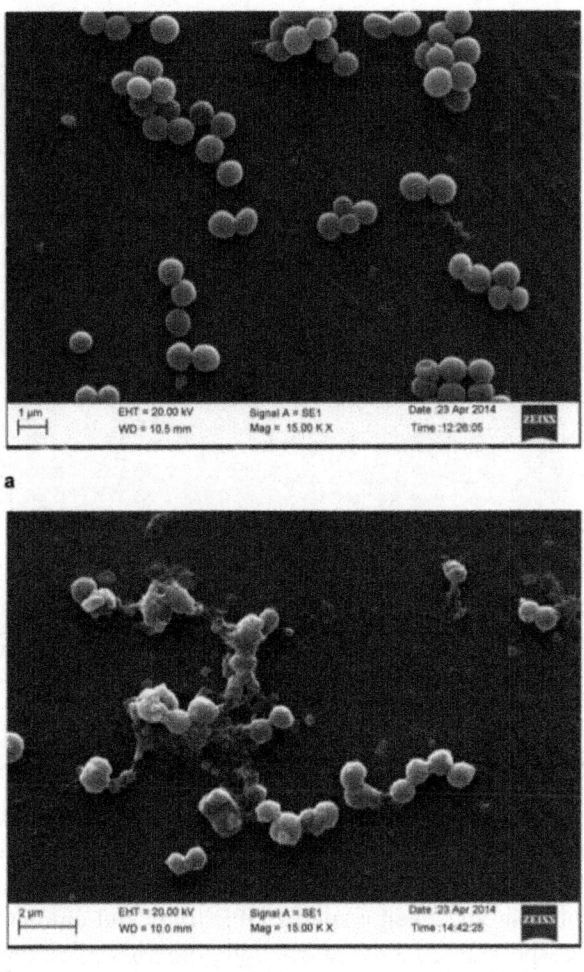

Figure 6: The scanning electron micrographs of *S. aureus* **cells**

Bioactive Polyethylene Film

The low density polyethylene film (LDPE), coated with sonorensin and nisin, showed inhibitory activity against *S. aureus* (Fig. 7). Untreated film did not show any antimicrobial activity. As shown in Fig. 7, the superficial growth of indicator strain was limited to the area surrounding the activated film that could clearly inhibit the development of the *S. aureus* in contrast, it could grew homogeneously on the surface of the plate and underneath the untreated film used as control.

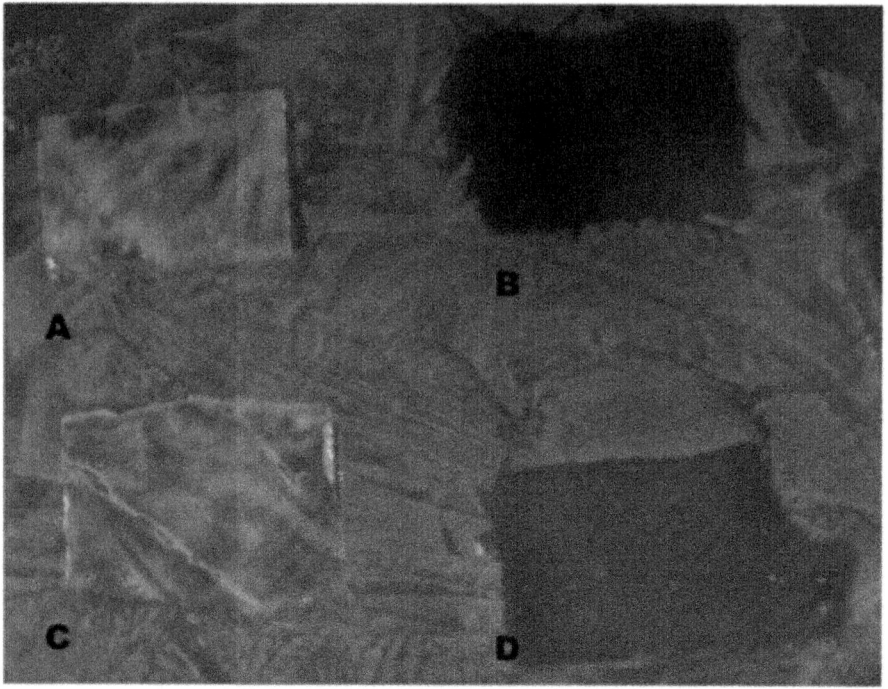

Figure 7: Inhibitory activity of coated LDPE films against *S. aureus*.

The sonorensin and nisin coated LDPE films were checked for their efficacy to inhibit the growth of food spoiling bacteria such as *S. aureus* and *L. monocytogenes*. Fresh meat spiked with these organisms and tomatoes were packed in sonoresin and nisin coated LDPE films and untreated LDPE films (control). The spoilage of both meat and tomatoes was observed in case of untreated packaging films after 4 days and 7 days of incubation at 4 °C respectively (Fig. 8). However, no signs of spoilage were seen in meat and tomatoes packed with sonorensin and nisin coated films (Fig. 8) even after 15 days of storage at refrigerated conditions. Moreover foul/stinky smell that was observed in meat samples packed with untreated films was absent in meat packed with sonorensin and nisin coated films. This suggested that like nisin, sonorensin could also be used as a food bio-preservative.

Figure 8: Preservative effect of coated LDPE film during the storage of (a) meat (b) tomatoes. (a) Meat samples were spiked with *L. monocytogenes* (1–3) and *S. aureus* (4–6). Spoilage of meat is visible in meat samples packaged in control LDPE films ...

DISCUSSION

The growth of biofilms is a significant problem within the healthcare and food industries. The characteristic resistance offered by biofilm-associated communities of microorganisms leading to their persistent survival is an important challenge to address. Biofilms have been attributed to food-borne illnesses[19,20] and can also cause premature biofouling in dairy and other processed foods. Though, many antibiotics are effective against planktonic cells, fewer are active against biofilms[21]. Furthermore, while it is suggested that bacteriocins may inhibit the development of biofilms[22], their effect upon microbial cells in a biofilm is not fully understood. When applied on preformed *S. aureus* biofilms, sonorensin was able to significantly reduce biofilm cell viability even at lower concentrations as indicated by XTT assay (Fig. 1c). SEM analysis of the biofilms also indicated that sonorensin not only inhibited biofilm formation but also caused thinning of mature biofilms (Fig. 2a,b).

The fall in the emergence of new antimicrobials in the market during the past two decades is worrying, particularly in view of the rise in bacterial resistance against many of the currently used antibiotics. Shifting the aim of drug development from multiplying to non-multiplying bacteria is expected to generate a new set of prospectives for antibiotic development and could result in the development of the drugs that would shorten the duration of therapy[8,23]. In addition to possessing a broad spectrum of antimicrobial activity, sonorensin also showed promising antimicrobial action against non-multiplying bacteria, which constitute a major cause of recurrence of chronic infections. The activity of sonorensin against non-multiplying cells of *S. aureus* was comparable to nisin. However, in case of non-multiplying cells of *E. coli*, sonorensin was effective at a comparatively higher concentration. The effectiveness of sonorensin against non-multiplying cells of *E. coli* was increased in combination with 20 mM EDTA that indicated the effects of sonorensin in reducing population being facilitated by the chelation of Mg^{2+} ions, present in the outer membrane of *E. coli*, by EDTA. This is in accordance with the reports that the removal of Mg^{2+} ions from the lipopolysaccharide layer of the outer membrane results in the loss of lipopolysaccharide and an increase in cell permeability[24]. This increase in outer membrane permeability is proposed to facilitate inactivation of the cell by sonorensin action at the cytoplasmic membrane. Similar results of the effect of EDTA on the activity of nisin against gram negative bacteria *Salmonella* spp. have been reported by Stevens *et al.*[25]. The activity of sonorensin against non-multiplying cells suggested that the bacterial resistance would not develop against sonorensin as all the bacterial cells (multiplying as well as non-multiplying) would be killed and no bacteria would survive that could evolve and develop resistance. Hu *et al.* reported an antibiotic, called HT61, active against non-multiplying bacteria, including methicillin resistant and sensitive *S. aureus*[23].

The membrane is the main barrier that limits the distribution and entry of antibiotics[18]. In addition to antimicrobial activities, bacteriocins serve as an anti-resistance compound to classic antibiotics as they are able to interact with bacterial membranes, create ion permeable channels leading to increased cytoplasmic membrane permeability and hence, bacterial cell death[26]. In case of sonorensin, it showed permeabilization effect on *S. aureus* membrane and increased the plasma membrane permeability for influx of ONPG into cells. Moreover the investigation of sonorensin treated cells stained with PI also revealed the influx of PI into the cells indicating that the cytoplasmic membrane could be the most probable target of action of sonorensin. In this sense, sonorensin could be the potential candidate for therapeutics as it does not target the cell components like nucleic acids and proteins which are the major targets of most antibiotics, leading to the targeted bacteria developing resistance against such antibiotics.

Antimicrobial packaging is a promising form of active food packaging and an emerging technology. Antimicrobial packaging film prevents microbial growth on food surface by direct contact of the package with the surface of foods. For this reason, the antimicrobial packaging film must be in contact with the surface of the food so that bacteriocins can diffuse to the surface[13]. When the food product is packaged with such films the antimicrobial substance is released slowly onto the surface of the food product thus providing protection for extended periods[27].

The results of present study of antimicrobial packaging applications on meat are quite promising. The bioactive packaging film of sonorensin showed its biopreservative effect during the storage of meat for long duration. Similar results of coating of LDPE films with bacteriocins have been previously reported[15,28,29]. Dawson et al. evaluated the effect of lauric acid and nisin impregnated soy based films on the growth of *L. monocytogenes* on turkey Bologna[30]. Ming et al.developed pediocin-coated casings that showed useful in controlling the growth of *L. monocytogenes* in meat and poultry products[31]. Sonorensin active packaged film also showed preservative effect upon storage of vegetables. These results indicated sonorensin to be a promising biopreservative agent and its incorporation in films may control the growth of undesirable bacteria, thereby extending the shelf life and enhancing the microbial safety of food products.

MATERIALS AND METHODS

Reagents and Media

Nisin and Histopaque-1077 were procured from Sigma- Aldrich Inc. (St. Louis, MO, USA) and microtitre plates were procured from Nunc (Nalge Nunc International, Denmark). BD vacutainer was obtained from BD biosciences (BD biosciences, CA, USA). All other reagents and media components used were either of analytical grade or of highest purity grade available in India.

Bacterial Strains

Bacillus sonorensis MT93 (accession numberHF944961.1) was isolated from marine soil sample collected from Parangipettai, India as reported previously[16]. Indicator strains: *B. subtilis* (MTCC 121), *S. aureus* (MTCC 1430), *E. coli* (MTCC 1610) and *L. monocytogenes* (MTCC 839) were procured from Microbial Type Culture Collection (MTCC), Chandigarh, India.

Minimal Inhibitory Concentration (MIC)

The MIC of sonorensin against *E. coli*, *S. aureus* and *L. monocytogenes*was determined as reported previously[16].

Biofilm Formation

The overnight culture of *S. aureus* was diluted to $\sim 2 \times 10^7$ CFU/ml. 200 μl of this suspension and 1% sucrose was added to wells of 96-well plates and incubated at 37 °C for 24 h. The negative control was un-inoculated media processed similarly. After incubation, the spent media was aspirated gently and wells were washed with 250 μl of PBS to remove planktonic bacteria and air-dried. 200 μl of 99% (v/v) methanol was added and incubated for 15 min for fixation and aspirated, and plates were allowed to dry. Wells were stained with 200 μl of 0.1% (v/v) crystal violet for 5 min. Excess stain was gently rinsed off and plates were air-dried. Stain was resolubilized in 200 μl of 95% (v/v) ethanol and cell concentration was measured at OD_{595} nm[32].

Biofilm Attachment Assay and Inhibition of Biofilm Formation

Biofilm attachment assay and inhibition of biofilm formation were performed as described previously[33,34]. The overnight culture of *S. aureus* (diluted to $\sim 2 \times 10^7$ CFU/ml) was added to wells of 96-well plates with different concentrations of sonorensin (in triplicate). The plates were incubated at 37 °C for 4 h and 24 h for biofilm attachment assay and inhibition of biofilm formation, respectively. The positive control was *S. aureus* in Brain heart infusion (BHI) -sucrose medium without sonorensin. After the incubation, wells were washed with PBS and the cell concentration was measured at OD_{595} nm[34].

After sonorensin treatment and incubation period of biofilm formation, XTT reduction assay was performed as a measure of metabolic activity in order to estimate viable cells. The wells were washed with PBS, and 20 μl of XTT (500 mg/ml) was added and incubated for 2 h at 37 °C. The color developed was read at 495 nm in a plate reader.

SEM

To examine the anti-biofilm activity of sonorensin by microscopy, *S. aureus* biofilm was developed on poly (L-lysine) coated cover slips. Following addition of sonorensin (50 μg/ml), cover slips were incubated for 24 h. The samples were then fixed with Karnovsky's fixative[35] for 2 h at 4 °C, washed with phosphate buffer, dehydrated with a graded ethanol series and finally by tert- butyl alcohol. Then, the cover slips were dipped in tert- butyl alcohol, kept at −20 °C followed by freeze drying and platinum coating. The samples were observed using Zeiss EVO 40 instrument (Ukraine).

EFFECTIVENESS OF SONORENSIN AGAINST NON-MULTIPLYING BACTERIA

Non-Multiplying Cell Preparation

S. aureus and *E. coli* cells were grown in 50 ml LB medium at 37 °C and 200 rpm for 7 days[23]. The 7-day-old cultures were centrifuged and washed twice with physiological buffered saline (PBS, pH 7.2), and following resuspension in same buffer were incubated at 37 °C and 200 rpm for 7 days. Cell viability was checked by counting colony-forming units (CFUs) every 24 h. The culture showed a decrease in CFU/ml in the first 4 days, and remained constant thereafter, at 3.18×10^5 and 4.2×10^6 CFU/ml for *S. aureus* and *E. coli*, respectively. The cells after incubation for 4 days were analysed for dormancy state by overnight treatment with lethal dose of ampicillin (2 mg/ml) and by comparison of regrowth curves with that of vegetative cells and were used as stocks of non-multiplying cells. Regrowth was performed in 200 μl LB medium in 96 well plate and OD_{600} nm was monitored.

Antimicrobial Activity Assay Against Non-Multiplying Cells

Different antimicrobial agents (ampicillin, nisin and sonorensin) at concentration range of 50 μg/ml–200 μg/ml were added to 100 μl of non-multiplying cell suspension of *S. aureus* and *E. coli* in 96 well plate. Cells were incubated at 37 °C and 200 rpm overnight, and then washed with PBS to remove excess antimicrobial agents and then plated in triplicate on BHI agar plates for estimation of CFU. An untreated sample, taken as control, was processed in the same way.

Effect of EDTA on Sonorensin Activity

Effect of EDTA on sonorensin activity against *E. coli* was tested in the presence of 20 mM EDTA, as described previously[25] (seeSupplementary methods online).

Sonorensin, 20 mM EDTA, buffer and sonorensin in combination of 20 mM EDTA were assayed for antimicrobial activity.

Haemolytic Activity Assay

The haemolytic activity was measured on human red blood cells (RBCs) as reported previously[36] and the protocol was approved by the Ethics Committee of the Institute. Complete lysis was measured by suspending RBC's in 1% Triton X-100[37] (see Supplementary methodsonline).

MEMBRANE DAMAGE OF BACTERIAL CELLS

Cytoplasmic Membrane Permeability

The cytoplasmic membrane permeabilization by sonorensin and nisin was investigated by using ONPG and measuring β-galactosidase activity in cells as described previously[38] (see Supplementary methods online). The hydrolysis of ONPG to O-nitrophenol over time was monitored at 405 nm with a microplate reader (Biotek, USA).

Flow Cytometry

The experiment was performed according to Joshi et al.with some modifications[37]. S. aureus cells were collected in mid exponential phase, washed three times with phosphate buffered saline, and resuspended at a concentration of 1×10^6 CFU/ml in the same buffer. This was followed by addition of 50 µg/ml of sonorensin and nisin and incubation at 37 °C for 1 h. Then, the mixtures were incubated with PI solution (10 µg/ml) for 30 min at 4 °C in the dark. PBS was used as negative control. Flow cytometry analysis was conducted using Accuri C6 Flow Cytometer in FL2 channel. For each sample 10^4 cells were analysed. Data was analyzed by C Flow Plus software (Becton Dickinson, San Jose, CA, USA).

SEM

To examine the bactericidal activity, S. aureus cells were grown in NB to an exponential phase and harvested by centrifugation. The pellet obtained was resuspended in fresh NB and aliquots of 5 ml containing about 1×10^6 cells/ml were incubated at 37 °C for 4 h with sonorensin (50 µg/ml). Samples withdrawn at 4 h were centrifuged and pellets were resuspended in 500 µl of phosphate buffer. Each sample was spread on a poly (L-lysine)-coated glass slides, fixed and observed with a Zeiss EVO 40 instrument.

Antimicrobial Polyethylene Films Preparation

Sonorensin and nisin were coated on LDPE as described previously[39]. The antimicrobial coating mixtures were prepared by adding 2 g of methyl cellulose to 10 ml of sonorensin and nisin (500 μg/ml) separately. The mixtures were then homogenised at 16,000 rpm for 2 min in a homogenizer (Silent crusher M, Heidolph, Germany). Then, 10 ml of ethanol and 4 ml of poly(ethylene glycol) (average Mn400) were added to these mixtures and re-homogenized for another 5 min. A control coating mix (without sonorensin/nisin) was prepared in a similar manner. The LDPE films (15 cm * 10 cm) were fixed on glass plates and 15 ml of the sonorensin and nisin mixtures, separately prepared, were then applied on the prepared film to have sonorensin/nisin at 20 μg/cm². The control mixture was also casted on LDPE in a similar manner. Plates were then allowed to dry at 37 °C for 2 h. After 2 h, the LDPE was detached from the glass plate and analyzed for its antimicrobial activity by placing it on BHI soft agar (0.8%) previously spread with *S. aureus* as indicator strain. The treated face of the LDPE film was in contact with agar. The control film was also tested similarly. The plates were then incubated at 30 °C overnight and observed for the presence or absence of zone of inhibition around the films in a lawn of bacterial cells.

Antimicrobial Activity of Sonorensin Coated Films during the Storage of Meat Products

The developed packaging films were used in challenge tests of control of *L. monocytogenes* and *S. aureus* growth during the storage of meat products. Following superficial spiking of chicken meat pieces with 2 ml suspension of *L. monocytogenes* and *S. aureus* at 1.5×10^6 CFU/ml, the pieces were packed with the active films and stored at 4 °C. Meat pieces packed with untreated films were included in the analysis as controls. After regular intervals of storage, the pieces were observed for the appearance of visible growth of bacteria and obnoxious smell. In another set of experiments the developed packaging films were used for preventing the spoilage during the storage of vegetables. Fresh tomatoes were packed with the active films and stored at 4 °C. Tomatoes packed with untreated films were included in the analysis as controls. After regular intervals of storage, the tomatoes were observed for the appearance of visible signs of spoilage and rottenness.

Statistical Analysis

All experiments were performed in triplicate and repeated in three independent experiments. The results were presented as the mean±standard deviations.

Statistical significance of difference between the control and the test samples was determined using ANOVA- test and $p$value<0.005 was considered as significant.

CONCLUSIONS

Sonorensin, a bacteriocin from *B. sonorensis* MT93, was found to be effective against non-multiplying cells of *S. aureus* indicating its potential for antimicrobial therapy and it did not show activity against normal mammalian cells. Sonorensin also effected the inhibition of biofilm formation. The mode of action of sonorensin as revealed by flow cytometry and SEM is the damage of bacterial membrane unlike most of the antibiotics that target cellular components and hence more prone to bacterial resistance. Furthermore, sonorensin, predicted to be the first bacteriocin of subfamily of heterocycloanthracin, demonstrated its efficacy as food biopreservative as the packaging films activated with sonorensin showed preservative effect on food products. Thus, sonorensin could prove to be promising antibiofilm agent as well as natural food biopreservative.

ACKNOWLEDGMENTS

A special thanks to Mr. Anil Theophilus for his help in Electron microscopy. LC and GS acknowledge their fellowships from DST and CSIR, Government of India, respectively.

REFERENCES

1. Monroe D. Looking for chinks in the armor of bacterial biofilms.PLoS Biol. 5, e307 (2007).

2. Hall-Stoodley L., Costerton J. W. & Stoodley P. Bacterial biofilms: from the natural environment to infectious diseases.Nature Rev. Microbiol. 2, 95–108 (2004).

3. Stewart P. S. & Costerton J. W. Antibiotic resistance of bacteria in biofilms. Lancet 358, 135–138 (2001).

4. Di Lorenzo A. *et al.* Characterization and performance of a toluene-degrading biofilm developed on pumice stones. Microb. Cell Fact. 4, 4 (2005).

5. Giaouris E. D. & Nychas G. J. E. The adherence of *Salmonella enteritidis* PT4 to stainless steel: the importance of the air-liquid interface and nutrient availability. Food Microbiol. 23, 747–752 (2006).

6. Meyer B. Approaches to prevention, removal and killing of biofilms. Int. Biodeter. Biodegr. 51, 249–253 (2003).

7. Park S. C., Park Y. & Hahm K. S. The role of antimicrobial peptides in preventing multidrug-resistant bacterial infections and biofilm formation. Int. J. Mol. Sci. 12, 5971–5992 (2011).

8. Coates A. R., Hu Y., Bax R. & Page C. The future challenges facing the development of new antimicrobial drugs. Nat. Rev. Drug Discov. 1, 895–910 (2002).

9. Coates A. R. & Hu Y. New strategies for antibacterial drug design: targeting non-multiplying latent bacteria. Drugs R D 7, 133–51 (2006).

10. Deegan L. H., Cotter P. D., Colin, H. & Ross P. Bacteriocins: biological tools for bio-preservation and shelf-life extension.Internat. Dairy 16, 1058–1071 (2006).

11. Jack R. W., Tagg J. R. & Ray B. Bacteriocins of gram-positive bacteria. Microbiol. Rev. 59, 171–200 (1995).

12. Han J. H. Antimicrobial food packaging. Food Technol. 54, 56–65 (2000).

13. Appendini P. & Hotchkiss J. H. Review of antimicrobial food packaging. Innov. Food Sci. and Emerg. Technol. 3, 113–126 (2002).

14. Padgett T., Han I. Y. & Dawson P. L. Incorporation of food-grade antimicrobial compounds into biodegradable packaging films. J. Food Prot. 61, 1330–1335 (1998).

15. An D. S., Kim Y. M., Lee S. B., Paik H. D. & Lee D. S.Antimicrobial low density polyethylene film coated with bacteriocins in binder medium. Food Sci. Biotechnol. 9, 14–20 (2000).

16. Chopra L., Singh G., Choudhary V. & Sahoo D. K. Sonorensin: an antimicrobial peptide, belonging to heterocycloanthracin subfamily of bacteriocins, from a new marine isolate *Bacillus sonorensis* MT93. Appl. Environ. Microbiol. 80, 2981–2990 (2014).

17. Chopra L., Singh G., Jena K. K., Verma H. & Sahoo D. K.Bioprocess development for the production of sonorensin by*Bacillus sonorensis* MT93 and its application as a food preservative. Biores. Technol. 175, 358–366 (2015).

18. Li L., Shi Y., Cheserek M. J., Su G. & Le G. Antibacterial activity and dual mechanisms of peptide analog derived from cell-penetrating peptide against *Salmonella typhimurium* and*Streptococcus pyogenes*. Appl. Microbiol. Biotechnol. 97, 1711–1723 (2013).

19. Hetrick E. M. & Schoenfisch M. H. Reducing implant-related infections: active release strategies. Chem. Soc. Rev. 35, 780–789 (2006).

20. Kumar C. G. & Anand S. K. Significance of microbial biofilms in food industry: a review. Int. J. Food Microbiol. 42, 9–27 (1998).

21. Shanks R. M. Q., Dashiff A., Alster J. S. & Kadouri D. E.Isolation and identification of a bacteriocin with antibacterial and antibiofilm activity from *Citrobacter freundii*. Arch. Microbiol.194, 575–587 (2012).

22. Hancock V., Dahl M. & Klemm P. Probiotic *Escherichia coli*strain Nissle 1917 outcompetes intestinal pathogens during biofilm formation. J. Med. Microbiol. 59, 392–399 (2010).

23. Hu Y., Tousi A. S., Liu Y. & Coates A. A new approach for the discovery of antibiotics by targeting non-multiplying bacteria: A novel topical antibiotic for *Staphylococcal* infections. PLoS One5, e11818 (2010).

24. Nikaido H. & Vaara M. Outer membrane. in Escherichia coli and Salmonella S. typhimurium: cellular and molecular biology (ed. Neidhardt F. C.) American Society for Microbiology, Washington, DC. 1, 7–22 (1987).

25. Stevens K. A., Sheldon B. W., Klapes N. A. & Klaenhammer T. R. Nisin treatment for inactivation of *Salmonella* species and other Gram-negative bacteria. Appl. Environ. Microbiol. 57, 3613–3615 (1991).

26. Yeaman M. R. & Yount N. Y. Mechanisms of antimicrobial peptide action and resistance. Pharmacol. Rev. 55, 27–55 (2003).

27. Kim H. J., Lee N. K., Paik H. D. & Lee D. S. Migration of bacteriocin from bacteriocin -coated film and its antimicrobial activity. Food Sci. and Biotechnol. 9, 325–329 (2000).

28. Lee C. H., An D. S., Lee S. C., Park H. J. & Lee D. S. A coating for use as an antimicrobial and antioxidative packaging material incorporating nisin and α-tocopherol. J. Food Eng. 62, 323–329 (2004).

29. Quintavalla S. & Vicini L. Antimicrobial food packaging in meat industry. Meat Sci. 62, 373–380 (2002).

30. Dawson, P. L., Carl, G. D., Acton, J. C. & Han, I. Y. Effect of lauric acid and nisin impregnated soy-based films on the growth of *Listeria monocytogenes* on turkey bologna. Poult. Sci. 81, 721–726 (2002).

31. Ming X., Weber G. H., Ayres J. W. & Sandine W. E. Bacteriocins applied to food packaging materials to inhibit *Listeria monocytogenes* on meats. J. Food Sci. 62, 413–415 (1997).

32. Durham-Colleran M. W., Verhoeven A. B. & van Hoek M. L.*Francisella novicida* forms *in vitro* biofilms mediated by an orphan response regulator. Microb. Ecol. 59, 457–465 (2010).

33. Overhage J. *et al.* Human host defense peptide LL-37 prevents bacterial biofilm formation. Infect. Immun. 76, 4176–4182 (2008).

34. Mataraci E. & Dosler S. *In vitro* activities of antibiotics and antimicrobial cationic peptides alone and in combination against methicillin resistance *Staphylococcus aureus* biofilms.Antimicrob. Agents Chemother. 56, 6366–6371 (2012).

35. David G. F., Herbert J. & Wright G. D. The ultrastructure of the pineal ganglion in the ferret. J. Anat. 115, 79–97 (1973).

36. Mangoni M. L. *et al.* Structure–activity relationship, conformational and biological studies of temporin L analogues. J Med. Chem. 54, 1298–1307 (2011).

37. Joshi S. *et al.* Interaction studies of novel cell selective antimicrobial peptides with model membranes and *E. coli* ATCC 11775. Biochim. Biophys. Acta 1798, 1864–1875 (2010).

38. Ibrahim H. R., Sugimoto Y. & Aoki T. Ovotransferrin antimicrobial peptide (OTAP-92) kills bacteria through a membrane damage mechanism. Biochim. Biophys. Acta 1523, 196–205 (2000).

39. Franklin N. B., Cooksey K. D. & Getty K. J. Inhibition of*Listeria monocytogenes* on the surface of individually packaged hot dogs with a packaging film coating containing nisin. J. Food Prot. 67, 480–485 (2004).

Chapter 4

BIOPRESERVATION OF TOMATO PASTE AND SAUCE WITH LEUCONOSTOC SPP. METABOLITES

Djadouni Fatima, Kihal Mebrouk and Heddadji Miloud

Laboratory of Applied Microbiology, Department of Biology, Faculty of Sciences, Es-Senia University, Oran, Algeria

ABSTRACT

The aim of this study was to evaluate 40 lactic acid bacteria (LAB) and 20 Bacillus strains isolated from the fermented tomato (Solanum lycopersicum) for their capacity to produce antimicrobial activities against several bacteria and fungi. The strain designed LBc03 has been selected for advanced studies. The supernatant culture of this strain inhibits the growth of Escherichia coli, Staphylococcus aureus and Aspergilus sp. Based on the cultural, morphological, physiological and biochemical characteristics, LBc03 was identified as Leuconeustoc spp. Its antimicrobial compound was determined as a proteinaceous substance, but it is possible that the bacteriocin may also be bound to other molecules like a lipid or a carbohydrate moiety. Metabolite extracts from selected LAB were more effective in preserving tomato paste and sauce stored at 4°C against spoilage bacteria like E. coli and the application of bio-preservative should be encouraged in food processing industries.

INTRODUCTION

Microbial spoilage of fruits and vegetable is known as rot, which manifests as loss of texture (soft rot), changes in color (black or grey) and often off odor (Trias et al., 2008). Also, the high water content in tomatoes makes it very susceptible to spoilage bacteria and fungi during storage, harvesting and transportation (Spadaro and Gullino, 2004). Fresh food like fruits and vegetables, are normal part of the human diet and are consumed in large quantities in most countries. These products are rich in carbohydrates and poor in proteins with pH value from slightly acidic to 7.0 and provide a suitable niche to several bacteria, yeasts and moulds (Wiessinger et al., 2000; Trias et al., 2008 ; Ogunbanwo et al., 2014). Tomato (Solanum lycopersicum) is one of the highly nutritious food ingredient used in the preparation of food all overthe world (Ogunniyi and Oladejo, 2011; Ogunbanwo et al., 2014). Its utilization as an ingredient in vegetable salads, other dishes and its processing into different products like puree, ketchups and juiceis well documented. Nutritionally, it contains a large amount of water, niacin, calcium and vitamins especially A, C, E which are important in the metabolic activities of man and protects the body against diseases (Taylor, 1987). Lycopene (acarotene) an essential component of tomato contributes in the prevention of cardiovascular disease and cancer of the prostrate (Clinton, 1998; Bernard et al., 1999).

Table 1: S01, S02, S03, and S04 tomatoes Solanum lycopersicum samples

Samples	Origin	Tomato Variety	Characteristics	Additives added	Temperature of storage (°C)	Storage time
S01[C1]	Mascara, Algeria	*Solanum lycopersicum*	Fresh-cut tomato	Absence	25	24 h
S02[C2]	Mascara, Algeria	*Solanum lycopersicum*	Tomato paste was boiled for 90°C for 02 h	Absence	25	24 h
S03	Mascara, Algeria	*Solanum lycopersicum*	Fresh-cut tomato	Absence	-20	06 months
S04	Mascara, Algeria	*Solanum lycopersicum*	Tomato paste was boiled for 90°C for 02 h	05% NaCl and oil	+4	06 months

C1and C2 are control samples.

The characteristic flavor of tomato is produced by the complex interaction of the volatiles and non-volatile components (Petro-Turza, 1987; Buttery, 1993). The nutritional value of tomato products is a topic attracting much attention, particularly regarding the effects resulting from food processing and storage treatments (Capanoglu et al., 2010). Among the common post-harvest fungal pathogens of tomatoes are Pencillium expansum, Monilinia laxa and Rhizopus stolinifer (Ogawa et al., 1995; Pla et al., 2005). Many LAB strains are able to produce protein compounds with efficient antimicrobial effect, which are known as bacteriocins (Davidson and Harrison, 2002). In recent times, the understanding of the preservation mechanisms of LAB is being exploited for industrial production of foods (Trias et al., 2008) because of their natural acceptance as Generally Recognized as Safe (GRAS) for human consumption and exhibit antimicrobial property (Aguirre and Collins, 1993). There is a complimentary effect by the production of acid and antimicrobial compounds that increases inhibition of both pathogen and spoilage bacteria (Edwards et al., 1983; Artés et al., 1999). Although many efforts have been made to develop bio-protective lactic acid bacteria strains, the application of these strains in fresh fruits and vegetables have not been developed yet (Toivonen and DeEll, 2002). This work is designed to investigate the effectiveness of lactic acid bacteria and Bacillus metabolites in preserving tomato paste and sauce against S. aureus, E. coli, Clostridium sp., Aspergillus sp., and Penicillium sp.

MATERIALS AND METHODS

Samples Collection

Fresh tomato samples Solanum lycopersicum (Table 1) were purchased from markets in Mascara, Algeria. The samples were collected in separate sterile polythene bags and immediately transported to the laboratory for analysis. Determination of the physical-chemical characteristics of the tomato samples (S01, S02, S03, and S04)

Moisture Content

The moisture content of the samples (S01, S02, S03, and S04) was determined before storage time by weighing into moisture cans then weighed and then placed in an oven at 80°C for 24 h to dry to a constant weight. Then, brought out and allowed to cool in desiccators and then reweighed (A.O.A.C., 1995).

$$\text{Thus. moisture content} = \frac{(\text{initial weight} - \text{final weight})}{\text{initial weight}} \times 100$$

Dry Matter Content

Five grams of each sample were obtained and placed into preweighed crucibles and dried in at 100°C for 12 h. The dried samples were weighed after cooling in the desiccators (A.O.A.C., 1984).

Determination of Lactic Acid Produced By Lab Isolates

This was achieved based on the methods described (Ogunbanwo et al., 2008; Bamidele et al., 2011). For these measurements lactic acid was determined by transferring 25 ml of the tomato juice (S01, S02, S03, and S04) into conical flasks and 3 drops of phenolphthalein were added as indicator. From a burette, 0.1 M NaOH was slowly added to the samples until a pink color appeared. Each ml of 0.1 M NaOH is equivalent to 90.08 mg of lactic acid.

Determination of the Ph

pH was determined after and before fermentation of the tomato samples (S01, S02, S03, and S04) by the used of pH meter (Inolab MLM) (Lehninger, 1981).

Detection of Pathogens, Contaminants, LAB, and Bacillus

The main objective of this experiment was to detect microbial contaminants in tomato samples (S01, S02, S03, and S04) in addition to other related hygiene tests. Total aerobic bacterial count, total coliform count, Total anaerobic bacteria count, Salmonella, S. aureus, Clostrudium, lactique acid bacteria, and Bacillus were detected (Delarras et al., 2006).

Isolation and Selection of LAB Strains

LAB strains were isolated from the tomato samples (S01, S02, S03, and S04). The samples were plated directly on MRS as detailed by De Man, Rogosa and Sharpe (Merck, Germany) and M17 agar (Merck, Germany) at 30°C for 2-3 days (pH6.5) under aerobiosis and anaerobiosis conditions. They were routinely propagated and stored at − 20°C supplemented with glycerol (20%, v/v, final concentration). Working cultures were sub-cultured twice (1 inoculum, 24 h, 30°C) prior to use (Djadouni and Kihel, 2012; Ogunbanwo et al., 2014).

Identification of the LAB Isolates

The pure isolate selected as a potential bacteriocin - producer was identified on the basis of its cultural, morphological, physiological and biochemical characteristics. The selected LAB isolates were characterized by Gram stain, absence of spores and catalase test. Gram+, catalase and spores negative strains were maintained frozen until needed for the antimicrobial activity testing. Confirmation of the identification was based on the use of Bergey's manual of systems bacteriology (Sneath et al., 1986).

Isolation and Identification of Bacillus Strains

Samples of tomato were weighed as 1 g portions and thoroughly homogenized insterile distilled water; serial dilutions were plated on LB agar (Luria–Bertani media, Merck, Germany) plates were incubated at 30°C for 2-3 days (Kalil et al., 2009). The pure isolate was examined macroscopically and microscopically and identified with reference to Holt et al., (1994). Isolates were identified by colonial appearance, gram positive, and presence of spores, the presence or absence of β-haemolysis, lecithinase activity, motility, penicillin susceptibility and biochemistry (Abriouel et al., 2010).

The Indicator Strains

The indicator strains (S. aureus, E. coli, Clostriduim sp., Aspergillus sp., and Penicillium sp.) used in this work was provided by the Laboratory of Bacteriology, Microbiology Department at the Faculty of Sciences, Es-Senia, Oran University, Algeria. For the antimicrobial assay, the pathogenic cultures were grown in the nutrient agar media (NA) at pH, 7.4 (Ogunbanwo et al., 2014). For antifungal activities determination, Aspergillus sp. and Penicillium sp. were grown in potato dextrose agar (PDA, Merck, Germany) for 7 days at 30°C. Spores were collected in sterile distilled water and then concentrated to 104 spores ml-1 (Smaoui et al., 2010).

Preparation of Cell-Free Filtrate

MRS broth (1000 µl) were inoculated separately with isolates (LAB or Bacillus) previously characterized and incubated at 30°C for 72 h. After incubation, a cell free supernatant was obtained by centrifuging (Spectrafuge 24D, Labnet, USA) the bacterial culture at 10.000 rpm for 45 min, followed by filtration of the supernatant through 0.2 mm pore size filter paper thus obtaining cell free filtrate (Khalil et al., 2009; Djadouni and Kihel, 2013).

Antagonistic Activity of LAB and Bacillus Metabolites against Spoilage Microorganisms

Sixty isolates (40 LAB and 20 Bacillus) were grown in MRS broth for 72 h at 30°C and the broth cultures were centrifuged at 10.000 rpm for 30 min and the supernatant containing the metabolites were obtained and 100 µL of the supernatant was transferred into wells (6 mm diameter) bored in Muller Hinton and potato dextrose agar previously seeded with the spoilage bacteria cells and fungi spores. The culture plates were incubated at 30 °C for 48 h and 7 days respectively and observed for zones of inhibition (Ogunbanwo et al., 2014).

Characterization of the Antimicrobial Substance

The isolated crude antimicrobial substance was characterized with respect to the effect of proteolytic enzymes on the bacteriocin activity. Selected enzymes were tested on the cell free supernatant (Bizani and Brandelli, 2002). Proteolytic enzymes including trypsin, pepsin, and papain were dissolved in 40 mM Tris-HCl (pH, 8.2), 0.002 M HCl (pH, 7), and 0.05 M sodium phosphate (pH, 7.0) respectively to a final concentration of 0.1 mg ml-1 . Other enzymes such as lipase and α-amylase were dissolved in 0.1 M potassium phosphate (pH, 6.0), and 0.1 M potassium phosphate (pH, 7.0) respectively to a final concentration of 0.1 mg/ml. Equal aliquots of both filter sterilized of each test strain and each enzyme solution were mixed, incubated at 30°C for each enzyme for 2 h and heated in boiling water for 5 min to inactivate the enzymes. These sample mixtures and the controls (without enzyme treatment) were inoculated with the indicator strains as previously mentioned and tested for antimicrobial activity by the optical density method (ODM) (Smaoui et al., 2010; Djadouni and Kihel, 2013).

Shelf Life Study of Tomato Paste and Sauce

The tomato paste and sauce were boiled for 10 min and dispensed in 20 g amount separately into three pre-sterilized containers. Twenty milliliters of the crude bacteriocin-like substance of Leuconostoc spp. were added (v/v) differently to the tomato paste and sauce inoculated with E. coli 107 CFU g-1 (2 ml) and stored at 4°C for 72 h. Microbial load of each treatment was monitored by determining the colony forming unit (CFU ml-1) of E. coli on the hektoen medium (Safdar et al., 2010; Cottaz et al., 2008).

Statistical Analysis

Data were expressed as mean ± standard deviation. Statistical significance was determined using one-way analysis of variance on the replicates, where a p-value of ≤ 0.05 was considered significant.

RESULTS AND DISCUSSION

The results of the physical and chemical analyses of tomato samples are shown in Table 2. The percentage of moisture contents decreased in S02 and S04 samples to 60.30 and 62.50%; this may be due to the boiling at 90°C for 2 h and the production process steps of tomato paste, that is made by cooking tomatoes for several hours to reduce moisture, straining them to remove the seeds and skin (Adam, 1998; Ife Fitz and Bas uipers 00 r gera et al., 2007; Souci et al., 2008). Dry matter content was

Table 2: Physical-chemical characteristics of tomatoes samples (S01, S02, S03, and S04)

Samples	Moisture content %	Dry matter content (g)	lactic acid content (°D)	pH	Firmness, taste and color
S01	94.20 ± 0.1	5.9 ± 0.1	0.99 ± 0.02	04.20±0.03	No change
S02	60.30 ± 0.1	5.7 ± 0.2	0.75 ± 0.02	04.43±0.03	Change in color
S03	94.20 ± 0.1	5.8 ± 0.1	0.89 ± 0.02	04.28±0.02	No change
S04	62.50 ± 0.1	5.9 ± 0.1	01.58 ± 0.02	03.70±0.01	Change in color, taste, and production of CO_2

stable along the different time and temperature of storage in S01, S02, S03 and S04 from 5.9 g to 5.7 g, but the pH values decreased to 03.70 in sample S04 that was stored at 4°C for six months, and lactic acid concentration increased to 01.58 D° during this storage period with an important change in the color, taste, and CO2 production; this changes may be due to the thermal treatment, fermentation process and organic acids (citric acid, ascorbic acid, and lactic acid) produced by LAB during storage in anaerobic conditions in presence of oil and NaCl (Jean-Louis, 2007). The fresh-cut tomato S03 was not affected during cold storage and shelf-life. On the other hand, S01 and S02 did not show significant differences among them. Color of fresh-cut tomatoes was significantly affected by the antimicrobial treatments and the storage time (AyalaZavala et al., 2008). Previous experiments demonstrated that firmness losses of fresh-cut tomato slices were probably linked to the ripening stage of whole fruit at processing and storage temperature. In addition, processing operations may have triggered important losses in membrane integrity due to mechanical stressing of plant tissues. In this regard, diminishing membrane integrity could cause a loss of texture related enzymes and their substrates, leading to a rise in Xuids and solute exchanges as well as an increase in the enzymatic activity (Ayala-Zavala et al., 2008). Results obtained in this study show that transformation process, temperature and storage period can affect the tomato components and quality (Goodman et al., 2002; Chong et al., 2009; Ife Fitz and Bas Kuipers, 2003). During transport and storage, intactfruitsandvegetablesarepronetodeleterious changes induced by respiratory, metabolic and enzymatic activities, as well as by desiccation, pests, microbial spoilage, and temperature-induced injury. Many of these changes adversely affect the antioxidant status of the tomato products (Lindley, 1998). Capanoglu et al. (2010), worked closely with a Turkish tomato paste factory and obtained samples from each step during the paste production process, from the tomatoes arriving at the factory gate to the final canned product. Both targeted (for carotenoids) and untargeted (LC-MS of semi-polar, hydrophilic extracts) metabolic approaches were used to follow biochemical changes after each step. Results show that a multitude of modifications took place, involving both increases and decreases in individual components. Those steps causing the greatest changes were identified and predictions were made as to how these steps could be tackled in a modified processing strategy to improve the antioxidant capacity of the end product. Gahler et al. (2003) investigated home preparation methods such as peeling, tomato soup preparation, etc., and three different steps of tomato juice production including sieving, homogenization, sterilization, filling, and pasteurization. Results suggested that homogenization increased the hydrophilic antioxidant capacity of the different tomato products. However, the exact mechanism still remains unclear. Similarly, in industrial processing, Capanoglu et al. (010) also showed the "breaker" or homogenization step most significantly altered the biochemical composition of tomato paste. Abushita et al. (2000) also analyzed samples taken from three steps of paste processing (raw tomato, crushed sieved puree, and pasteurized paste) which were obtained from a canning factory in Hungary. The results show that the contents of ascorbic acid and tocopherols decreased during processing while carotenoids either remained unchanged or were found to increase. Absence of Salmonella, S. aureus, and Clostridium in the tomatoes samples (Table 3) showed that thermal treatment and cold storage of tomato inhibit the growth and developments of pathogens and spoilage microorganisms. However, thermal treatments applied during industrial preparation of tomato products may involve various chemical reactions leading to the degradation of these antioxidants. Besides, addition of vegetable oil for the preparation of tomato sauce may lead to lipid oxidation contributing to the micro-constituent instability (Chanforan et al., 2010a and b). The presence of some aerobic mesophilic microorganisms, coliforms, and anaerobic bacteria compared to control samples was related to the averments conditions and preparations methods of tomatoes. Ayala-Zavala et al. (2008) suggest that mechanical damage during minimal processing enhance contamination by epithelial microflora, and promote leaking of nutrients, which are rich substrates to microorganisms, supporting thefast microbial development in contrast with intact tissue. The highest count of LAB and Bacillus in S04 sample was explained by the storage of tomato paste for 06 month at 4°C in the presence of oil and NaCl, which favor the fermentation and organics acids production that

Table 3: Detection of pathogens, contaminants, LAB, and Bacillus in the tomato samples (S01, S02, S03, and S04)

Samples	Total aerobic bacterial count (CFU ml^{-1})	total coliform count (CFU ml^{-1})	Total anaerobic bacteria count (CFU ml^{-1})	Salmonella (CFU ml^{-1})	Clostrudium (CFU ml^{-1})	S. aureus (CFU ml^{-1})	LAB (CFU ml^{-1})	Bacillus (CFU ml^{-1})
S01	86.10^5	36.10^5	30.10^5	ABS	ABS	ABS	ABS	20.10^5
S02	55.10^5	17.10^5	34.10^5	ABS	ABS	ABS	20.10^5	25.10^5
S03	76.10^5	ABS	ABS	ABS	ABS	ABS	47.10^5	29.10^2
S04	ABS	25.10^5	36.10^5	ABS	ABS	ABS	60.10^6	72.10^5

* ABC, absence.

decreased the pH of products and increased also the CO_2 production; these conditions inhibit the growth of contaminants in the tomato paste (Buta and Moline, 1998; Benkeblia, 2004). Also, this LAB and Bacillus have antibacterial and antifungal activities against a variety of Gram-negative and Gram-positive bacteria and fungi. Related to fungal development, the highest counts were observed in control samples (Siboukeur, 2011). A total of 60 isolates were obtained from fermented tomato paste and screened for antimicrobial spectrum against the Gram-positive, Gram-negative bacteria, and fungi using the well diffusionmethod. The average diameter of the inhibition zone measured ranged from 2 to 4 mm size. One strain of LAB was selected for further studies, because

LBc03 contained antimicrobial compound with wide spectrum that inhibited the growth of three indicator strains E. coli, S. aureus, and Aspergilus sp but did not inhibited Penicillium sp. and Clostridium sp. On the basis on their positive Gram reaction, non-motility, absence of catalase activity and of spore formation, the rod or coccal shape, physiological and biochemical characters as well as sugar utilization pattern, LBc03 was identified as Leuconostoc spp. (Table 4). The isolation of LAB species from healthy tomato fruits is in accordance with findings of Sajur et al. (2007) and Settanni and Corsetti, (2008). The presence of LAB in tomato fruit is attributed to their high survival in post harvest conditions of tomatoes (Trias et al., 2008; Ogunbanwo et al., 2014). The LAB ability to produce antimicrobial compounds is due the absence of true catalyses to break down hydrogen peroxide generated which accumulates and becomes inhibitory to some organisms. The antagonistic activity of LAB metabolites against the spoilage bacteria and fungi agrees with the findings of Trias et al. (2008). The inhibitory effect of lactic acid is due to undissociated forms of the acids which penetrates the pathogen's membrane and liberate hydrogen ion in the neutral cytoplasm thus inhibiting vital cell functions (Corleh and Brown, 1980; Adeniyi et al., 2006). Diacetyl is known to be very effective against fungi and this is due to the interference with the utilization of arginine (De Vyust and Vandamme, 1994) and in addition to a strong oxidizing effect on the organisms cell especially bacteria (Condon, 1987; Sangorrin et al., 2014). The Leuconostoc spp. (LBc03) bacteriocin-like substance produced was inactivated by the tested enzymes (Table 5). However, since the activity of the filtrate was not completely inhibited, it is possible that the bacteriocin may also be bound to other molecules like a lipid or a carbohydrate moiety. These data clearly show that the bacteriocin-like substance was of a proteinaceous nature. Similar results were obtained (Joshi et al., 2006; Diop et al., 2008; khalil et al., 2009; Chanforan et al., 2010a and b; Rakshita, 2011). Bacteriocin-like substance produced by LBc03 resulted in the decrease of E. coli from an initial population of 107 to 05.103CFU mL-1 and 1.103 CFU mL-1 in tomato paste and sauce respectively (Figure 1). This reveals a possible potential of the LAB metabolites in the retardation of food spoilage which agrees with the findings of Ogunbanwo et al. (2008). The bio-preservative potential of LAB metabolites has been tested on other food product like suya (Adesokan et al., 2008) and chicken meat (Ogunbanwo and Okanlawon, 2006). A major advantage in the use of lactic acid bacteria and their metabolites is that they are considered as generally recognized as safe (GRAS) and comply often with the recommendations for food products (Stiles and Holzapfel, 1997). Unlike some chemical preservatives, LAB metabolites have not been reported to have residual effect on the food product or the consumer's health.

CONCLUSION

Our bacteriocin-like substance revealed interesting properties that justifies it importance regarding food safety and protection.

Table 4: Morphological, physiological and biochemical properties of LAB isolate (LBc03)

Test	Isolate LBc03
Morphology	Rod, circular and white colonies 3.0mm
Growth at temperature (°C)	
10	+
15	+
30	+
37	+
45	+
Growth at pH	
3.5	-
4.5	+
5.5	+
6.5	+
7.5	+
8.5	+
9.5	w
Growth at NaCl %	
4	+
6.5	+
10	+
15	+
CO_2 from glucose	+
CO_2 from gluconate	+
ADH	+
Citrate	+
Thermoresistance at 60°C for 30 min at 45°C	+
Fermentation Type	He
Fermentation of: glucose, saccharose, and lactose.	+
Lait Sherman (1% BM) at 42°C	+
Lait Sherman (3% BM) at 42°C	-
Hydrolyse of caseine	+
Fructose	+
Mannane	+
Maltose	+
Trehalose	+
Manose	+
Melibiose	-
Palmitine	-
Raffinose	+
Xylose	-
Mannitol	+

+, Growth; -, no growth; w, weak growth; He, hetero-fermentation.

Table 5: Effect of enzymes treatment (1 mg/mL) on bacteriocin activity against S. aureus Results are expressed as % of means values of growth reduction (n= 3) ± standard deviations

Enzymes	Enzyme concentration (1 mg mL⁻¹)
Pepsin	38.50 ± 0.5
Trypsin	40.30 ± 0.5
α- Amylase	41.80 ± 0.3
Lipase	40.80 ± 0.5

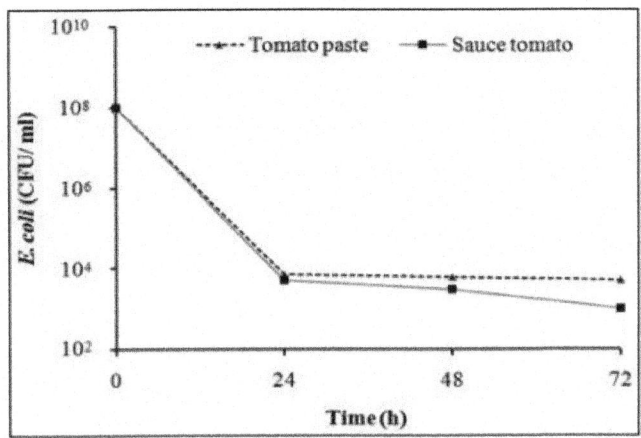

Figure 1: Reduction of E. coli population in paste and sauce tomato treated with natural antimicrobial of LBc03 isolate and stored at 4°C for 03 days.

ACKNOWLEDGEMENT

This research work was supported by Microbiology Research Laboratory. Faculty of Sciences, Es-Senia University, Oran, Algeria.

REFERENCES

1.   A.O.A.C. (1984). Official methods of analysis, 14th revised edition, Association of Official Analytical Chemists. Washington. D.C. USA G. Eason B. Noble and I. N. Sneddon "On certain integrals of

2.   Lipschitz-Hankel type involving products of Bessel functions "

3.   A.O.A.C. International (1995). Official Methods of Analysis of AOAC International. 16th revised edn. Association of Official Analytical Communities. Arlington, VA, USA.

4.   Abriouel H, Franz C, Ben Omar N, Galvez A (2010). Diversity and applications of Bacillus bacteriocins. FEMS Microbiol. Rev. 35:201-232.

5.   Abushita AA, Daood HG, Biacs PA (2000). Change in carotenoids and

antioxidant vitamins in tomato as a function of varietal and technological factors. J. Agric. Food Chem. 48: 2075–2081.

6.   Adam E (1998). Solar drying of sliced onion and quality attributes as affected by the drying process and storage conditions. Dissertation.

7.   Stuttgart: Universitat Hohenheim, Institut für Agrartechnik in den Tropen und Subtropen.

8.   Adeniyi BA, Ayeni FA, Ogunbanwo ST (2006). Antagonistic activities of lactic acid bacteria isolated from Nigerian fermented dairy food against organisms implicated in urinary tract infection. Biotechnology5(2): 183-188.

9.   Adesokan IA, Odetoyinbo BB, Olubamiwa AO (2008). Biopreservative activity of lactic acid bacteria on suya produced from poultry meat. Afr. J. Biotechnol. 7(20): 3799-3803.

10. Aguirre M, Collins MD (1993). Lactic acid bacteria and human clinical infection. J. Appl. Bacteriol. 75(2): 95-107.

11. Artés F, Conesa MA, Hernandez S, Gil MI (1999). Keeping quality of fresh-cut tomato. Postharvest Biol. Technol. 1: 153-162.

12. Ayala-Zavala JF, Oms-Oliu G, Serrano IO, González AGA, ÁlvarezParrilla E, Martin-Belloso O (2008). Biopreservation of fresh-cut tomatoes using natural antimicrobials. Eur. Food. Res. Technol. 226:1047-1055.

13. Bamidele TA, Adeniyi BA, Ogunbanwo ST, Smith SI, Omonigbehin EA (2011). Antibacterial activities of lactic acid bacteria isolated from selected vegetables grown in Nigeria: A preliminary report SierraLeone. J. Biomed. Res. 3(3): 128-132.

14. Benkeblia N (2004). Antimicrobial activityof essential oil extracts of various onions (Allium cepa) and garlic (Allium sativum) Lebensm. Wiss. Technol. 37(2):263-268.

15. Bernard OE, Obioma EN, Fabian JO (1999). Growth inhibition of tomato-rot fungi by phenolic acids and essential oil extracts of pepper fruit (Dennetiatripetala). Food Res. Int. 32: 395-399.

16. Bizani D, Brandelli A (2002). Characterization of a bacteriocin produced by a newly isolated Bacillus sp. strain 8 A. J. Appl. Microbiol. 93:512-519.

17. Buta IT, Moline HE (1998). Methyl jasmonate extends shelf life and reduces microbial contamination of fresh-cut celery and peppers. J. Agric. Food Chem. 46: 1253-1256.

18. Buttery RG (1993). Quantitative and sensory aspects of flavor of tomato and other vegetables and fruits. In: Flavor Science: Sensible Principles

And Techniques. Acree, T. and Teranishi, R. (eds.). Am.Chem. Soc. Books, Washington. D.C. pp. 259-286.

19. Capanoglu E, Beekwilder J, Boyacioglu DR, Hall RD (2010). The Effect of industrial food processing on potential ly health-beneficial tomato antioxidants. Crit. Rev. Food. Sci. Nutr. 50: 919–930.

20. Chanforan C, Caris-Veyrat C, Dufour C (2010b). Evaluation of the stability of tomato and sunflower oil microconstituents in systems modelling the processing of tomato paste into tomato sauce. UMR 408 UAPV-INRA d'Avignon.

21. Chanforan C, Caris-Veyrat C, Dufour C(2010a). Stability of tomato microconstituents (phenolic compounds, carotenoids, vitamins C and E) during processing: studies in model systems, development of a stoichio-kinetic model and validation for the unit step of tomato sauce preparation. UMR 408 UAPV-INRA d'Avignon.

22. Chong HH, Simsek S, Reuhs BL (2009). Analysis of cell-wall pectin from hot and cold break tomato preparations. Food Res. Int. 42: 770-772.

23. Clinton SK (1998). Lycopene; chemistry, biology and implications for human health and diseases. Nutr. Rev. 56(2): 35-51.

24. Condon S (1987). Responses of lactic acid bacteria to oxygen. FEMS Microbiol. Lett. 46(3): 269-280.

25. Corleh DA, Brown, MH (1980). pH and activity In: (Col) Microbial Ecology of Foods, Silkier. J. H. Academic Press, New York.

26. Cottaz A, Oulahal N, Sebti I (2008). Effet des ultra-violets sur les microorganismes de contamination de la tomate. Laboratoire d›InfoRmatique en Image et Systèmes d›information Davidson PM, Harrison MA (2002). Resistance and adaptation to food antimicrobials, sanitizers, and other process controls. Food Technol.56(11):69-78.

27. De Vyust L, Vandamme EJ (1994). Bacteriocins of lactic acid bacteria: Microbiology, genetics and applications. Blackie Academic and Professional, London. pp. 151-221.

28. Delarras C (006). Microbiologie pratique pour le laboratoire d'analyses ou de contrôle sanitaire. ISBN: 978-2-7430-0945-8, Lavoisier 2007, Paris.

29. Diop MB, Dauphin DR, Dortu C, Destain J, Tine E, Thonart P (2008). In vitro detection and characterization of bacteriocin-like inhibitory activity of lactic acid bacteria (LAB) isolated from Senegalese local food products. Afr. J. Microbiol. Res. 2: 206-216.

30. Djadouni F, Kihal M (2013). Characterization and determination of the

factors affecting anti-listerial bacteriocins from Lactobacillus plantarum and Pediococcus pentosaceus isolated from dairy milk products. Afr. J. Food. Sci. 7(2): 35-44.

31. Djadouni, F, Kihal M (2012). Antimicrobial Activity of Lactic Acid Bacteria and the Spectrum of their Biopeptides against Spoiling Germs in Foods. Braz. Arch. Biol. Technol. 5 (3): 435-443.

32. Edwards J, Saltveit M E, Henderson WR (1983). Inhibition of lycopene synthesis in tomato pericarp tissue by inhibitors of ethylene biosynthesis and reversal with applied ethylene. J. Am. Soc. Hort. Sci. 108: 512-514.

33. Sangorrín MP, Lopes C A, Vero S, Wisniewsk, M (2014). Cold-adapted yeasts as biocontrol agents: Biodiversity, adaptation strategies and biocontrol potential. Book Part. pp. 441-464.

34. Gahler S, Otto K, Bohm V (2003). Alterations of vitamin C, total phenolics, and antioxidant capacity as affected by processing tomatoes to different products. J. Agric. Food Chem. 51: 7962-7968.

35. Goodman CL, Fawcett S, Barringer SA (2002). Flavor, viscosity, and color analyses of hot and cold break tomato juices. J. Food. Sci. 67: 404-408.

36. Holt JG, Krieg NR, Sneath PHA, Staley JT, Williams ST (1994). Editors. In: Bergey's Manual of Determinative Bacteriology. 9th ed. Baltimore: Williams and Wilkins. p. 559.

37. Jean-Louis I (2007). Les relations aliments, micro-organismes, et consommateurs. Microbiologie alimentaire-STIA 2, Université Montpellier II.

38. Joshi VK, Sharma S, Rana N (2006). Production, purification, stability and efficacy of bacteriocin from isolates of natural Lactic acid fermentation of vegetables. Ind. J. Food Technol. Biotechnol. 3 (44):435-439.

39. Khali R, El Bahloul YA, Djadouni F, Hamed Omar S (2009). Isolation and characterization of a bacteriocin produced by a newly isolated Bacillus megaterium 19 strains. Pak. J. Nutr. 03 (8): 242-250.

40. Lehninger AL (1981). Biochimie bases moléculaires de la structure et des fonctions cellulaires. 2ème Ed, Flammarion médicine-science.

41. Lindley MG (1998). The impact of food processing on antioxidants in vegetable oils, fruits and vegetables. Trend Food Sci. Technol. 9:336–340.

42. Ogawa JM, Zehr EI, Bird GW, Ritchie DF, Uriu K, Uyemoto JK (1995). Compendium of stone fruit diseases. APSS press. St. Paul, Minessota. p. 57.

43. Ogunbanwo ST, Adebayo AA, Ayodele MA, Okanlawon B M, Edema MO (2008). Effects of lactic acid bacteria and Saccharomyces cerevisae co-culture starters on the nutritional contents and shelf life of cassava-wheat bread. J. Appl. Biosci. 12: 612-622.

44. Ogunbanwo ST, Fadahunsi IF, Molokwu A J (2014). Thermal stability of lactic acid bacteria metabolites and its application in preservation of tomato pastes. Malays. J. Microbiol. 10 (1):15-23.

45. Ogunbanwo ST, Okanlawon BM (2006). Microbial and sensory changes during the cold storage of chicken meat treated with bacteriocin from L. brevis OG1. Pak. J. Nutr. 5(6): 601-605.

46. Ogunniyi LT, Oladejo JA (2011). Technical efficiency of tomato ,production in Oyo State Nigeria. Agric. Sci. Res. J. 1(4):84-91.

47. Petro-Turza M (1987). Flavor of tomato and tomato products. Food Rev. Int. 2(3):309-351.

48. Pla M, Rodriguez-Lazaro D, Badosa E, Montesinis E (2005). Measuring microbiological contamination in fruit and vegetables. In: Improving The safety of fresh fruit and vegetables. Jongen, W. (ed.). CRC Press, Abington. pp. 147-155.

49. Rakshita PK (2011). Antimicrobial activity of lactic acid bacteria and partial characterization of its bacteriocins. Master thesis, New Delhi, Mysore. pp. 14-20.

50. Safdar MN, Mumtaz A, Amjad M, Siddiqui N, Hameed T (2010).

51. Development and quality characteristics studies of tomato paste stored at different temperatures. Pak. J. Nutr. 9(3): 265-268.

52. Sajur SA, Saguir FM, Manca de Nadra MC (2007). Effect of dominant species of lactic acid bacteria from tomato on the natural development in tomato puree. Food Control 18(5): 594-600.

53. Settanni L, Corsetti A (2008). Application of bacteriocins in vegetable food biopreservation. Int. J. Food. Microbiol. 121:123-138.

54. Siboukeur A (011). Etude de l'activité antibactérienne des bactériocines (type nisine) produites par Lactococcus lactis subsp lactis, isolée à partir du lait camelin. Magister en Biologie. Option: Microbiologie Appliquée. Ouargla. pp. 12-15.

55. Smaoui S, Elleuch L, Bejar W, Karray-Rebai I, Ayadi I, Jaouadi B, Mathieu F, Chouayekh H, Bejar S, Mellouli L (2010). Inhibition of Fungi and Gram-Negative Bacteria by Bacteriocin BacTN635

56. Produced by Lactobacillus plantarum sp. TN635. Appl. Biochem. Biotechnol. 9: 88217.

57. Sneath PH Mair NS Sharpe ME Holts JG (1986). Bergey's Manual Systematic Bacteriology. Vol. 2. Willam and Wilkins, Baltimore.

58. Souci, Fachman, Kraut. (2008). La composition des aliments. Tableaux des valeurs nutritives, 7éme édition. MedPharm Scientific Publishers / Taylor & Francis, ISBN 978-3-8047-5038-8.

59. Spadaro D, Gullino ML (2004). State of the art and future prospects of the biological control of post-harvest fruit diseases. Int. J. Food. Microbiol. 91(2): 185-194.

60. Stiles EM, Holzapfel WH (1997). Lactic acid bacteria of foods and their current taxonomy. Int. J. Food. Microbiol. 36(1): 1-2.

61. Taylor JH (1987). Text of lectures delivered at the national workshop on fruit and vegetable seedlings production held at NIHORT 9-13

62. Technicon Instrument Corporation (1975). Industrial method. pp. 155-171.

63. Toivonen PMA, DeEll JR (2002). Physiology of fresh-cut fruits and vegetables. In: Lamikanra (ed) Fresh-cut fruits and vegetables: Science, technology and market. CRC Press, Boca Raton. pp. 91-1238.

64. r gera K, Oliver H, Andreas B (2007). Conservation of onion and tomato in Niger assessment of post-harvest losses and drying methods. Conf. Int. Agric. Res. pp.1-7.

65. rias R Ba eras L, Montesinos E, Badosa E (2008). Lactic acid bacteria from fresh fruit and vegetables as biocontrol agents of phytopathogenic bacteria and fungi. Int. J. Microbiol. 11 (4):231-236.

66. Wiessinger WR, Chantarapanont W, Beuchat LR (2000). Survival and growth of Salmonella baildonin shredded lettuce and diced tomatoes, and effectiveness of chlorinated water as a sanitizer. Int. J. Food.Microbiol. 62 (1-2):123-131.

Chapter 5

KILLER TOXIN OF KLUYVEROMYCES PHAFFII DBVPG 6076 AS A BIOPRESERVATIVE AGENT TO CONTROL APICULATE WINE YEASTS

M. Ciani[1] and f. Fatichenti[2]

[1]Dipartimento di Biotecnologie Agrarie e Ambientali, Universita' di Ancona, Ancona 60131, Italy

[2]Sez. Microbiologia Applicata, Dipartimento di Biologia Vegetale e Biotecnologie Agroambientali, Universita' di Perugia, Perugia 06121, Italy

ABSTRACT

The use of Kluyveromyces phaffii DBVPG 6076 killer toxin against apiculate wine yeasts has been investigated. The killer toxin of K. phaffii DBVPG 6076 showed extensive anti-Hanseniaspora activity against strains isolated from grape samples. The proteinaceous killer toxin was found to be active in the pH range of 3 to 5 and at temperatures lower than 40°C. These biochemical properties would allow the use of K. phaffii killer toxin in wine making. Fungicidal or fungistatic effects depend on the toxin concentration. Toxin concentrations present in the supernatant during optimal conditions of production (14.3 arbitrary units) exerted a fungicidal effect on a sensitive strain of Hanseniaspora uvarum. At subcritical concentrations (fungistatic effect) the saturation kinetics observed with the increased ratio of killer toxin to H. uvarum cells suggest the presence of a toxin receptor. The inhibitory activity exerted by the killer toxin present in grape juice was comparable to that of sulfur dioxide. The findings presented suggest that the K. phaffii DBVPG 6076 killer toxin has potential as a biopreservative agent in wine making

Since Bevan and Makower (3) discovered the killer phenomenon in strains of *Saccharomyces cerevisiae*, several other yeast species have been found to produce a toxic proteinaceous factor that kills sensitive strains (17, 20, 21, 30, 31). Killer strains of *S. cerevisiae* show an antiyeast spectrum restricted to sensitive *Saccharomyces* strains except for a report on the killing of *Torulopsis glabrata* (37). Unlike the genus*Saccharomyces* a wide spectrum of intergeneric activity was reported for killer toxins from other genera such as *Pichia* (13, 14, 25),*Hansenula* (1, 18), *Williopsis* (34), and *Kluyveromyces* (22, 37).

Several potential applications for the killer phenomenon have been suggested since it was determined and studied. In fermentation industries, the killer character can be used to combat wild, contaminating *Saccharomyces* strains (33, 36). In the food industry, killer yeasts have been proposed to control spoilage yeasts in the preservation of food (16). In the medical field, killer yeasts have been used in the biotyping of pathogenic yeasts (15, 18), and the killer toxins of *Pichia anomala* (25, 26) and *Williopsis mrakii* (34) have been proposed as antimycotic agents.

In wine making, killer yeasts belonging to *S. cerevisiae* are currently used to initiate wine fermentation to improve the process of wine making and wine quality (28, 33). However, the main limit of the killer toxin of *S. cerevisiae* wine yeast (K2 type) resides in its narrow antiyeast spectrum which, being restricted to sensitive *Saccharomyces*strains, does not affect wild yeasts, such as those belonging to the genera *Hanseniaspora/Kloeckera, Pichia,* and *Saccharomycodes*.

Several ecological studies (9, 11, 12, 23) have clearly demonstrated that apiculate yeasts (*Hanseniaspora/Kloeckera*) predominate on grape surfaces and in freshly pressed juice. The control of the growth of apiculate yeasts in a nonsterile environment such as grape must is generally carried out by sulfur dioxide. However, several institutions, such as the World Health Organization and the European Economic Community, have highlighted the need to reduce the use of this antimicrobial agent in food products because of its toxicity. In this context, the use of a killer toxin as a control agent for apiculate yeasts in the prefermentative stage and during the fermentation of grape must has to be encouraged in order to reduce or eliminate the use of SO_2. The findings of this study indicate that the killer toxin produced by*Kluyveromyces phaffii* DBVPG 6076 may be used as a biological agent against apiculate yeasts, which are usually present in freshly pressed juice and during the first stage of alcoholic fermentation.

MATERIALS AND METHODS

Cultures and Media.

The strains used in the present study were from the Industrial Yeasts Collection of the University of Perugia (DBVPG): *K. phaffii* DBVPG 6076, a killer yeast; *S. cerevisiae* 6500 (NCYC 1006; National Collection of Yeast Cultures, Norwich, England), a sensitive strain; *S. cerevisiae* DBVPG 6664 (commercial strain Prise de Mousse; Red Star, Milwaukee, Wis.), a selected starter culture resistant to the DBVPG 6076 toxin; and *Hanseniaspora uvarum* DBVPG 3037. Strains isolated from grape berries from various vineyards of the Umbria region, classified as *Hanseniaspora* spp. by the methods of Kurtzman and Fell (10), were used to evaluate the frequency of the killer activity of DBVPG 6076. All yeast strains were subcultured at 6-month intervals on malt agar and maintained at 6°C. DBVPG 6076 was grown in yeast extract-peptone-dextrose (YPD) medium. The composition of YPD buffered at pH 4.5 (0.1 M citrate-phosphate buffer) was as follows (per liter): Bacto yeast extract, 10 g; Bacto Peptone, 10 g; and glucose, 50 g. Grape juice obtained from Grechetto grapes (pH 3.29; sugar, 209 g liter^{-1}) was used for trials in natural medium.

Assessment of DBVPG 6076 Killer Toxin (KPKT) Activity.

The frequency of killer toxin activity was evaluated by streak plate agar diffusion assay (21). Approximately 10^5 cells ml^{-1} (final concentration) of the strain to be tested for sensitivity to the killer toxin were uniformly suspended in 20 ml of Bacto malt agar (Difco Laboratories, Detroit, Mich.) buffered at pH 4.5 (0.1 M citrate-phosphate buffer), maintained at 45°C in a water bath, and poured immediately into sterile petri dishes. DBVPG 6076 was streaked on the agar surface, and the plates were incubated at 20°C for 72 h. Killer activity was evident as a clear zone of inhibition surrounding the streak.

Toxin preparations were assayed by the well test method of Somers and Bevan (29). Toxin samples were filter sterilized though 0.45-μm-pore-size membrane filters (Millipore Corp., Bedford, Mass.). Seventy microliters of toxin sample was placed in wells (7-mm diameter) cut in the malt agar medium buffered at pH 4.5 (0.1 M citrate-phosphate buffer) (Difco). Malt agar plates had been previously seeded with a sensitive indicator strain. The killing activity of each sample was measured and defined as the mean zone of inhibition of replicate wells after incubation for 72 h at 20°C.

Kpkt Production.

DBVPG 6076 was grown in YPD broth in a 2-liter flask with 1 liter of working volume for 72 h at 25°C on a rotary shaker (150 rpm^{-1}). After centrifugation (3,000 × *g* for 10 min at 4°C), the supernatant was filtered though 0.45-μm-pore-size membrane filters (Millipore). Ice-cold ethanol was added to the filtrate to a final volume of 50% (vol/vol). After 2 h at −18°C, the resulting precipitate was collected by centrifugation (10,000 × *g* for 30 min) and resuspended in a reduced volume of 10 mM citrate-phosphate buffer (pH 4.5). For further concentration, Amicon YM10 (10-kDa-cutoff membrane) (Pharmacia, Uppsala, Sweden) was used.

Measurement of KPKT Activity.

Under the experimental conditions used, a linear relationship was observed between the logarithm of killer toxin concentration and the diameter of the inhibition zone assayed by the well test method. One arbitrary unit (AU) is defined as the toxin concentration in the supernatant that caused a clear zone of 8.0 mm (actual diameter, 15 mm, minus diameter of the 7-mm well), using *S. cerevisiae* DBVPG 6500 as a sensitive indicator strain and 70 μl of sample.

Effects of Proteolitic Enzymes and Temperature Stability.

Five hundred microliters of concentrated killer toxin was mixed with 125 μl of type IV papain solution (10 mg ml^{-1}, 10 to 15 U mg^{-1}). The toxin and protease solution was added to 4.5 ml of YPD broth inoculated with a sensitive *H. uvarum* DBVPG 3037 strain (initial inoculum, 10^6 cells ml^{-1}). After an incubation period of 20 h at 20°C, the effects of the proteolytic enzyme were evaluated by the well test assay and a viable-cell count. Similar procedures were carried out in order to evaluate the temperature stability of the killer toxin. Samples of killer toxin were incubated at 20, 25, 30, 35, 40, 45, or 50°C for 2 h, mixed with YPD broth, and then inoculated with a sensitive *H. uvarum* DBVPG 3037 strain (initial inoculum, 10^6 cells ml^{-1}). Killer toxin was also subjected to heat treatment (100°C for 10 min). Killer activity was determined by the well test assay and a viable-cell count after incubation for 24 h at 20°C.

pH Stability.

In order to determine the pH range of activity, samples of killer toxin were tested by the well test assay using malt agar plates buffered at pH values of 2.8, 3.0, 3.5, 4.0, 4.5, 5.0, and 6.0.

Fungistatic and Fungicidal Effects.

Experiments were carried out in 300-ml Erlenmeyer flasks containing 100 ml of YPD broth buffered at pH 4.5 at 20°C. A 48-h preculture of *H. uvarum* DBVPG 3037, also grown at 20°C in the same medium as the test, was inoculated to obtain an initial count of 10^5 cells ml^{-1}. Different toxin concentrations were added, and cell growth was evaluated by measuring optical density at 580 nm (OD$_{580}$) using a DU 640 spectrophotometer (Beckman, Fullerton, Calif.) or by counting viable cells on plates. A control test (without toxin) was included in all assays.

Growth rate Reduction Assay.

The reduction of the growth rate was carried out according to the procedures of Sawant et al. (25), evaluating toxin concentrations lower than that which caused fungicidal activity. Cells of *H. uvarum* DBVPG 3037 in log phase were inoculated at an initial OD_{580} of 0.1 in 300-ml Erlenmeyer flasks containing 100 ml of YPD broth and incubated at 25°C in a rotary shaker (150 rpm). Killer toxin was added to the medium at the following concentrations (AU): 0.14, 0.26, 0.71, 1.43, 2.86, 5.72, 11.44, and 14.3. Optical densities were evaluated at 0 h and at 1-h intervals up to 9 h. Growth rates were determined by linear regression analyses of optical densities from 4 to 9 h of growth.

Activity of KPKT in Grape Juice.

Trials in grape juice were carried out in 300-ml Erlenmeyer flasks containing 100 ml of pasteurized grape juice (100°C for 10 min). Concentrated Kpkt was added, and the procedure was standardized to provide the following final concentrations (AU ml^{-1}): 5.14, 7.15, and 14.3. Sulfur dioxide at 37.5 (free SO_2, 3.20 ml liter^{-1}), 75.0 (free SO_2, 6.08 ml liter^{-1}), and 150 (free SO_2, 10.28 ml liter^{-1}) ml liter^{-1} was added 24 h before the inoculation in order to obtain the binding equilibrium. Precultures of *H. uvarum* DBVPG 3037 were grown at 20°C in the same medium as the test for 48 h and inoculated in order to obtain an initial count of 10^5 cells ml^{-1}. Experiments were carried out in static conditions at 20°C by fitting each flask with a glass valve containing sulfuric acid, which allowed only CO_2 to escape the system (5). An inoculum of *S. cerevisiae* DBVPG 6664 starter culture (5×10^6 cells ml^{-1}) was added 48 h after the beginning of fermentation. *S. cerevisiae* was preincubated by the procedures used for the inoculum of *H. uvarum* DBVPG 3037. Trial fermentations with pure cultures of *S. cerevisiae* DBVPG 6664 and *H. uvarum* DBVPG 3037 were also included as control tests. The progress of fermentation was monitored by the amount of weight lost due to the carbon dioxide evolved. When the weight of the apparatus became constant, the fermentation was considered to be finished and samples were collected by filtration (0.45-μm-pore-size membrane; Millipore) for chemical analysis. Volatile acidity (expressed as grams of acetic acid per liter) was quantified by steam distillation according to official analytical procedures (6). Acetaldehyde and ethyl acetate were detected by gas-liquid chromatographic analysis as described by Bertuccioli (2).

RESULTS

Killer Activity of *K. phaffii* on Apiculate Wine Yeasts

The evaluation of the spectrum of activity of DBVPG 6076 is the first step toward the practical application of this killer toxin in the control of apiculate yeasts in wine making. Thus, 298 strains of apiculate yeasts belonging to*Hanseniaspora* spp., isolated from 52 different grapes and sampled in the course of four years, were tested for their sensitivity to the *K. phaffii*DBVPG 6076 killer strain. Interestingly, 94.9% of the strains of apiculate wine yeasts were sensitive to the killer activity of *K. phaffii*(Table (Table1).1).

TABLE 1: Frequency of killer activity of *K. phaffii*DBVPG 6076 among *Hanseniaspora*spp. strains isolated from grape berries from different vineyards of the Umbria region

Yr of isolation	No. of sensitive strains	No. of resistant strains	No. of samples[a]	Sensitive strains (%)
1994	80	1	20	98.8
1995	90	1	12	98.9
1996	98	14	12	87.5
1997	30	0	8	100.0
Total	298	16	52	94.9

[a] Number of bunches from which the apiculate strains were isolated.

Characterization of KPKT.

Studies were carried out in order to elucidate the biochemical properties of Kpkt, particularly in relation to its possible use in wine making. The toxin was concentrated 10-fold by ethanol precipitation and this or more-concentrated preparations were used for successive characterization of the killer factor. The activity of Kpkt in the supernatant and after precipitation in ethanol are shown in Table Table2.2.

TABLE 2: Killer activity of Kpkt in the supernatant and after ethanol precipitation[a]

Activity measured	Toxin concn factor	AU ml^{-1}	Killer activity against:	
			S. cerevisiae (diam [mm])	*H. uvarum* (diam [mm])
In the supernatant	1	14.3	8.0	4.0
After ethanol precipitation	10	143.3	16.0	8.0

[a] Mean of eight different experiments. The variation was less than 5%.

Since the killer toxins known so far are proteins or glycoproteins, the effects of protease treatment on the killer factor of *K. phaffii* were assayed. Papain, which breaks sulfide bonds, inactivated the killer toxin (Table).

TABLE 3: Effect of papain on Kpkt[a] activity[b]

Time (h)	CFU (10^6 ml^{-1})		Killer toxin and papain
	Control		
	No killer toxin	Killer toxin	
0	0.3	0.2	0.4
24	58.0	0.2	18.0

Further characterization of the DBVPG 6076 toxin was carried out by evaluating the effects of pH and temperature on the activity of the killer factor. Results reported in Fig. Fig.1A1A show that Kpkt was active in the pH range of 3.0 to 5.0, but a 30% reduction in activity was observed at pH 3.0 against the sensitive *H. uvarum* strain.

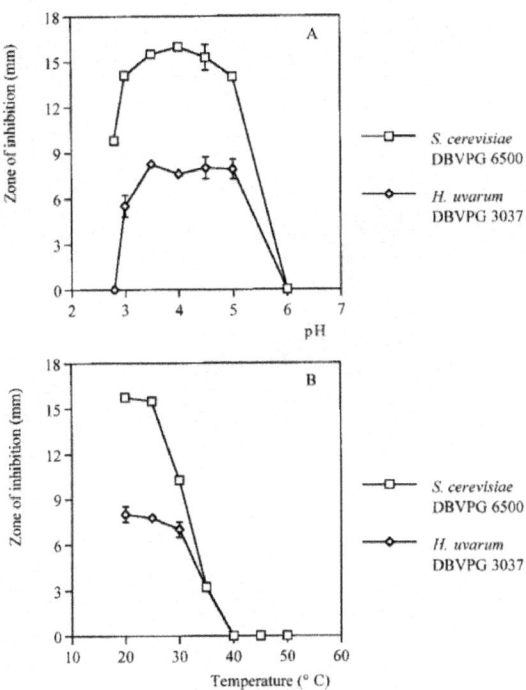

Figure. 1: Effects of pH (A) and temperature (B) on the activity of Kpkt. pH effects were evaluated by the well test assay. Malt agar was buffered from pH 2.8 to 6.0 10 mM citrate-phosphate buffer. The toxin concentration was 14.3 AU ml^{-1}. After 2 h of incubation ...

The killer activity observed for both sensitive strains was completely lost at pH 6. At pH 2.8 the toxin showed reduced activity against *S. cerevisiae*, whereas no activity was observed against *H. uvarum*. Regarding the influence of temperature, the killer activity was stable up to 25°C (Fig. (Fig.1B),1B), whereas higher temperatures caused a progressive decrease of the activity of Kpkt. No activity was found after 2 h of incubation at 40°C. Heat treatment (100°C for 10 min) caused loss of activity. These results were confirmed by a viable-cell count assay (data not shown).

Mode of Action of KPKT Against *H. Uvarum* DBVPG 3037.

The effects of Kpkt on *H. uvarum* DBVPG 3037 growth were tested in liquid media at different concentrations of the killer factor. The growth of *H. uvarum*(assayed by measuring the OD_{580} after 24 h of incubation at 20°C) showed a progressive reduction with increased toxin concentrations. Interestingly, at the toxin concentration corresponding to 14.3 AU in the supernatant (Fig. (Fig.2A),2A), no growth was observed. In another trial, the evaluation of the viable cells during the first 8 h of growth confirmed that the critical concentration necessary for a fungicidal effect was 14.3 AU (Fig. (Fig.2B).2B). Lower concentrations appeared fungistatic for *H. uvarum*, extending lag phase for 2 h before growth resumed.

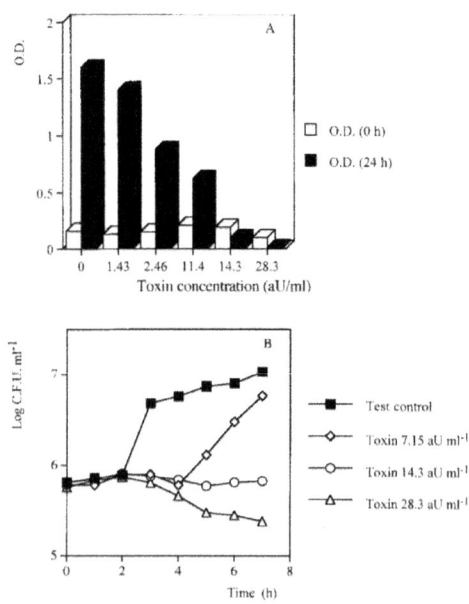

Figure. 2: Fungistatic and fungicidial activity of Kpkt evaluated by OD unit (A) and log CFU ml⁻¹ (B). Cells were grown in YPD broth buffered at pH 4.5. The initial

inoculum contained 10^5 *H. uvarum* cells ml⁻¹. Each sampling point represents the mean ...

In order to assess the modality of action of Kpkt, the growth rate reduction assay (25) was carried out. The plotting of increased concentrations of the killer toxin against the growth rate (Fig. (Fig.3)3) showed an exponential relationship with typical saturation kinetics.

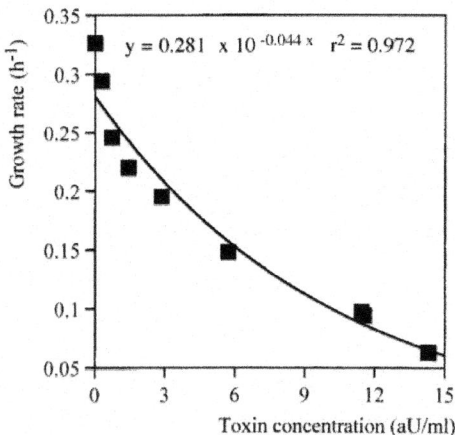

Figure. 3: Effects of increasing concentrations of Kpkt on the growth rate of *H. uvar-um*DBVPG 3037. Procedures are described in Materials and Methods. Each sampling point represents the mean of duplicate experiments. The variation was less than 10%.

Activity of KPKT in Grape Juice.

In order to verify the potential of DBVPG 6076 as a biological antiyeast agent in wine making, DBVPG 6076 toxin activity was assayed in natural grape juice. Moreover, the activities of Kpkt and SO_2, the chemical antiseptic agent universally used in winemaking, were compared. Figure Figure44 shows the development of viable *H. uvarum* cells in the presence of different concentrations of Kpkt and SO_2 during the first 48 h of growth at the stage of fermentation when apiculate yeasts naturally dominate the process. The positive control (inoculum of *H. uvarum* without antimicrobial agents) showed 45×10^6 cells ml⁻¹ after 48 h of fermentation. As expected, the presence of toxin caused reduced growth of *H. uvarum* (Fig. (Fig.4A).4A). A toxin concentration of 14.3 AU ml⁻¹ resulted in a complete inhibition of the apiculate strain after 48 h of incubation. Absence of growth was obtained with 7.15 AU of toxin ml⁻¹ within 24 h. After this time, reduced growth was exhibited by an apiculate yeast showing approximately 10×10^6 cells ml⁻¹ after 48 h. Lower concentrations (5.14 AU ml⁻¹) caused only reduced growth of *H.*

uvarum without a prolonged lag phase. The effects of sulfur dioxide on the growth of *H. uvarum* were similar to those of the DBVPG 6076 toxin (Fig. B). After 48 h of incubation, no growth was exhibited in the presence of 150 mg of SO_2 liter^{-1}, whereas 37.5 and 75 mg of this antiseptic agent liter^{-1} caused lag phases of 24 and 30 h, respectively, before growth resumed.

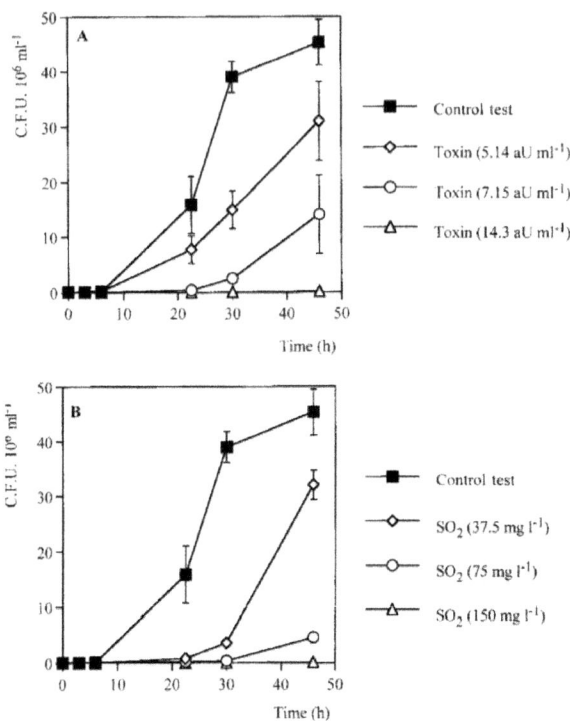

Figure. 4: Comparison of Kpkt (A) and SO_2 (B) activities against *H. uvarum* DBVPG 3037 in natural grape juice. The initial inoculum of *H. uvarum* was 10^5 cells per ml. The grape juice had a pH of 3.29 and a sugar concentration of 209 g liter^{-1}. Data are given ...

The evaluation of some fermentation products obtained after the inoculation of the *S. cerevisiae* starter culture (Table (Table4)4) showed that the control of *H. uvarum* is accompanied by a progressive reduction of ethyl acetate. Amounts of ethyl acetate similar to those produced by the *S. cerevisiae* control test were exhibited by the trials containing 14.3 aU of toxin ml^{-1} and 150 mg of SO_2 liter^{-1}. Increased amounts of ethyl acetate were caused by lower concentrations of two antiseptic agents without reaching the taste threshold (150 mg liter^{-1}) (8). No relevant differences were found among the trials for volatile acidity, whereas the trials with SO_2 showed a small increase of acetaldehyde compared to the trial tests.

Table 4: Influence of Kpkt and SO$_2$ activity on some fermentation products of wines

Organism(s)	Volatile acidity (g liter^{-1})	Acetaldehyde (mg liter^{-1})	Ethyl acetate (mg liter^{-1})
H. uvarum	0.47	21.0	307
S. cerevisiae	0.33	29.5	43
H. uvarum and S. cerevisiae	0.42	12.3	150
H. uvarum, S. cerevisiae, and toxin (5.14 AU)	0.36	15.4	138
H. uvarum, S. cerevisiae, and toxin (7.15 AU)	0.36	21.0	128
H. uvarum, S. cerevisiae, and toxin (14.3 AU)	0.34	15.2	45
H. uvarum, S. cerevisiae, and SO$_2$ (37.5 mg liter^{-1})	0.36	24.4	130
H. uvarum, S. cerevisiae, and SO$_2$ (75 mg liter^{-1})	0.32	27.0	104
H. uvarum, S. cerevisiae, and SO$_2$ (150 mg liter^{-1})	0.38	36.0	38

DISCUSSION

The use of *S. cerevisiae* killer yeasts in vinification prevents stuck fermentations caused by wild killer yeasts (33). However, antiyeast activity restricted to sensitive *Saccharomyces* strains does not permit the control of wild non-*Saccharomyces*, particularly apiculate yeasts which are present in freshly pressed grape juice. *K. phaffii* exhibits killer activity against apiculate yeasts (*Kloeckera apiculata/H. uvarum*) (22) and other spoilage yeasts (16). In order to assess the potential use of DBVPG-6076 in wine making we verified the following features: (i) the diffusion of killer activity among apiculate wine yeasts, (ii) the killer toxin activity at the conditions used in vinification, and (iii) the antiseptic effect compared with that of a chemical preservative agent used in wine making (SO$_2$).

Results showed that Kpkt exhibits widespread killer activity against apiculate wine yeasts. Evidence from protease treatments suggests that Kpkt is a protein with sulfide bonds like several other killer factors (4,35), since papain has been shown to destroy the toxin and its activity. Like most toxins (28), Kpkt is unstable at both high temperatures and high pH values. In contrast to the toxin of *Kluyveromyces lactis* (32,35), Kpkt has a low pH range of activity (up to pH 3.0). These findings are in agreement with the use of Kpkt at the majority of pHs and temperatures in wine making conditions.

The characterization of Kpkt activity against *H. uvarum* indicates that fungistatic or fungicidal effects depend on the toxin concentration. A toxin concentration of 14.3 aU ml^{-1} exerted zymocidal activity against *H. uvarum* DBVPG 3037. At subcritical concentrations of toxin (fungistatic effect), the saturation kinetics observed with an increased ratio of Kpkt to *H. uvarum* cells suggest the presence of a toxin receptor, probably on the cell wall. Cell wall receptors are known to mediate toxin action in the killer systems of several yeast species (7,19, 24, 27).

The activity of Kpkt in grape juice is comparable to that of sulfur dioxide. The toxin concentration normally present in the supernatant (14.3 AU ml^{-1}) is capable of controlling apiculate yeasts for 48 h at cell densities generally found in grape juice at the beginning of wine fermentation (9). The inhibition exerted by Kpkt on apiculate fermentation activity is reflected by the decrease in by-products such as ethyl acetate, the main compound responsible for the vinegary odor in wines, and acetaldehyde (linked to SO_2 addition), an undesirable compound due to its capacity to combine with SO_2, which is added for the preservation of wine. In conclusion, Kpkt, the sole toxin known to exhibit killer activity against apiculate yeasts, has great potential as a biopreservative agent in wine making and can profitably be substituted for SO_2 at the prefermentation stage. Moreover, this killer toxin could be also used in the food industry since broad-spectrum activity against spoilage yeasts was shown (16).

Further studies are in progress to acquire additional information on the biochemical properties of Kpkt, contributing to the understanding and development of a novel biopreservative agent to combat wild microflora in winemaking.

ACKNOWLEDGMENTS

We thank Fausto Maccarelli for technical assistance and two anonymous reviewers for helpful comments on the manuscript.

This work was supported in part by the Ministero dell'Università e della Ricerca Scientifica e Tecnologica (MURST) and the University of Ancona (progetto di Ateneo).

REFERENCES

1.   Ashida S, Shimazaki S T, Kitano K, Hara S. New killer toxin of *Hansenula mrakii*. Agric Biol Chem. 1983;47:2953–2955.

2.   Bertuccioli M. Determinazione gas-cromatografica diretta di alcuni composti volatili nei vini. Vini d'Italia. 1982;24:149–156.

3.   Bevan E A, Makower M. The physiological basis of the killer character in yeast. In: Geerts S J, editor. Genetics today, XIth International Congress on Genetics vol. 1. Oxford, England: Pergamon Press; 1963. pp. 202–203.

4.   Chen W-B, Han Y F, Jong S-C, Chang S-C. Isolation, purification, and characterization of a killer protein from *Schwanniomyces occidentalis*. Appl Environ Microbiol. 2000;66:5348–5352

5.   Ciani M, Rosini G. Definizione dell'indice di moltiplicazione della CO_2 nella valutazione, per via ponderale della capacita alcoligena di un lievito. Ann Fac Agrar Univ Stud Perugia. 1987;41:753–762.

6.   European Economic Community. Methods of community analyses of wines. Regulation no. 2676/90 of the Commission, 17 September 1990.

Brussels, Belgium: European Economic Community; 1990.

7.   Hutchins K, Bussey H. Cell wall receptor for yeast killer toxin: involvement of $(1 \to 6)$ β-D-glucan. J Bacteriol. 1983;154:161–169

8.   Jackson R S. Sensory perception and wine assessment. In: Jackson R S, editor. Wine science. San Diego, Calif: Academic Press, Inc.; 1994. p. 447.

9.   Kunkee R E, Goswell R W. Table wines. In: Rose A H, editor.Alcoholic beverages. London, England: Academic Press; 1977. pp. 315–386.

10. Kurtzman C P, Fell J W, editors. The yeasts, a taxonomic study. 4th ed. Amsterdam, The Netherlands: Elsevier Science B. V.; 1998.

11. Martini A. Origin and domestication of the wine yeast*Saccharomyces cerevisiae*. J Wine Res. 1993;4:165–176.

12. Martini A, Ciani M, Scorzetti G. Direct enumeration and isolation of wine yeasts from grape surfaces. Am J Enol Vitic. 1996;47:435–440.

13. Middelbeek E J, Hermans J M H, Stumm C. Production, purification and properties of a *Pichia kluyveri* killer toxin. Antonie Leeuwenhoek. 1979;45:437–450.

14. Middelbeek E J, van de Laar H H A M, Hermans J M H, Stumm C, Vogels G D. Physiological conditions affecting the sensitivity of*Saccharomyces cerevisiae* to a *Pichia kluyveri* killer toxin and energy requirement for toxin action. Antonie Leeuwenhoek. 1980;46:483–497.

15. Morace G, Manzara S, Dettori G, Fanti F, Campana L, Polonelli L, Chezzi C. Biotyping of bacterial isolates using the killer system. Eur J Epidemiol. 1989;5:85–90.

16. Palpacelli V, Ciani M, Rosini G. Activity of different "killer" yeasts on strains of yeast species undesiderable in food industry. FEMS Microbiol Lett. 1991;84:75–78.

17. Phillskirk G, Young T W. The occurence of killer character in yeasts of various genera. Antonie Leeuwenhoek. 1975;41:147–151.

18. Polonelli L, Archibusacci C, Sestito M, Morace G. Killer system: a simple method for differentiating *Candida albicans* strains. J Clin Microbiol. 1983;17:774–780.

19. Radler F, Schmitt M J, Meyer B. Killer toxin of *Hanseniaspora uvarum*. Arch Microbiol. 1990;154:175–178.

20. Rosini G. Interaction between killer strains of *Hansenula anomala*var. anomala and *Saccharomyces cerevisiae* yeast species. Can J Microbiol. 1985;31:300–302.

21. Rosini G. The occurence of killer characters in yeasts. Can J Microbiol. 1983;29:1462–1464.

22. Rosini G, Cantini M. Killer character in *Kluyveromyces* yeasts: activity on *Kloeckera apiculata*. FEMS Microbiol Lett. 1987;44:81–84.

23. Rosini G, Federici F, Martini A. Yeast flora of grape berries during ripening. Microb Ecol. 1982;8:83–89.

24. Santos A, Marquina D, Leal J A, Peinado J M. (1 → 6)-β-D-Glucan as cell wall receptor for *Pichia membranifaciens* killer toxin. Appl Environ Microbiol. 2000;66:1809–1813.

25. Sawant A D, Abdelal A T, Ahearn D G. Anti-*Candida albicans*activity of *Pichia anomala* as determined by a growth rate reduction assay. Appl Environ Microbiol. 1988;54:1099–1103.

26. Sawant A D, Abdelal A T, Ahearn D G. Purification and characterization of the anti-*Candida* toxin of *Pichia anomala* WC 65.Antimicrob Agents Chemother. 1989;33:48–52.

27. Schmitt M, Radler F. Molecular structure of the cell wall receptor for killer toxin KT28 in *Saccharomyces cerevisiae*. J Bacteriol. 1988;170:2192–2196.

28. Shimizu K. Killer yeasts. In: Fleet G H, editor. Wine microbiology and biotechnology. Chur, Switzerland: Harwood Academic; 1993. pp. 243–264.

29. Somers J M, Bevan E A. The inheritance of the killer character in yeast. Genet Res Commun. 1969;13:71–83.

30. Starmer W T, Ganter P F, Aberdeen V, Lachance M A, Phaff H J. The ecological role of killer yeasts in natural communities of yeasts.Can J Microbiol. 1987;33:783–796.

31. Stumm C, Hermans J M H, Middelbeek E J, Croes A F, De Vries G J M L. Killer-sensitive relationships in yeasts from natural habitats.Antonie Leeuwenhoek. 1977;43:125–128.

32. Sugisaki Y, Gunge N, Sakaguchi K, Kamasaki M, Tamura G. Characterization of novel killer toxin encoded by a double-stranded linear DNA plasmid of *Kluyveromyces lactis*. Eur J Biochem.1984;141:241–245.

33. van Vuuren H J J, Jacobs C J. Killer yeasts in the wine industry: a review. Am J Enol Vitic. 1992;43:119–128.

34. Walker G M, McLeod A H, Hodgson V J. Interactions between killer yeasts and pathogenic fungi. FEMS Microbiol Lett.1995;127:213–222.

35. Wilson   C,   Whittaker   P   A.   Factors   affecting   activity   and stability   of   the   *Kluyveromyces   lactis*   killer   toxin.   Appl   Environ

Microbiol.1989;55:695–699.

36. Young T W. Killer yeasts. In: Rose A H, Harrison J S, editors. The yeasts. 2nd ed. Vol. 2. London, England: Academic Press; 1987. pp. 131–164.

37. Young T W, Yagiu M. A comparison of the killer character in different yeasts and its classification. Antonie Leeuwenhoek.1978;44:59–77.

Chapter 6

NATURAL ANTIMICROBIALS FOR BIOPRESERVATION OF SPROUTS

Antonio Gálvez, Hikmate Abriouel, Antonio Cobo and Rubén Pérez Pulido

Área de Microbiología, Departamento de Ciencias de la Salud, Facultad de Ciencias Experimentales, Universidad de Jaén, Campus Las Lagunillas s/n. 23071-Jaén Spain

INTRODUCTION

In recent years there has been an increase in consumers demands for mungbean, alfalfa, soybean, radish and other seed sprouts (Rosas and Escartin, 2000) that are usually eaten raw in salads or in sandwiches. Seed sprouts have been part of the human diet since old times in countries such as Japan where they are widely consumed. Tthe interest in consuming fresh green sprouts has extended all over the world because they are considered to provide health benefits (Rosas and Escartin, 2000). A great variety of seed sprouts can be found at present in the market, such as adzuki bean (Phaseolus angularis), alfalfa (Medicago sativa), broccoli (Brassica oleracea convar. botrytis), cress (Lepidium sativum), lentil (Lens culinaris), mung bean (Phaseolus aureus), soybean (Glycine max), white mustard (Sinapis alba), green and yellow pea (Pisum sativum), onion (Allium cepa), radish (Raphanus sativus), rice (Oryza sativa L.), rye (Secale cereale), sesame (Sesamum indicum), sunflower (Helianthus annuus) and wheat (Triticum aestivum), although the most popular are alfalfa, soybeans, mung beans and raddish (Taormina et al., 1999). Seed sprouts are usually eaten raw in salads or in sandwiches, and concerns for the safety of these raw foods have increased lately. Sprouts are grown from seeds placed in environmentally controlled, hydroponic conditions and incubated in warm, moist, nutrient-rich conditions, which are ideal environments for microbial growth. The seeds usually carry microbial loads comprised between 3 and 6 log CFU/g, including pseudomonads and enterobacteriaceae as main components (Andrews et al., 1982; Prokopowich and Blank, 1991; Robertson et al., 2002; Splittstoesser et al., 1983). The bacterial load increases rapidly during the sprouting process, reaching from 6 to 8 log CFU/g after two days in one study (Fu et al., 2001) and between 7.8 and 8.8 in another (Weiss et al., 2007). Other reports have indicated final counts of up to 8-9 log CFU/g in commercial sprouts (Patterson and Woodburn, 1980; Prokopowich and Blank, 1991). In addition, the pathogenic bacteria can survive on sprouts through the typical refrigerated shelf life of the products (Harris et al., 2003). Recent studies indicate that pathogenic bacteria can survive both on and in plant tissues (Lynch et al., 2009). For example, when alfalfa seeds contaminated with Escherichia coli O157 or with Salmonella are sprouted, the bacteria enter the growing sprout, and appear throughout the deep tissues of the young plants (Itoh et al., 1998; Charkowski et al., 2002). The bacterial cells located inside plant tissue will be refractile to inactivation by common disinfection methods. A wide variety of pathogens have been isolated from sprouted seeds (including alfalfa, mung bean, cress, soybean, and mustard): Aeromonas hydrophila, Salmonella spp., Listeria monocytogenes, Staphylococcus aureus, Bacillus cereus, Yersinia enterocolitica and Vibrio cholerae (Beuchat, 1996; Harris et al., 2003). Outbreaks have been associated primarily with Salmonella serotypes but have also been attributed to B. cereus, E. coli O157:H7, and Y. enterocolitica (Harris et al., 2003). Alfalfa sprouts rank in the first place of associated outbreaks, followed by far by clover, radish, and mung bean sprouts. This is due to the fact that alfalfa sprouts are the most popular type of seed sprouts that are commonly eaten raw, while others such as mung beans are often cooked before consumption. Therefore, it is necessary to plan intervention strategies to decrease the risks for transmission of human pathogenic bacteria through sprouted seeds. In this chapter we discuss how different biocontrol strategies can be applied either alone or in combination with other intervention strategies to improve the control of pathogenic bacteria most frequently found in sprouts (Fig. 1).

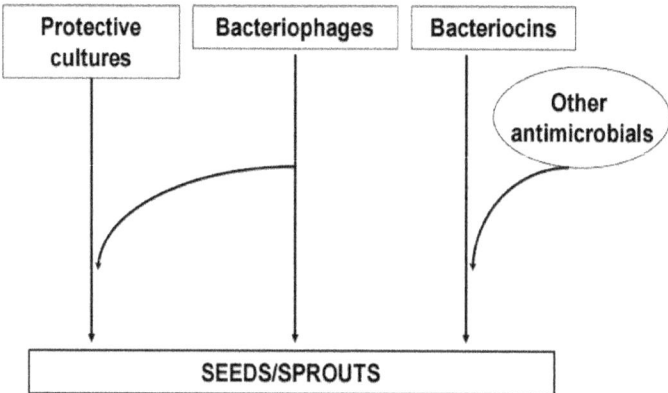

Figure. 1: Intervention strategies based on biopreservation to decrease the risk of transmission of human pathogenic bacteria in sprouts.

INOCULATION WITH PROTECTIVE CULTURES

One of the oldest approaches for biopreservation of foods is based on the natural interactions that microbes play in Nature. Such interactions can be redirected and exploited to selectively inhibit microbial populations of concern, such as pathogenic or spoilage bacteria in food ecosystems. Microbial antagonism is usually based on a combination of mechanisms which include competition for space and nutrients, and selectively changing the environmental conditions, eg. by acidification, or production of an array of antimicrobial substances that include organic acids, oxygen radicals, siderophores, antibiotics, lytic enzymes, and bacteriocins. The bacteriocins were defined by Jack et al. (1995) as ribosomally synthesized antimicrobial peptides or proteins. When using live microbial antagonists as protective cultures in food biopreservation, there is a number of limitations and rules that need to be taken into consideration (Galvez et al., 2007): i) the protective cultures must not be pathogenic or produce toxins against humans or animals, and must not modify the organoleptic properties of the food product; ii) they must be able to multiply in the food and produce antagonistic activity at the food storage temperature; ii) they must be efficient in controlling the bacterial pathogen of concern under the common food storage and processing conditions. There has been a great interest in the application of bacteriocins and bacteriocin-producing strains (especially those produced by the LAB) on the preservation of foods of animal origin, but to a much less extent on vegetable foods such as sprouted seeds (Gálvez et al., 2008; Abriouel et al., 2010). Cai et al. (1997) isolated a collection of bacteria from mung- and soy bean-sprouts from retail stores and tested them for anti-listerial activity. Ten strains showed highest inhibitory activity, among which one strain was identified as a nisin-producing Lactococcus lactis subsp. lactis. This strain was able to reduce the levels of L. monocytogenes by 1.0-1.4 logs in Caesar salad after storage for 10 days at 7 or 10°C. Similar results were reported when the bacteriocin-producing Enterococcus faecium ATCC 19434 strain was tested on the salad for comparison (Cai et al., 1997). However, none of the strains were tested on sprouts. The enterococcal strain Enterococcus faecalis A-48-32 (which produces the cyclic peptide bacteriocin AS-48) was inoculated as a protective culture against B. cereus on soybean sprouts stored at 15 and 22 °C simulating temperature-abuse events (Cobo Molinos et al., 2008a). Enterococci multiplied rapidly on sprouts at the two temperatures tested. In both cases, bacteriocin activity could be recovered from the sprouts during days 1 to 3 of storage, but not after more prolonged periods. In cocultures at 15 °C, growth of B. cereus was completely inhibited for the whole storage period, with viable cell counts being significantly lower than control cultures for the first 5 days of storage. By contrast, cocultivation with the A-48-32 strain at 22 °C only produced some growth inhibition of B. cereus during the first 3 days of storage. Bennik et al. (1999) tested two bacteriocin-producing Pediococcus parvulus strains (isolated from minimally-processed vegetables and one Enterocococcus mundtii ATO6 strain (isolated from chicory endive) that produced the pediocin-like mundticin ATO6 (Bennik et al., 1998). Only E. mundtii was capable of bacteriocin production at 4-8°C, a reason why this strain was selected for experiments to control L. monocytogenes on mungbean sprouts, which were stored under modified atmosphere (MA) of 1.5%O2/20%CO2/78.5%N2 at 8°C. However, in spite of its capacity for growth and bacteriocin production under MA, this strain failed to inhibit the growth of L. monocytogenes on sprouts and for this reason was not likely to be feasible for application as a protective culture (Bennik et al., 1999). More recently, the bacteriocin-producing strain E. mundtii CUGF08 isolated from alfalfa sprouts was shown to produce an anti-listerial bacteriocin called mundticin L (Feng et al., 2009). This bacteriocin is highly similar to mundticin ATO6, differing in only by one amino acid residue in the YYGNGV (YYGNGL) motif and the last two amino acid residues, which are reversed as in the almost identical mundticin SK (Kawamoto et al., 2002). The antagonistic capacity of this strain on sprouts has not been reported so far. In one study, approximately 120 strains of indigenous microbiota from raw vegetables (including alfalfa and clover sprouting seeds, Romaine lettuce, and pre-peeled baby carrots) were tested for their ability to inhibit the growth of Salmonella Chester, L. monocytogenes, E. coli, or Erwinia carotovora subsp. carotovora (Liao and Fett, 2001). Six isolates capable of inhibiting the growth of at least one pathogen were isolated and identified: Bacillus spp. (three strains), Pseudomonas aeruginosa (one strain), Pseudomonas fluorescens (strain A3), and yeast (strain D1). When strains A3 and D1 were tested on green pepper, the growth of Salmonella Chester and L. monocytogenes was reduced by 1 and 2 logs, respectively, over a period of 3 days. The authors stressed the potential of these two strains as biocontrol agents for reducing the growth of these two pathogens on fresh produce products, although their inhibitory effects on sprouts were not determined. Matos and Garland (2004) tested different potential biological control inoculants against Salmonella in alfalfa. Inocula included raw bacterial suspensions derived from market sprouts or laboratory-grown alfalfa sprouts (obtained by sonication of sprouts in a buffered peptone solution) and the P. fluorescens 2-79 strain. This strain was originally isolated from the rhizosphere of wheat and shown to suppress take-all, a major fungal root and crown disease of wheat and barley. A cocktail of four different serovars of Salmonella were used, all of them isolated from sprout-related outbreaks (Salmonella Newport H1275, Salmonella Anatum F4317, Salmonella Stanley

H0558, and Salmonella Infantis F4319). According to results of the competitive exclusion bioassays, P. fluorescens 2-79 was able to reduce the growth of Salmonella on alfalfa sprouts by 4.22, 4.24, and 1.81 logs on days 1, 3, and 7, respectively. The laboratory-grown sprout inoculum also had its highest inhibitory effect on Salmonella at the beginning of the growth period with log reductions of 2.56 on day 1; however, viable cell reductions then decreased to 0.21 on day 3 and 0.71 logs on day 7. The best results were observed for the market sprout inoculum, which reduced the number of Salmonella cells by 1.61, 2.80, and 5.48 log on days 1, 3 and 7, respectively. The market sprouts inoculum achieved the highest log reduction of Salmonella on day 7 than any of the other experimental treatments. Apparently, the natural populations of bacteria found in commercial sprouts were better adapted to utilization of an array of nutrient substrates, suggesting competition with the Salmonella for available resources as a potential mode of action. Unfortunately, the authors did not characterize the microbial composition of this inoculum or how it could be propagated and standardized as a commercial inoculant. The potential of P. fluorescens 2-79 as a biocontrol agent on sprouts was further investigated by Fett (2006). The addition of a P. fluorescens 2-79 inoculum to the seed soak water prior to the germination of alfalfa seeds previously inoculated with a cocktail of Salmonella enterica strains led to outgrowth inhibition of the salmonellae through the complete 6-days sprouting period, without an adverse effect on sprout yield or appearance. At day 6, the counts of S. enterica were between 4 and 5 log CFU/g lower compared to seeds not treated with the antagonist. Final cell numbers of S. enterica on the sprouted seeds were shown to be dependant on the cell density of the inoculum and on the application or not of a sanitation treatment (Liao, 2008). Remarkably, the population of salmonellae on sanitized seeds was approx. 1 log higher than that on nonsanitized seeds which contained 2 logs higher number of native microflora. In another study, a Pseudomonas jessenii strain isolated from raddish sprouts showed high inhibitory activity against a variety of target bacteria (L. innocua, S. aureus, Bacillus subtilis, S. Typhimurium, E. cloacae and other Pseudomonas strains). When tested as a protective culture in challenge experiments against S. Senftenberg on hydroponically grown mung bean seeds inoculated simultaneously with the two bacteria, Salmonella showed a reduced growth on the sprouts, which resulted in cell counts exceeding 3 log and 2 log CFU/g below the control after 24 and 48 h incubation, respectively (Weiss et al., 2007). The best results were obtained when the seeds were preinoculated with P. jenssenii on day 0 and challenged with Salmonella Senftenberg on day 1 of sprouting. Pre-inoculation with P. jenssenii completely suppressed growth of S. Senftenberg during germination of sprouts. After 7 days, while the salmonellae reached 8 log CFU/g in the control samples, counts on sprout samples with the pre-treated seeds were below 1 log CFU/g.

Enterobacter asburiae is commonly associated with plants and has been used as a biocontrol strain for inhibiting the growth of enteric pathogens such as Salmonella and E. coli O157:H7 (Cooley et al., 2003). E. asburiae JX1 isolated from mungbean sprouts exhibited stable antagonistic activity against a broad range of Salmonella serovars (Agona, Berta, Enteritidis, Hadar, Heidelberg, Javiana, Montevideo, Muenchen, Newport, Saint Paul, and Typhimurium DT104) (Ye et al., 2010). Mung beans inoculated with a cocktail of the Salmonella serovars in combination with E. asburiae JX1 attained much lower levels of salmonellae (1.16 log CFU/g) than the single cocktail inoculated samples (6.72 log CFU/g) after 4 days of sprouting. The inhibitory activity of E. asburiae JX1 was attributed partially to nutrient competition, but also to the production of inhibitory substance(s) released in solid medium but not in liquid cultures.

APPLICATION OF LYTIC BACTERIOPHAGES

Lytic bacteriophages are gaining interest as biocontrol agents because they are highly species-specific, and therefore can target the pathogenic bacteria of concern without affecting other bacteria that may be of interest in the particular food system where they are being applied. The self-propagating capacity of lytic bacteriophages is also an advantage concerning the amount of inoculant to be added, decreasing in principle the processing costs. Phages have been shown to control the growth of pathogens such as L. monocytogenes, Salmonella and Campylobacter jejuni, as well as spoilage organisms in fruit, dairy products, poultry and red meats (Greer, 2005; Hudson et al., 2005; Rees and Dodd, 2006). Nevertheless, due to their high strain specificity, it is often necessary to apply complex phage mixtures, making the phage selection procedures rather complicated. Factors such as strain sensitivity, transient resistance, and host range specificity of the bacteriophages clearly seem to limit their application as biocontrol agents on sprouts. Pao et al. (2004) tested the potential of lytic phages (the Myoviridae Phage-A, targeting Salmonella Typhimurium and Salmonella Enteritidis, and the Siphoviridae Phage-B, specific for Salmonella Montevideo) in experimentally contaminated broccoli and mustard seeds. Salmonella counts increased to a greater extent in mustard seeds than in broccoli seeds during soaking. Application of Phage-A achieved a 1.37 log suppression of Salmonella growth on mustard seeds. The mixture of Phage-A and Phage-B caused a 1.50 log suppression of Salmonella growth in the soaking water of broccoli seeds. Because of the observed host specificity of phages, the authors stressed the importance of developing phage mixtures that can control a broad range of potential contaminants. However, experiments with two Salmonella bacteriophages isolated from enrichment of sewage samples (SSP5 and SSP6, belonging to the Myoviridae and Siphoviridae families, respectively) against a Salmonella Oranienburg strain (isolated from an outbreak of salmonellosis associated with alfalfa sprouts in Australia) provided unsatisfactory results. In liquid cultures, maximum reductions of Salmonella counts of ca. 1 log unit were achieved after 3 h of incubation. However, the remaining viable population was refractile to a second bacteriophage addition, entering a temporary non-specific phage resistance stage (Kocharunchitt et al., 2009). The temporary character of resistance was further demonstrated because bacterial colonies isolated from agar plates after the phage challenge showed the same degree of phage sensitivity as the original strains. On alfalfa seeds experimentally contaminated with S. Oranienburg, phage treatment (SSP6) caused a still lower reduction of Salmonella levels after 3 h of presoaking (as compared to liquid cultures) and did not inhibit growth of the Salmonella population thereafter. The phage survived on both treated and control seeds as well as in their treatment solutions, indicating that the phage could survive throughout the sprouting process. In a more recent study, a cocktail of six bacteriophages (F01, P01, P102, P700, P800, and FL 41) isolated from manure effluent sampled from pig or feedlot farms was prepared and evaluated for controlling a cocktail of Salmonella serovars in sprouting seeds (Ye et al., 2010). Mung beans inoculated with Salmonella and sprouted over a 4-day period attained levels of 6.72 log CFU/g. Levels of Salmonella were reduced to 3.31 log CFU/g when the pathogen was coinoculated with bacteriophages. However, by using a combination of E. asburiae JX1 and bacteriophages, the levels of Salmonella associated with mung bean sprouts were only detected by enrichment. With alfalfa sprouts derived from seeds inoculated with Salmonella alone, the counts of the enteric pathogen attained levels in the order of 7.62 log CFU/g. However, in the presence of E. asburiae JX1 and bacteriophages, no Salmonella was recovered even when samples (25-g batches) were enriched. The biocontrol preparation was effective at controlling the growth of Salmonella under a range of sprouting temperatures (20 to 30 °C). The combination of E. asburiae JX1 and bacteriophages represents a promising, chemical-free approach for controlling the growth of Salmonella on sprouting seeds.

TREATMENT WITH BACTERIOCIN PREPARATIONS

Bacteriocins can be concentrated from cultured broths of their producer strains and used as partially-purified preparations in food systems. Nisin and pediocin PA-1/Ach are available on the market for application in foods. These convenient, natural preservatives, can be added to foods in a more flexible way than protective cultures or phage preparations, and the bacteriocin dose can be adjusted to maximize its bactericidal effects. Very often, bacteriocins are used in food systems as part of hurdle technology, whereby their bactericidal effects are enhanced by other antimicrobial factors with which they may act synergistically. Compared to meat and dairy products, where the application of bacteriocins has been studied extensively, much less work has been done on vegetable foods. In mungbean sprouts dipped in a solution of purified mundticin ATO6 (200 U/ml) or coated with an alginate film containing the bacteriocin (200 U/ml), a 2 log-decline of listerial counts was observed after treatments (Bennik et al., 1999). Although outgrowth of the listeria was not inhibited during storage of samples, counts obtained for the mundticin-treated samples were always lower compared to the untreated control. The mundticin-alginate coating provided the best results, especially within days 5 and 10 of incubation. Application of bacteriocins immobilized in edible coatings has been investigated in greater details for other types of foods. Bacteriocin immobilization has several advantages like a more regular delivery of bacteriocin molecules to the food substrate, lower inactivation of bacteriocin by interaction with food components or enzymes, and protection from cross contamination. Nisin and pediocin solutions were tested as possible sanitizer treatments on cabbage, broccoli, and mung bean sprouts against a bacterial cocktail of five L. monocytogenes strains (Bari et al., 2005). Two hours after inoculation of the listeria, the samples were washed vigorously with agitation for 1 min with different antimicrobial solutions (0.02 M EDTA, 2% sodium lactate, 0.02% potassium sorbate, 0.02% phytic acid, 10 mM citric acid, and 50 mg/l of nisin, 48 mg/l of pediocin, or combinations of bacteriocins and chemical preservatives). Viable cell count determinations after treatments indicated an overall lower efficacy for all treatments on mung bean sprouts compared to the other vegetable substrates, resulting in reductions of 0.5 log CFU/g or lower compared to samples washed simply with distilled water. Antilisterial activity of pediocin in sprouts was potentiated by citric acid, while the activity of nisin was by phytic acid. The most effective antimicrobial treatment combination on mung bean sprouts was nisin plus pediocin plus phytic acid, which caused a 1.2-log CFU/g reduction compared to washing treatment with distilled water. This combined treatment also reduced the native microflora by ca 1.4 log CFU/g. However, the fate of survivors after treatment was not investigated. It would be interesting to know whether the surviving bacterial populations were able to multiply or decreased during further storage of the treated samples. Enterocin AS-48 is a broad-spectrum cyclic antimicrobial peptide produced by E. faecalis and E. faecium strains (reviewed by Maqueda et al., 2004). This bacteriocin has been widely investigated in food systems against foodborne pathogenic bacteria such as L. monocytogenes, S. aureus, B. cereus, E. coli or S.

enterica (Abriouel et al., 2010; Ananou et al., 2005; Cobo Molinos et al., 2005, 2009a; Grande et al., 2006, 2007a, b; Martínez Viedma et al., 2008, 2009a, 2009b; Muñoz et al., 2007). Partially-purified preparations of enterocin AS-48 can be produced easy on semi-synthetic media (Abriouel et al., 2003) and on whey-based substrates (Ananou et al., 2010), which makes this bacteriocin an amenable antimicrobial for application in foods. Because of its broad spectrum of inhibition and increased stability due to its cyclic structure, enterocin AS-48 can be a sound candidate for decontamination of vegetable foods containing L. monocytogenes and other foodborne bacteria sensitive to this bacteriocin. In experiments carried out for enterocin AS-48 on sprouts, L. monocytogenes was able to grow without bacteriocin on alfalfa and soybean sprouts and in green asparagus at temperatures of 6 to 22°C and to reach high cell numbers (up to 6 log CFU/g, depending on temperature) during storage (Cobo Molinos et al., 2005). Sprouts inoculated with L. monocytogenes were treated by immersion for 5 min in distilled water or in bacteriocin solutions of 5, 12.5, and 25 µg/ml. The effect of bacteriocin treatment was directly proportional to bacteriocin concentration and inversely proportional to storage temperature of samples after treatment. In alfalfa and soybean sprouts, the concentration of viable listeria was reduced by approx. 2 log units by the 25 µg/ml bacteriocin immersion treatment. Washing treatments with 12.5 µg/ml bacteriocin caused apparently much lower reductions of listeria populations in soybean sprouts compared to alfalfa sprouts (1-1.5 log cycles). Nevertheless, for both alfalfa and soybean sprouts, the bacteriocin concentrations of 12.5 as well as 25 µg/ml reduced the concentrations of viable listeria below the detection levels from day 1 to day 7 of storage at temperatures of 6°C as well as 15°C. These results indicated that residual bacteriocin adsorbed to the treated sprouts was able to provide a protective effect after treatment for samples stored under a broad interval of refrigeration temperatures (Cobo Molinos et al., 2005). Nevertheless, incubation of the treated samples at a higher temperature of 22°C seriously compromised the protective effect of the bacteriocin. In soybean sprouts, the concentration of viable cells was reduced much more slowly, and in alfalfa sprouts the listeria were able to multiply even in the samples treated with 25 µg/ml bacteriocin. For green asparagus, no viable listeria were detected during storage at 15°C of samples treated with bacteriocin concentrations of 12.5 and 25 µg/ml. The bacteriocin apparently had a lower effect at 6°C (reducing the viable concentration of listeria below detection levels only for 25 µg/ml bacteriocin after day 3 of storage) and did not avoid growth of the listeria in the samples stored at 22°C.

In order to improve the bactericidal effects against listeria, the bacteriocin (25 µg/ml) was applied on green asparagus in washing treatments in combinations with a variety of antimicrobials: acetic acid, citric acid, lactic acid, potassium lactate, sodium propionate, potassium sorbate, sodium nitrite and nitrate, trisodium-tri-metaphosphate, potassium thiosulfate, potassium permanganate, n-propyl p-hydroxybenzoate, p-hydoxybenzoic acid methyl esther, hexadecylpyridinium chloride, peracetic acid, and sodium hypochlorite (Cobo Molinos et al., 2005; Table 1). The combinations of AS-48, acetic acid, citric acid, sodium propionate, potassium sorbate or sodium nitrite had limited or no effect on viability of listeria compared to the effect of AS-48 alone. Remarkably, solutions containing AS-48 plus lactic acid (0.1% and 0.5%), sodium lactate (0.1% and 0.5%), n-propyl phydroxybenzoate (0.1% and 0.5%), p-hydroxybenzoic acid methyl esther (0.5%), peracetic acid (80 ppm), sodium hypochlorite (100 ppm), potassium nitrate (100 ppm), or tri-sodium tri-metaphosphate (0.5%) reduced viable counts of listeria below detection limits upon application of the immersion treatment and/or further storage for 24 h. A lower increase of AS-48 activity was noticed for hexadecylpyridinium chloride (0.5%), sodium thiosulphate (0.01 N), tri-sodium phosphate (1.5%) and potassium permanganate (25 ppm). This study provided a variety of combinations of enterocin AS-48 and sanitizers that improved the efficacy for decontamination of Listeria while at the same time decreasing the effective concentration of sanitizers to be added. Since most foods (including sprouts) are complex ecosystems where mixed microbial populations coexist, the impact of bacteriocin treatment was investigated on soybean sprouts during storage at 10°C. Changes in microbial populations during storage of bacteriocin-treated foods may provide insights on the global impact of bacteriocin treatment in addition to the selective inhibition of the bacterial pathogen being studied by selective counting methods. For doing these studies, culture-independent methods based on the total DNA of the bacterial community are frequently used, and the results of amplification of species-specific DNA regions are analysed by denaturing gel electrophoresis (DGGE). Results from sprout samples treated with enterocin AS-48 revealed modifications of the microbial populations during storage, apparently increasing the proportion of Enterococcus and Leuconostoc bacteria and decreasing the levels of Gram-negative bacteria (such as Pantoea, Escherichia and Enterobacter) on the sprouts during storage (Cobo Molinos et al., 2009b). These results indicate that bacteriocin treatment with enterocin AS-48 has additional effects besides inhibition of L. monocytogenes on sprouts, disturbing the microbial balance. The consequences of these changes for the biopreservation of sprouts and for the survival of pathogenic and spoilage bacteria need to be investigated in deeper details. Bacillus cereus is a toxin-producing common soil inhabitant that is often present in a variety of foods, including those of vegetable origin (Granum, 2001). In soybean sprouts, endospore-forming bacteria were found in the order of 2 log CFU/g, and 53 out of 55 B. cereus isolates were found to produce diarrheic enterotoxins (Kim et al., 2004). A study was carried out on application of a washing treatment with enterocin AS-48 (25 µg/ml) for decontamination of sprouts and green asparagus challenged with B. cereus and Bacillus weihenstephanensis and stored at temperatures of 6, 15 or 22°C (Cobo Molinos et al., 2008a). The best results were obtained for samples refrigerated at 6 °C, in which the washing treatments reduced the population of B. cereus by 1 to 1.6 log cycles, and the remaining viable population was reduced below detection levels after days 1 to 3 of storage. By contrast, reductions of viable cell counts obtained after treatments at 15 or 22 °C were much lower in the three types of food tested, and the remaining viable cells multiplied during storage of the treated samples. A similar trend was observed for samples challenged with B. weihenstephanensis, with reductions of viable counts up to 3.4 log cycles after treatment, complete inhibition of outgrowth during storage at 6°C for up to 7 days, and proliferation of survivors during storage at 15 and 22°C. In order to improve the efficacy of treatments, the bacteriocin (25 µg/ml) was tested in combination with a variety of chemicals (lactic acid, sodium lactate, sodium hypochlorite, tri-sodium tri-metaphosphate, hexadecyl pyridinium chloride, peracetic acid, polyphosphoric acid, carvacrol, hydrocinnamic acid, n-propyl phydroxybenzoate and p-hydroxybenzoic acid methyl esther) against B. cereus in alfalfa sprouts stored at 15°C (Cobo Molinos et al., 2008a; Table 1). The bactericidal effect of treatments was enhanced significantly by addition of AS-48 for carvacrol (0.3%), cinnamic acid (0.3%), hydrocinnamic acid (0.5%), polyphosphoric acid (at 0.1 and 0.5%), peracetic acid (40 ppm), hexadecylpyridinium chloride (0.5%), sodium hypochlorite (100 ppm), and tri-sodium tri-metaphosphate (0.5%). Interestingly, for some of the combined treatments with AS-49 (with sodium hypochlorite, peracetic acid, polyphosphoric acid, and hydrocinnamic acid) the levels of B. cereus remained below detection limit after 24 h of treatment, indicating that treatments were effective in preventing outgrowth of possible survivors or that no surviving cells were left by the treatments. The degree of protection afforded by these combined treatments was tested in alfalfa sprouts challenged with B. cereus and with B. weihenstephanensis, and stored at 15°C for one week. In samples treated with carvacrol, cinnamic and hydrocinnamic acids in combination with AS-48, viable cell counts of B. cereus remained significantly lower compared to each individual treatment during most part or the whole storage period. The best results were obtained for the combinations of AS-48 and sodium hypochlorite, hexadecylpyridinium chloride, peracetic acid and polyphosphoric acid, which reduced the population of B. cereus below detection limits for the whole or at least most of the storage period. The combinations of 25 µg/ml AS- 48 and hydrocinnamic acid, peracetic acid, sodium hypochlorite, polyphosphoric acid and hexadecylpyridinium chloride also reduced the population of B. weihenstephanensis below detection limits for one week when tested on alfalfa sprouts (Cobo Molinos et al., 2008a). Bacteriocins from Gram-positive bacteria are usually not active against Gram-negative bacteria, with some exceptions. However, Gram-negative cells can be rendered sensitive to these bacteriocins when exposed to treatments that damage the bacterial outer membrane, which acts as a barrier against diffusion of bacteriocin molecules to the cytoplasmic membrane where they exert their lethal action. Exposure to chelating agents, acids, or sublethal heat, are often used to sensitize Gram-negative bacteria to bacteriocins. Since most of the Gram-negative bacteria are resistant to enterocin AS-48, the bacteriocin was tested on sprouts in combination with several sensitizing treatments. Once the synergistic effects were demonstrated in washing treatments against S. enterica as test organism, they were corroborated on other Gram-negative species. Inactivation of S. enterica cells inoculated on soybean sprouts increased greatly when sprouts were heated for 5 min at 65°C in an alkaline solution (25 μg/ml, pH 9.0) of enterocin AS-48 (Cobo Molinos et al., 2008b). Washing treatments containing AS-48 (25 μg/ml) and 1.5% lactic acid, 1.5% trisodium phosphate, 0.5% trisodium tri-metaphosphate, 0.1% polyphosphoric acid, 80 ppm peracetic acid, 0.5% hexadecylpyridinium chloride, 100 ppm sodium hypochlorite, 0.5% n-propyl phydroxybenzoate, 0.5% p-hydroxybenzoic acid methyl esther and 2% hydrocinnamic acid significantly reduced the population of S. enterica to a much greater extent that the single treatments (Table 1). The best results were obtained for combinations of AS-48 and lactic acid, peracetic acid, as well as polyphosphoric acid.

The efficacy of enterocin AS-48-polyphosphoric acid treatment for decontamination of sprouts was tested on other Gram-negative bacteria, including species involved in human disease and in food spoilage E. coli O157:H7, Shigella flexneri, Shigella sonnei, Enterobacter aerogenes, Y. enterocolitica, A. hydrophila and P. fluorescens). Enterocin addition increased the bactericidal effects of treatments in all cases, although there were large differences in the concentrations of polyphosphoric acid required for a complete inactivation and prolonged protection from one bacterial species to another. For example, E. coli required at least 0.4% polyphosphoric acid for complete inactivation while Y. enterocolitica was very sensitive to low polyphosphoric acid treatment (0.1-0.2%).

Table 1: Antimicrobials that potentiate the efficacy of enterocin AS-48 on sprouts (X). Unmarked combinations were not tested or did not potentiate bacteriocin activity

	L. monocytogenes	*B. cereus*	*S. enterica*
Acetic acid	X		
Citric acid	X		
Propionate	X		
Sorbate	X		
Lactic acid	X	X	X
Lactate	X	X	X
Nitrite	X		
Nitrate	X		
EDTA			X
Tri-sodium phosphate	X		X
Tri-sodium tri-metaphosphate	X	X	X
Thiosulphate	X		
Permanganate	X		
Propyl-*p*-hydroxybenzoate	X	X	X
p-Hydoxybenzoic acid methyl esther	X	X	X
Peracetic acid	X	X	X
Hexadecylpyridinium chloride	X	X	X
Sodium hypochlorite		X	X
Polyphosphoric acid		X	X
Carvacrol		X	
Cinnamic acid		X	
Hydrocinnamic acid		X	X

A. hydrophila and E. aerogenes were very sensitive to combined treatments with enterocin AS-48 plus 0.5% and 1% polyphosphoric acid, respectively. P. fluorescens was also very sensitive to 1% polyphosphoric acid (with an average decrease of 5 log CFU/g after the combined treatment), but it required a higher concentration of at least 2% to avoid overgrowth during storage. The shigellae were the most resistant of all bacteria to the combined treatment. S. sonnei was not completely eliminated by treatments containing 2% polyphosphoric acid and S. flexneri was even more resistant, although the combined treatment with 2% polyphosphoric acid always reduced the surviving fraction and avoided overgrwth during storage of the samples. These results suggest that combined treatments with polyphosphoric acid at 1% are sufficient to suppress the most common Gram-negative pathogens found in sprouts (that is, S. enterica and E. coli) as well as others such as Y. enterocolitica and the spoilage Pseudomonas. Other bacteria which are much less common, such as the shigella, would require a polyphosphoric acid concentration of at least 2% for inactivation on sprouts. Therefore, the disinfection capacity of the combined washing treatments could be adjusted depending on the estimated risks. The bacteriocins from Gram-negative bacteria have seldom been exploited in food preservation, although they could be useful in the control of enteric pathogens. Many Gramnegative species have shown to produce bacteriocins, but those produced by E. coli strains (or colicins) have been studied in greater details (Riley and Gordon, 1996). A recently isolated E. coli strain Hu194 (from a human fecal sample) was capable of inhibiting 22 E. coli strains of serotype O157:H7 and it also exhibited some degree of antimicrobial activity against Salmonella. This inhibition was mediated by the production of a colicin named Hu194 (Nandiwada et al., 2004). Semi-crude colicin Hu194 was applied by mixing and drying on alfalfa seeds challenged with three different E. coli O157:H7 strains. The bacteriocin treatment caused variable effects depending on the strain. One of the strains was successfully inactivated (5 log CFU/g reduction) from inoculated alfalfa seeds, without any bacterial overgrowth being observed for up to 5 days of further incubation. By contrast, the other two strains required 20-fold higher colicin concentrations to achieve a reduction of 3 log cycles (Nandiwada et al., 2004). This reduction was slightly enhanced after soaking the inoculated seeds for a longer period of time, and the maximum bacterial reduction was typically observed during the first 2 days after soaking the seeds in the colicin extract suspension. The variations observed in strain sensitivity need to be further investigated by testing a larger number of strains in order to establish the minimum bacteriocin dose that is needed to achieve an acceptable reduction of viable counts for the target bacteria.

CONCLUSIONS AND PERSPECTIVES

Biocontrol strategies aimed at reducing the transmission of human pathogenic bacteria through the consumption of sprouts seem a promising alternative strategy to the use of chemical preservatives or at least to reduce the concentrations of preservatives to be used. Application of live bacterial cultures requires a careful selection of strains that are well adapted to the food substrate and storage temperature of sprouts, and is somehow limited by their capacity for an effective in situ production of antimicrobial substances and/ or competitive exclusion of the pathogens. Strains of Gram-negative bacteria belonging to Pseudomonas and Enterobacter genera seem to be promising for the control of Salmonella on sprouts. Intervention strategies based on lytic bacteriophages require a careful selection of complex phage preparations in order to overcome efficacy limitations due to strain specificity. Bacteriophages may act synergistically with antagonistic bacteria against selected human pathogenic bacteria. This is an interesting approach that has very seldom been exploited and where further studies are required. The lytic enzymes produced by bacteriophages could also be exploited as antimicrobial agents on sprouts (either alone or in combination with other treatments), minimizing the impact of phage specificity. However, this approach has not been investigated yet. Some bacteriocins (especially those produced by the lactic acid bacteria) have shown interesting results when tested on sprouts. The efficacy of these bacteriocins can be enhanced in combination with other antimicrobials, decreasing the concentrations of chemical preservatives required for disinfection of sprouts and protection from bacterial overgrowth during shelf life storage. The application of bacteriocins from Gram-negative bacteria for biopreservation of sprouts has been seldom investigated. This is an interesting field for new research as well. In addition, the application of cocktails based on protective cultures, bacteriophages and bacteriocins should also be investigated as broad-spectrum intervention strategies capable of simultaneous inhibition of mixed bacterial populations of human pathogenic bacteria on sprouts.

REFERENCES

1. Abriouel, H.; Valdivia, E.; Martínez-Bueno, M.; Maqueda, M. & Gálvez, A. (2003). A simple method for semi-preparative-scale production and recovery of enterocin AS-48

2. derived from Enterococcus faecalis subsp. liquefaciens A-48-32. Journal of Microbiological Methods, 55 (2), 599–605, ISSN 0167-7012

3. Abriouel, H.; Lucas, R.; Ben Omar, N.; Valdivia, E. & Gálvez, A. (2010). Potential applications of the cyclic peptide enterocin AS-48 in the preservation of vegetable foods and beverages. Probiotics and Antimicrobial Proteins, 2(2), 77-89, ISSN 1867-1306

4. Ananou, S.; Garriga, M.; Hugas, M.; Maqueda, M.; Martínez-Bueno, M.; Gálvez, A. & Valdivia, E. (2005) Control of Listeria monocytogenes in model sausages by enterocin AS-48. International Journal of Food Microbiology, 103 (2), 179-190, ISSN 0740-0020

5. Ananou, S.; Muñoz, A.; Martínez-Bueno, M.; González-Tello, P.; Gálvez, A.; Maqueda, M. & Valdivia, E. (2010). Evaluation of an enterocin AS-48 enriched bioactive powder obtained by spray drying. Food Microbiology, 27 (1) 58-63, ISSN 0740-0020

6. Andrews, W.H.; Mislivec, P.B.; Wilson, C.R.; Bruce, V.R.; Polema, P.L.; Gibson, R.; Trucksess, M.W. & Young, K. (1982). Microbial hazards associated with bean sprouting. Journal of Association of Official Analytical Chemists, 65 (2), 241–248, ISSN 1060-3271

7. Bari, M.L.; Ukuku, D.O.; Kawasaki, T.; Inatsu, Y.; Isshiki, K. & Kawamoto, S. (2005). Combined efficacy of nisin and pediocin with sodium lactate, citric acid, phytic acid, and potassium sorbate and EDTA

in reducing the Listeria monocytogenes population of inoculated fresh-cut produce. Journal of food protection, 68 (7), 1381-1387, ISSN 0362-028X

8.   Bennik, M.H.J.; Overbeek, W. van; Smid, E.J. & Gorris, L.G.M. (1999). Biopreservation in modified atmosphere stored mungbean sprouts: the use of vegetable-associated bacteriocinogenic lactic acid bacteria to control the growth of Listeria monocytogenes. Letters in Applied Microbiology, 28 (3), 226-232, ISSN 0266-8254

9.   Beuchat, L.R. (1996). Pathogenic microorganisms associated with fresh produce. Journal of food protection, 59 (2), 204-216, ISSN 0362-028X Cai, Y.; Ng, L.-K. & Farber, J. M. (1997). Isolation and characterization of nisin-producing Lactococcus lactis subsp. lactis from bean sprouts. Journal of Applied Microbiology, 83 (4), 499-507, ISSN 1364-5072

10.   Charkowski, A.O.; Barak, J.D.; Sarreal, C.Z. & Mandrell, RE. (2002). Differences in growth of Salmonella enterica and Escherichia coli O157:H7 on alfalfa sprouts. Applied and Environmental Microbiology, 68 (6), 3114–3120, ISSN 0099-2240

11.   Cobo Molinos, A.; Abriouel, H.; Ben Omar, N.; Valdivia, E.; Lucas, R.; Maqueda, M.; Martínez Cañamero, M. & Gálvez, A. (2005). Effect of immersion solutions containing enterocin AS-48 on Listeria monocytogenes in vegetable foods. Applied and Environmental Microbiology, 71 (12), 7781-7787, ISSN 0099-2240

12.   Cobo Molinos, A.; Abriouel, H.; Lucas, R.; Ben Omar, N.; Valdivia, E. & Gálvez, A. (2008a). Inhibition of Bacillus cereus and B. weihenstephanensis in raw vegetables by application of washing solutions containing enterocin AS-48 alone and in combination with other antimicrobials. Food Microbiology, 25 (6), 762-770, ISSN 0740-0020

13.   Cobo Molinos, A.; Abriouel, H.; Lucas, R.; Valdivia, E.; Ben Omar, N. & Gálvez, A. (2008b). Combined physico-chemical treatments based on enterocin AS-48 for inactivation of Gram-negative bacteria in soybean sprouts. Food and Chemical Toxicology, 46 (6), 2912-2921, ISSN 0278-6915

14.   Cobo Molinos, A.; Abriouel, H.; Ben Omar, N.; Lucas, R.; Valdivia, E. & Gálvez, A. (2009a). Enhanced bactericidal activity of enterocin AS-48 in combination with essential oils, natural bioactive compounds, and chemical preservatives against Listeria monocytogenes in ready-to-eat salads. Food and Chemical Toxicology, 47 (9), 2216-2223, ISSN 0278-6915

15. Cobo Molinos, A., Abriouel, H.; Ben Omar, N.; Lucas, R. & Gálvez, A. (2009b). Microbial diversity changes in soybean sprouts treated with enterocin AS-48. Food Microbiology, 26 (8) 922–926, ISSN 0740-0020

16. Cooley, M.B.; Miller, W.G. & Mandrell, R.E. (2003). Colonization of Arabidopsis thaliana with Salmonella enterica and enterohemorrhagic Escherichia coli O157:H7 and competition by Enterobacter asburiae. Applied and Environmental Microbiology, 69 (8), 4915–4926,ISSN 0099-2240

17. Feng, G.; Guron, G.K.; Churey, J.J. & Worobo, R.W. (2009). Characterization of mundticin L, a Class IIa anti-Listeria bacteriocin from Enterococcus mundtii CUGF08. Applied and Environmental Microbiology, 75 (17), 5708–5713, ISSN 0099-2240

18. Fett, W. F. (2006). Inhibition of Salmonella enterica by plant-associated pseudomonads in vitro and on sprouting alfalfa seed. Journal of Food Protection, 69 (4), 719–728, ISSN 0362-028X

19. Fu, T.; Stewart, D.; Reineke, K.; Ulaszek, J.; Schliesser, J. & Tortorello, M. (2001). Use of spent irrigation water for microbiological analysis of alfalfa sprouts. Journal of Food Protection, 64 (6), 802–806, ISSN 0362-028X

20. Gálvez, A.; Abriouel, H.; Lucas López, R. & Ben Omar, N. (2007). Bacteriocin-based strategies for food biopreservation. International Journal of Food Microbiology, 120 (1-2), 51-70, ISSN 0168-1605

21. Galvez, A.; Lopez, R.L.; Abriouel, H.; Valdivia, E. & Ben Omar, N. (2008). Application of bacteriocins in the control of foodborne pathogenic and spoilage bacteria. Critical Reviews Biotechnology, 28 (2), 125–152, ISSN 0738-8551

22. Grande, M.J.; Lucas, R.; Abriouel, H.; Valdivia, E.; Ben Omar, N.; Maqueda, M.; MartínezBueno, M.; Martínez-Cañamero, M. & Gálvez, A. (2006). Inhibition of toxicogenic Bacillus cereus in rice-based foods by enterocin AS-48. International Journal of Food Microbiology, 106 (2), 185-194, ISSN 0168-1605

23. Grande, M.J.; Lucas, R.; Abriouel, H.; Valdivia, E.; Ben Omar, N.; Maqueda, M.; MartínezCañamero, M. & Gálvez, A. (2007a). Treatment of vegetable sauces with enterocin AS-48 alone or in combination with phenolic compounds to inhibit proliferation of Staphylococcus aureus. Journal of Food Protection, 70 (2), 405-411, ISSN 0362-028X

24. Grande, M.J.; Abriouel, H.; Lucas, R.; Valdivia, E.; Ben Omar, N.; Martínez-Cañamero, M. & Gálvez, A. (2007b). Efficacy of enterocin AS-

48 against bacilli in ready-to-eat vegetable soups and purees. Journal of Food Protection, 70 (10), 2339-2345, ISSN 0362-028X

25. Granum, P.E. (2007). Bacillus cereus. In: Food Microbiology. Fundamentals and Frontiers, (third edn), Doyle, M.P. & Beuchat, L.R. (Eds.), pp. 445-455, ASM Press, ISBN 978-1-55581-407-6, Washington, DC

26. Greer, G.G. (2005). Bacteriophage control of foodborne bacteria. Journal of Food Protection, 68 (5), 1102–1111, ISSN 0362-028X

27. Harris, L.J.; Farber, J.N.; Beuchat, L.R.; Parish, M.E.; Suslow, T.V.; Garrett, E.H. & Busta, F.F. (2003). Outbreaks associated with fresh produce: incidence, growth, and survival of pathogens in fresh and fresh-cut produce. Comprehensive Reviews in Food Science and Food Safety, 2 (Supplement), 78-141, ISSN 1541-4337

28. Hudson, J.A.; Billington, C.; Carey-Smith, G. & Greening, G. (2005). Bacteriophages as biocontrol agents in food. Journal of Food Protection, 68 (2), 426–437, ISSN 0362-028X

29. Itoh, Y.; Sugita-Konishi, Y.; Kasuga, F.; Iwaki, M.; Hara-Kudo, Y.; Saito, N.; Noguchi, Y.; Konuma, H. & Kumagai, S. (1998). Enterohemorrhagic Escherichia coli O157:H7

30. present in radish sprouts. Applied and Environmental Microbiology, 64 (4), 1532–1535, ISSN 0099-2240

31. Jack, R.W.; Tagg, J.R. & Ray, B. (1995) Bacteriocins of Gram positive bacteria. Microbiology Reviews, 59 (2), 171–200, ISSN 0146-0749

32. Kawamoto, S.; Shima, J.; Sato, R.; Eguchi, T.; Ohmomo, S.; Shibato, J.; Horikoshi, N.; Takeshita, K. & Sameshima, T. (2002). Biochemical and genetic characterization of mundticin KS, an antilisterial peptide produced by Enterococcus mundtii NFRI 7393.

33. Applied and Environmental Microbiology, 68 (8), 3830–3840, ISSN 0099-2240 Kim, H.J.; Lee, D.S. & Paik, H.D. (2004). Characterization of Bacillus cereus isolates from raw soybean sprouts. Journal of Food Protection, 67 (5), 1031–1035, ISSN 0362-028X

34. Kocharunchitt, C.; Ross, T. & McNeil, D.L. (2009). Use of bacteriophages as biocontrol agents to control Salmonella associated with seed sprouts. International Journal of Food Microbiology, 128 (3), 453–459, ISSN 0168-1605

35. Liao, C.H. (2008). Growth of Salmonella on sprouting alfalfa seeds as affected by the inoculum size, native microbial load and Pseudomonas fluorescens 2-79. Letters in Applied Microbiology, 46 (2), 232-6, ISSN 0266-8254

36. Liao, C.-H. & Fett, W.F. (2001). Analysis of native microflora and selection of strains antagonistic to human pathogens on fresh produce. Journal of Food Protection, 64 (8), 1110–1115, ISSN 0362-028X

37. Lynch, M.F.; Tauxe, R.V. & Herberg, C.W. (2009). The growing burden of foodborne outbreaks due to contaminated fresh produce: risks and opportunities. Epidemiology and Infection, 137 (3), 307–315, ISSN 0950-2688

38. Maqueda, M.; Gálvez, A.; Martínez Bueno, M.; Sanchez-Barrena, M.J.; González, C.; Albert, A.; Rico, M. & Valdivia, E. (2004). Peptide AS-48: prototype of a new class of cyclic bacteriocins. Current Protein & Peptide Science, 5 (5), 399-416, ISSN 1389-2037

39. Martínez-Viedma, P.; Sobrino, A.; Ben Omar, N.; Abriouel, H.; Lucas López, R.; Valdivia, E.; Martín Belloso, O. & Gálvez, A. (2008). Enhanced bactericidal effect of HighIntensity Pulsed-Electric Field treatment in combination with enterocin AS-48

40. against Salmonella enterica in apple juice. International Journal of Food Microbiology, 128 (2), 244-249, ISSN 0168-1605

41. Martínez-Viedma, P.; Abriouel, H.; Ben Omar, N.; Lucas, R.; Valdivia, E. & Gálvez, A. (2009a). Inactivation of Geobacillus stearothermophilus in canned foods and drinks by addition of enterocin AS-48. Food Microbiology, 26 (3), 289-293, ISSN 0740-0020

42. Martínez-Viedma, P.; Abriouel, H.; Sobrino, A.; Ben Omar, N.; Lucas López, R.; Valdivia, E.; Martín Belloso, O. & Gálvez, A. (2009b). Effect of enterocin AS-48 in combination with High-Intensity Pulsed-Electric Field treatment against the spoilage bacterium Lactobacillus diolivorans in apple juice. Food Microbiology, 26 (5),491-496, ISSN 0740-0020

43. Matos, A. & Garland, J.L. (2005). Effects of Community Versus Single Strain Inoculants on the biocontrol of Salmonella and microbial community dynamics in alfalfa sprouts. Journal of Food Protection, 68 (1), 40–48, ISSN 0362-028X

44. Muñoz, A.; Ananou, S.; Gálvez, A.; Martínez-Bueno, M.; Rodríguez, A.; Maqueda, M. & Valdivia, E. (2007). Inhibition of Staphylococcus aureus in dairy products by enterocin AS-48 produced in situ and ex situ: bactericidal synergism with heat. International Dairy Journal, 17 (7), 760-769, ISSN 0958-6946

45. Nandiwada, L.S.; Schamberger, G.P.; Schafer, H.W. & Diez-Gonzalez, F. (2004). Characterization of an E2-type colicin and its application to treat alfalfa seeds to reduce Escherichia coli O157:H7. International Journal of Food Microbiology, 93 (3), 267–279, ISSN 0168-1605

46. Pao, S.; Rolph, S.P.; Westbrook, E.W. & Shen, H. (2004). Use of bacteriophages to control Salmonella in experimentally contaminated sprout seeds. Journal of Food Science, 69 (5), 127–130, ISSN 0022-1147

47. Patterson, J.E. & Woodburn, M.J. (1980). Klebsiella and other bacteria on alfalfa and bean sprouts at the retail level. Journal of Food Science, 45 (3), 492–495, ISSN 0022-1147

48. Prokopowich, D. & Blank, G. (1991). Microbiological evaluation of vegetable sprouts and seeds. Journal of Food Protection, 54 (7), 560–562, ISSN 0362-028X

49. Rees, C.E.D. & Dodd, C.E.R. (2006). Phage for rapid detection and control of bacterial pathogens in food. Advances in Applied Microbiology, 59, 159–186, ISSN 0065-2164

50. Riley, M.A. & Gordon, D.M. (1996). The ecology and evolution of bacteriocins. Journal of Industrial Microbiology, 17 (3-4), 151– 158, ISSN 1367-5435

51. Robertson, L.J.; Johannessen, G.S.; Gjerde, B.K. & Loncarevic, S. (2002). Microbiological analysis of seed sprouts in Norway. International Journal of Food Microbiology, 75 (1-2), 119–126, ISSN 0168-1605

52. Rosas, C.J. & E.F. Escartin. (2000). Survival and growth of Vibrio cholerae O1, Salmonella typhi and Escherichia coli O157:H7 in alfalfa sprouts. Journal of Food Science, 65 (1), 162–165, ISSN 0022-1147

53. Splittstoesser, D.F.; Queale, D.T. & Andaloro, B.W. (1983). The microbiology of vegetable sprouts during commercial production. Journal of Food Safety, 5 (2), 79–86, ISSN 0149-6085

54. Taormina, P.J.; Beuchat, L.R. & Slutsker, L. (1999). Infections associated with eating seed sprouts: an international concern. Emerging Infectious Diseases, 5 (5), 626– 634, ISSN 1080-6040

55. Weiss, A.; Hertel, C.; Grothe, S.; Ha, D. & Hammes, W.P. (2007). Characterization of the cultivable microbiota of sprouts and their potential for application as protective cultures. Systematic and Applied Microbiology, 30 (6), 483-93, ISSN 0723-2020

56. Ye, J.; Kostrzynska, M.; Dunfield, K. & Warriner, K. (2010). Control of Salmonella on sprouting mung bean and alfalfa seeds by using a biocontrol preparation based on antagonistic bacteria and lytic bacteriophages. Journal of Food Protection, 73 (1), 9–17, ISSN 0362-028X.

Chapter 7

EMERGING PRESERVATION TECHNIQUES FOR CONTROLLING SPOILAGE AND PATHOGENIC MICROORGANISMS IN FRUIT JUICES

Kamal Rai Aneja,[1] Romika Dhiman,[2] Neeraj Kumar Aggarwal,[2] and Ashish Aneja[3]

[1]Vaidyanath Research, Training and Diagnostic Centre, Kurukshetra 136118, India

[2]Department of Microbiology, Kurukshetra University, Kurukshetra 136119, India

[3]University Health Centre, Kurukshetra University, Kurukshetra 136119, India

ABSTRACT

Fruit juices are important commodities in the global market providing vast possibilities for new value added products to meet consumer demand for convenience, nutrition, and health. Fruit juices are spoiled primarily due to proliferation of acid tolerant and osmophilic microflora. There is also risk of food borne microbial infections which is associated with the consumption of fruit juices. In order to reduce the incidence of outbreaks, fruit juices are preserved by various techniques. Thermal pasteurization is used commercially by fruit juice industries for the preservation of fruit juices but results in losses of essential nutrients and changes in physicochemical and organoleptic properties. Nonthermal pasteurization methods such as high hydrostatic pressure, pulsed electric field, and ultrasound and irradiations have also been employed in fruit juices to overcome the negative effects of thermal pasteurization. Some of these techniques have already been commercialized. Some are still in research or pilot scale. Apart from these emerging techniques, preservatives from natural sources have also shown considerable promise for use in some food products. In this review article, spoilage, pathogenic microflora, and food borne outbreaks associated with fruit juices of last two decades are given in one section. In other sections various prevention methods to control the growth of spoilage and pathogenic microflora to increase the shelf life of fruit juices are discussed.

INTRODUCTION

Consumer demand for nutritious foods such as fresh cut fruits and unpasteurized fruit juices has increased in the last decades owing to their low content of sodium, cholesterol and fat and high concentration of vitamin C, polyphenols, and antioxidants that play important role in the prevention of heart diseases, cancer, and diabetes [1–4]. Juice is defined as unfermented but fermentable juice, intended for direct consumption, obtained by the mechanical process from sound, ripe fruits, and preserved exclusively by physical means. The addition of sugars or acids can be permitted but must be endorsed in the individual standard [5–7] Increased consumption of fruit juices has direct influence on economy in positive way but in negative way also when food borne disease outbreaks and spoilage problems occur [8]. Juices have become frequent vehicle for transmitting pathogens such as enterohaemorrhagic Escherichia coli O157, Salmonella, and Cryptosporidium [9]. Several emerging spoilage microorganisms also have great concern in fruit juice industry; for example, Alicyclobacillus acidoterrestris has been isolated from several juices and juice products with reported occurrence between 14.7% and 18.3%. Propionibacterium cyclohexanicum and heat resistant species of mycelial fungi such as Byssochlamys fulva, B. nivea, and Neosartorya fischeri and species of Talaromyces have also been reported to spoil fruit juices [10–13]. There is tremendous increase in the food borne disease outbreaks associated with the consumption of fruit juices [14].

Keeping in view the threat challenge posed by spoilage and pathogenic microorganisms to both the fruit juice industry and public health authorities, several guidelines have been published by national food standard agencies, such as HACCP and FDA, to control or reduce the incidence of food borne disease outbreaks or spoilage [8]. For prevention of these microorganisms in fruit juices, thermal treatment is effective method for microbial inactivation but it may produce some undesirable effects on foods such as loss of nutrients and reduction of fresh like flavor [15, 16]. New technologies such as high hydrostatic pressure (HHP), high pressure homogenization (HPH), pulsed electric field (PEF), ultrasound, and irradiations have been developed to maintain nutritional and sensory quality of fruit juices [17, 18].

Chemical preservatives, such as sodium benzoate and potassium sorbate, are commonly used in fruit juices and beverages to extend their shelf life [11]. However, consumer demand for fresh and safe foods without chemically synthesized preservatives leads to increase the interest in use of food preservatives from natural sources [19]. Natural preservatives such as bacteriocins, organic acids, essential oils, and phenolic compounds have been used in some food products [19, 20].

MICROORGANISMS INVOLVED IN SPOILAGE

Change in the appearance, smell or taste of a food that makes it unacceptable to the consumer is called food spoilage [21]. Spoilage of fruit and vegetable juices is primarily owing to the proliferation of their natural acid tolerant and osmophilic microflora [22]. Fresh fruit juices are more susceptible to spoilage because fluid contents are in touch with air and microorganisms from the environment during the time of handling [4]. Yeasts, heat sensitive moulds, and lactic acid bacteria are indicator for the quality of raw materials. Heat resistant fungi and other spore forming bacteria such as Clostridium pasteurianum and Bacillus coagulans are used as targets for fruit juice pasteurization processes [8].

YEASTS

Yeasts have the ability to grow at low pH, high sugar concentration, and low water activity. Fruit juices are generally rich in simple carbohydrates and complex nitrogen sources and hence are ideal substrates for yeasts [22]. Over 110 species of yeasts have been reported to be associated with food and food products [23] and are dominating contaminants in fruit juices ranging from 1.0 to 6.83 log 10 cfu/mL [24]. The presence of yeasts in fruit juices may result from failures in fruit juice pasteurization and failure in sanitation practices [8]. Spoilage by yeasts in fruit juices is characterized by formation of CO_2 and alcohol. Yeasts may also produce turbidity, flocculation, pellicles, and clumping. Yeasts also produced pectinesterases which degrade pectin causing spoilage, organic acids, and acetaldehyde, which contribute for a "fermented flavor," may also be formed [6, 25].

Pichia, Candida, Saccharomyces, and Rhodotorula are the genera mainly responsible for the spoilage of fruit juices; species frequently isolated are Pichia membranifaciens, Candida maltosa, C. sake, Saccharomyces bailii, S. bisporus, S. cerevisiae, S. rouxii, S. bayanus, Brettanomyces intermedius, Schizosaccharomyces pombe, Torulopsis holmii, Hanseniaspora guilliermondii, Schwanniomyces occidentalis, Dekkera bruxellensis, D. naardenensis Torulaspora delbrueckii, and Zygosaccharomyces microellipsoides [7]. Major yeast species found in citrus juices are Candida parapsilosis, C. stellata, Saccharomyces cerevisiae, Torulaspora delbrueckii, and Zygosaccharomyces rouxii [26]. Some of these species are sensitive to thermal pasteurization treatment applied to fruit juices [6].

Yeasts Resistant to Preservatives

Resistance to preservatives is a great threat to the stability of fruit juices [27]. Examples of yeasts resistant to preservatives include Zygosaccharomyces bailli, Candida krusei, Saccharomyces bisporus, Schizosaccharomyces pombe, and Pichia membranifaciens [27, 28]. Resistance to preservatives has been attributed to the ability of cells to tolerate chronic intracellular pH drops by phosphofructokinase enzyme [6]. P. membranifaciens is resistant to heat, moderate amount of salt, SO_2, sorbic, benzoic and acetic acid; hence, it is considered as target microorganism for optimization of thermal pasteurization [7].

MOULDS

Moulds are aerobic which grow at low pH and high sugar concentration. In response to thermal treatment, moulds are divided into two categories: heat labile and heat resistant [29, 30]. Former kinds produce mycelial mats in juice and adhere to the package interior, carton seams and produce musty and stale off-flavours. Juice cloud loss occurs through the activity of pectinesterases [6, 29]. The dominant moulds recorded in fruit juices belong to Penicillium sp., Clad-osporium sp., Aspergillus niger, A. fumigatus, Botrytis sp., and Aureobasidium pullulans [25]. Rhizopus and Mucor are also associated with spoilage of fresh fruits and vegetables [30]. Among these, some moulds produce mycotoxins which are of great threat to human health. Major mycotoxins associated with fruit juices are byssochlamic acid (Byssochlamys fulva, B. nivea), patulin (B. fulva, B. nivea, andP. expansum), ochratoxin (Aspergillus carbonarius), and citrinin (Penicillium expansum, P. citrinum) [29, 31]. Presence of patulin in fruit juices is indicator of poor quality of fruits used in processing of juices [32].

Heat Resistant Moulds

The moulds, which able to survive at 85°C for 4.5 minutes, low oxygen tension, low pH (3.0–4.5) and produce pectinolytic enzymes, have an influence on juice stability [24]. Some notable species are Byssochlamys fulva, B. nivea, Neosartorya fischeri, and Talaromyces [6, 24, 33]. These moulds survive commercial heat pasteurization treatment, usually applied to fruits and fruit products because of the presence of heat resistant ascospores [6,25, 34]. Heat resistance also depends upon the fruit product. As the concentration of sugar increases, heat resistance in microorganisms also increases [6, 24, 33].

The presence of heat resistant fungi such as Paecilomyces variotii, Aspergillus tamari, A. flavus, and A. ochraceushas been reported in sixty packaged Nigerian fruit juices consisting of mango, pineapple, orange, and tomato [35]. Chlamydospores, sclerotia, and aleurospores are the resistant structures/spores produced by these moulds [34, 36].

A pasteurization temperature for fruit and fruit products often tested is 90°C for 3 minutes. This treatment may not be adequate to inactivate ascospores of Byssochlamys fulva, Neosartorya fischeri, and Talaromyces species [24]. Salomão et al. [34] reported that the thermal tolerance level of the ascospores of heat resistant moulds varies from strain to strain and with the composition of the heating medium as explained in Table 1.

Table 1: Thermal tolerance level of heat resistant moulds

Heat resistant moulds	Thermal tolerance level	References
Talaromyces flavus	100°C for 5 to 12 minutes in many fruit syrups	[24]
Byssochlamys fulva	86° to 88°C for 30 minutes	[24]
Paecilomyces variotii, Fusarium sp.	95°C for 10–20 seconds	[90]

The sources of contamination of these ascospores of heat resistant fungi found in fruit juices are soil, especially in case of grapes, passion fruits, pineapples, mangoes, strawberries, and other berries [6]. Other sources of contamination are processing facilities, air, utensils, fields, and orchards [24].

BACTERIA

Bacteria are present in low numbers on fresh fruits and vegetables due to low pH. Acid tolerant bacteria such as heterofermentative lactic acid bacteria, acetic acid bacteria, Erwinia sp., Enterobacter sp., Clostridium, Alicyclobacillus acidoterrestris, Propionibacterium cyclohexanicum, Pseudomonas sp., and Bacillus sp. have been reported as deteriorative in fruit juices [7, 19] which are discussed here.

Lactic Acid Bacteria (LAB)

Lactic acid bacteria are gram positive, rod shaped, and catalase negative. Heterofermentative lactic acid bacteria were reported as the most important group of spoilage microorganisms in fruit juices [25]. Lactobacillus and Leuconostoc are the two taxa frequently isolated from fruits and spoiled fruit juices [6]. They produce lactic acids in fruit juices along with lesser amount of acetic and gluconic acids, ethanol and CO_2, but some species of lactic acid bacteria such as L. mesenteroides ssp. cremoris, Leuconostoc paramesenteroides, and Leuconostoc dextranicum are more prominent as they produce diacetyl and acetoin as metabolites in spoiled fruit juices, contributing to buttery or butter milk off flavor to citrus juices [6, 13, 25].

Acetic Acid Bacteria

Acetic acid bacteria belong to three taxa, namely, Acetobacter, Gluconobacter, and Gluconacetobacter. These are gram negative or gram variable, aerobic, ellipsoidal to rod shaped cells that can occur in chains, single or in pairs and are among the main spoilage bacteria because they have the ability to grow at relatively low pH and low nutrient levels. Production of sour and vinegar like flavours in fruit juices is due to the formation of acetic acid by these bacteria [6, 25, 37].

Alicyclobacilli

In recent years, Alicyclobacillus a thermoacidophilic, endospore producing bacterium has emerged as major concern to the beverage industry worldwide as many high concentrated fruit products which are valuable semiprepared food components to the bakery, dairy, canning, baby foods, and distilling and beverage industries have been found to be contaminated with these spoilage microbes [13]. The thermoacidophilic nature and presence of highly resistant endospores is responsible for their survival during the production of concentrated fruit products. Soil is considered to be the main source of contamination of fresh fruits during harvesting. Some alicyclobacilli are soil borne microbes [38–42]. This bacterium was first classified as Bacillus acidocaldarius followed by B. acidoterrestris and now assigned to new genus Alicyclobacillus [6, 10, 12, 13, 33,43, 44].

Of the over 20 species of Alicyclobacillus isolated from different environments. Some of them areAlicyclobacillus acidocaldarius, A. hesperidium, A. acidophilus, A. cyclohaptanicus, A. fastidious, and A. pomorum have been implicated in spoilage incidents in high acid fruit and vegetable products [45].Alicyclobacillus acidoterrestris has emerged as new spoilage bacterium for commercialized fruit juices that can survive pasteurization at 95°C for 2 minutes and can spoil heat treated fruit juices by the formation of taint chemicals (guaiacol and halophenolic) [13, 46].

Pathogenicity of Alicyclobacillus acidoterrestris and A. caldarious strains have extensively been studied [38].Alicyclobacillus contains -alicyclic fatty acids (-cyclohexane and -cycloheptane fatty acids) in their cell membrane that are responsible for heat resistance of Alicyclobacillus by forming a protective coating with strong hydrophobic bonds. These hydrophobic bonds stabilize reduced membrane permeability in extreme and high temperature environments [43, 44]. Another factor contributing to the heat stability of Alicyclobacillus is its endospores along with presence of heat stable proteins and mineralization by divalent cations especially calcium-dipicolinate complex [44, 47, 48].

Contamination of Alicyclobacillus in fruit juices results from sources like soil, water, and processing facilities [44]. Spoilage of fruit juices by Alicyclobacillus is difficult to detect because it does not produce any visible changes such as gas during growth and incipient swelling of containers does not occur so that spoilage in retail products cannot be noticed [33, 49]. It produces a smoky, medicinal, and antiseptic off-odour associated with guaiacol. Other compounds such as 2,6-dibromophenol and 2,6-dichlorophenol have also been found [46, 50,51]. Endospores of Alicyclobacillus have D values in the range of 16–23 minutes at 90°C, greater than the pasteurization treatments applied in fruit juice processing [12]. Hence, Silva and Gibbs [33] suggested thatAlicyclobacillus is to be designated as the target microbe in the design of pasteurization processes for acidic foods and beverages.

Propionibacterium cyclohexanicum

Propionibacterium cyclohexanicum, a gram positive, acid tolerant, heat resistant, nonmotile pleomorphic rod shaped bacterium first isolated from spoiled orange juice in 1993 [52]. It possesses -cyclohexyl undecanoic acid in cell membrane as Alicyclobacillus genus but lacks the production of endospores [11]. Walker and Phillips [10] reported that Propionibacterium cyclohexanicum survives at 95°C for 10 minutes in orange juice and hence would survive treatments commonly used in pasteurization process used in fruit juice industry.

Streptomyces

Streptomyces is a gram positive, filamentous rod shaped soil bacterium, possesses spores in chains. Streptomyces griseus is frequently found in spoiled apple juice. This bacterium enters into fruit juices through poorly washed fruits that are contaminated with soil [53]. It produces earthy like off-flavour in juices attributable to the presence of compounds like geosmin, 2-methyl isoborneol, and 2-isobutyl-3-methoxy pyrazine [8].

Bacillus

Bacillus coagulans, B. marcesens, and B. polymyxa are gram positive, rod shaped, and endospore producers often spoil various fruit juices [54]. Bacillus coagulans spoils canned tomato juice and vegetable products. It causes flat sour spoilage in juice [6, 13, 33, 55].

Clostridium

Two species of Clostridium mainly C. pasteurianum and C. butyricum, gram positive, anaerobic endospore forming bacterium, have been isolated at low pH of fruit juices [54].

PATHOGENIC MICROORGANISMS

Fruit surfaces can be contaminated with feces and feces present on fruit surfaces can contaminate the washing water and permit the internalization of food borne pathogens which help in their survival under the acidic conditions of fruit juices [8, 56]. Some strains of Escherichia coli, Shigella, and Salmonella may survive for several days and even weeks in acidic environment by regulating their internal pH that maintained at neutral pH by combination of passive and active mechanisms [9].

Shigella flexneri and S. sonnei survive in apple juice (pH 3.3) and tomato juice (pH 4.0) at 7°C for at least 14 days [57]. Sospedra et al. [14] reported the presence of Salmonella sp. and Staphylococcus aureus in orange juice extracted by squeezing machine used in restaurants. Several studies carried out by researchers on pathogenic microorganisms associated with street vended unpasteurized fruit juices [58–62] as summarized in Table 2.

Table 2: Human pathogens isolated from street sold unpasteurized fruit juices

Place	Fruit juice	Pathogens	Reference
Vishakhapatnam (India)	Orange, pomegranate, mango, pine apple, and grape	Faecal coliforms and faecal streptococci	[58]
Mumbai (India)	Sugarcane, lime, and carrot	Vibrio cholerae, Escherichia coli, and Staphylococcus aureus	[59]
Jimma town (South west Ethiopia)	Avocado, papaya, and pine apple	Klebsiella, Enterobacter, and Serratia	[60]
Nagpur (India)	Pine apple, sweet lime, and carrot juice	Salmonella, Coliforms, and S. aureus	[61]
Amravati (India)	Apple, orange, pineapple, pomegranate, sweet lemon, and mix fruit	Salmonella, Coliforms, S. aureus, Pseudomonas, and Proteus	[62]

Because of the presence of pathogens in fruit juices, the food borne outbreaks associated with consumption of fruit juices have been increased [9, 14, 19, 63, 64]. Fruit juice borne outbreaks of last two decades from 1991–2010 are summarized in Table 3. Several outbreaks associated with consumption of fruit juices have been reported maximum in year 1999 (5) [59, 65–67] and 1996 (4) [68–70]. E. coli O157:H7 is associated with large number of outbreaks attributable to consumption of unpasteurized apple juice. Salmonella is main causal organism for outbreaks related to unpasteurized orange juice. Clostridium botulinum is reported from homemade as well as pasteurized carrot juice. Vibrio cholorae has been reported for outbreak in India by the consumption of unpasteurized sugarcane juice (Table 3). In year 1999, 423 people in the USA and Canada and 405 people in Australia were affected by consuming unpasteurized orange juice [59].

Table 3: Fruit juice borne outbreaks caused by pathogenic bacteria from 1991 to 2010

Type of fruit juices	Pathogens	Year	Country	Venue	Number of cases (deaths)	Reference
Unpasteurized apple juice	Escherichia coli O157:H7	1991	USA	Small cider mill	23 (0)	[91]
Unpasteurized orange juice	Enterotoxigenic E. coli	1992	India	Roadside vendor	6 (0)	[92]
Unpasteurized apple juice	Cryptosporidium	1993	USA	School	213 (0)	[93]
Carrot homemade juice	Clostridium botulinum	1993	USA	Home	1 (0)	[94]
Unpasteurized orange juice	Salmonella gaminara, S. hartford, and S. rubislaw	1995	USA	Retail	63 (0)	[95–97]
Unpasteurized orange juice	Shigella flexneri	1995	South Africa	Restaurant	14 (0)	[98]
Unpasteurized apple juice	C. parvum	1996	USA	Small cider mill	31 (0)	[68]
Unpasteurized apple juice	E. coli O157:H7	1996	USA	Small cider mill	14 (0)	[68]
Unpasteurized apple juice	E. coli O157:H7	1996	USA	Small cider mill	6 (0)	[70]
Unpasteurized apple juice	E. coli O157:H7	1996	Canada, USA	Retail	70 (1)	[69, 99]
Unpasteurized apple juice	E. coli O157:H7	1998	Canada	Farm/home	14 (0)	[100]
Unpasteurized apple juice	E. coli O157:H7	1999	USA	Not reported	25 (0)	[67]
Unpasteurized orange juice	S. muenchen	1999	Canada, USA	Restaurant	423 (1)	[65]
Unpasteurized orange juice	S. anatum	1999	USA	Roadside stand	6 (0)	[66]
Unpasteurized orange juice	S. typhimurium	1999	Australia	Retail	405 (0)	[59]
Unpasteurized sugar cane juice	Vibrio cholerae	1999	India	Not reported	—	[59]
Unpasteurized orange juice	S. enteritidis	2000	USA	Retail and food service	88 (0)	[101]
Unpasteurized apple juice	C. parvum	2003	USA	Farm/retail	144 (0)	[102]
Unpasteurized apple juice	E. coli O111 and C. parvum	2004	USA	Farm/home	212 (0)	[102]
Unpasteurized orange juice	S. typhimurium and S. saintpaul	2005	USA	Retail and food service	152 (0)	[103]
Unpasteurized sugar cane juice	Trypanosoma cruzi	2005	Brazil	Roadside kiosk	25 (3)	[104]
Pasteurized carrot juice	C. botulinum	2006	USA	Retail	4 (0)	[105]
Unpasteurized apple juice	E. coli O157:H7	2007	USA	Not reported	9 (0)	[67]
Unpasteurized apple juice	E. coli O157:H7	2008	USA	Retail	7	[67]
Unpasteurized orange juice	S. panama	2008	Netherlands	Not reported	33 (0)	[7]
Unpasteurized apple juice	E. coli O157:H7	2010	USA	Fair	7 (0)	[106]

ISOLATION OF SPOILAGE MICROORGANISMS FROM FRUIT JUICES

Microbiological examination of fruit juices is done by serial dilution pour plate methods [71]. For total aciduric microbial populations, orange serum agar (pH—5.5, Orange Serum 200 mL/L; Yeast Extract—3 g/L; Enzymatic Digest of Casein—10 g/L; Dextrose—4 g/L; Potassium Phosphate—2.5 g/L; Agar—17 g/L) is used. Potato dextrose agar (PDA) (pH—3.5, potato—200 g/L; dextrose—20 g/L, agar—15 g/L), PDA with antibiotics (pH—5.6), malt extract agar (MEA) (pH—5.5; Malt extract—30 g/L; Mycological peptone—5 g/L; Agar 15 g/L) supplemented with chloramphenicol (100 mg/L, added before autoclaving) are used for enumeration of yeasts and moulds at incubation temperature of 25°C for 3 and 7 days for yeasts and moulds, respectively [25].

Isolation of yeasts resistant to preservatives from fruit juices is done by plating on MEA with 0.5% acetic acid and tryptone glucose yeast extract agar (TGY) (pH—7.0, Casein enzymatic hydrolysate—5 g/L; Yeast extract—3 g/L;

Glucose—1 g/L; Agar—15 g/L) with 0.5% acetic acid at 25–30°C for 3–5 days. For the isolation of heat resistant fungi, fruit sample must be treated in water bath at 75°C to 80°C for 1.5 hour and isolated by using PDA, OSA, and MEA media [25].

PDA at pH 3.7 and OSA at pH 5.5 at an incubation temperature of 50°C for 3 days are the best isolation media for the detection and isolation of Alicyclobacillus sp. from fruit juices and concentrates [46]. Further identification of moulds and yeasts is done by using methods and media described in Samson et al. [72] and identification of bacterial isolates is done by using methods described in Bergey's Manual of Systematic Bacteriology [73] as well as by using advanced DNA-sequencing techniques.

PREVENTION OF SPOILAGE AND PATHOGENIC MICROORGANISMS IN FRUIT JUICES

There are various techniques to prevent pathogenic as well as nonpathogenic microflora such as chilling, freezing, water activity, modified atmosphere packaging, pasteurization, nonthermal physical techniques, and by addition of natural antimicrobials [19]. The most common method to inactivate microorganisms and enzymes for increasing the shelf life of fruit juices is by thermal processing; however, loss of original taste and flavor compounds occur in fruit juices. These negative effects have motivated a great interest in the development of new technology that offer advantages of using low processing temperatures, low energy consumption, and high retention of nutritional and sensory properties of the food and improving its microbiological quality [17].

THERMAL PASTEURIZATION

FDA has recommended 5 $\log_{10}$ reduction of infection pathogens in a fruit juice which can be achieved by pasteurization at 90–95°C for 4–10 seconds [4, 6]. This pasteurization temperature is effective against E. coli and Salmonella [50] but is not effective against ascospores of heat resistant fungi [34, 74] and heat resistant bacteria [13, 44, 49, 50]. In addition, the thermal pasteurization damages nutritional and physiochemical properties of fruit juices [4].

Nonthermal preservative methods are receiving good attention because of their potential for quality and safety improvement of food [15]. Some of the nonthermal processes used in food industries are high intensity pulsed electric field (HIPEF), high hydrostatic pressure (HHP), high pressure homogenization (HPH), ultraviolet (UV), ultrasound, and irradiation. These novel nonthermal technologies have the ability to inactivate microorganisms at ambient or near ambient temperatures, thus avoiding the deleterious effect of heat has on flavor, colour, and nutrient value of foods mean effect of thermal treatment on fruit juices [84]. Apart from these methods, sodium benzoate and potassium sorbate are two of the most commonly used chemical preservatives to increase the shelf life of fruit juices. But consumer demand for safe, natural, and environmental friendly food preservatives has been increasing. Natural antimicrobials such as bacteriocins, lactoperoxidase, herbs, leaves, oils, and spices have shown potential for use in some food products [18]. Some of the common nonthermal methods used in fruit juice industry are described below.

HIGH HYDROSTATIC PRESSURE (HHP)

High hydrostatic pressure is commercially used worldwide for a variety of foods such as cooked meat, shellfish, fruit, and vegetable juices, sauces, and dips [75]. In this process, fruit juices are subjected to 400 MPa pressure for a few minutes at 20°C or below which is sufficient to reduce the numbers of spoilage microorganisms such as yeasts, moulds, and lactic acid bacteria [75, 76]. The HHP treatment has a lethal effect on microorganisms by affecting their cell membrane along with inactivation of some key enzymes which are involved in DNA replication and transcription processes [15, 75]. Bacterial spores are resistant to HHP treatment and they can survive up to pressure of 1000 MPa [76]. This process has great potential to reduce the microbial load of fruit juices and increase the shelf life of fruit juices (Table 4).

Table 4: Effect of high hydrostatic pressure on microorganisms in fruit juices

Fruit juice(s)	Target microorganisms	Treatment parameters	Log reduction	References
Orange juice	Escherichia coli O157:H7	550 MPa, 30°C, 5 min	6	[107]
Apple juice	E. coli 29055	400 MPa, 25°C	>5	[108]
Apricot, sour cherry, and apple juices	Staphylococcus aureus E. coli O157:H7, Salmonella enteritidis	350 MPa, 30°C, 5 minutes	>5	[109]
Apple juice	E. coli, Listeria innocua, and Salmonella	545 MPa, 1 min	5	[110]
Orange juice	E. coli, L. innocua	241 MPa, 3 min	5	[111]

HIGH PRESSURE HOMOGENIZATION (HPH)

It involves the pumping of liquid through homogenizing valve at high pressure over 100 MPa. It produces high turbulence and shear along with compression, acceleration, and pressure drop result in the breakdown of particles and dispersion throughout the product. After homogenization, the particles of uniform size in the range from 0.2 μm to 2 μm are obtained [16]. In the past, HPH was purposed as a suitable method for the stabilization of dairy products but in last decades it has been suggested for its use for prolongation of the shelf life of fruit juices [22]. HPH inactivates microorganisms by damaging their structural integrity coupled with sudden rise in temperature produced in this process [75]. Table 5 summarizes the effect of HPH on the reduction of microbial load of fruit juices.

Table 5: Effect of high pressure homogenization on microorganisms in fruit juices

Fruit juice(s)	Target microorganisms	Treatment parameters	Log reduction	References
Orange juice	Escherichia coli O58:H21 ATCC 10536 E. coli O157:H7 CCUG 44857	300 MPa	4	[112]
Orange juice	Saccharomyces cerevisiae, Lactobacillus plantarum	>250 MPa	>5	[113]
Apple juice	Saccharomyces, Penicillium, Aureobasidium, and Aspergillus	300 MPa	>5	[114]
Apricot juice	S. cerevisiae	100 MPa (4 times)	2.2	[115]
Carrot juice	S. cerevisiae	100 MPa (3 times)	3	[115]
Apricot juice	Zygosaccharomyces bailii	100 MPa (8 times)	2.5	[116]
Carrot juice	Z. bailii	100 MPa (8 times)	2.5	[116]

PULSED ELECTRIC FIELD (PEF)

PEF involves the application of short duration and high intensity electric field pulses. The fluid foods are placed between two electrodes in batch and continuous flow treatment [17]. This process inactivates microorganisms and enzymes with only small increase in temperature [77], affects the cell membrane of microorganisms by electroporation which leads to leakage of cytoplasmic content from cells [78, 79]. The work carried out on the inactivation of microorganisms in fruit juices by PEF treatment is summarized in Table 6.

Table 6: Effect of pulsed electric field on microorganisms in fruit juices

Fruit juice(s)	Target microorganisms	Treatment parameters	Log reduction	Reference
Apple juice	Escherichia coli 8739, E. coli O157:H7	30 kV/cm, 172 µs, <35°C	5	[117]
Cranberry juice	Total aerobic count, moulds, and yeasts	40 kV/cm, 150 µs, <25°C	4	[118]
Apple juice	E. coli O157:H7	34 kV/cm, 166 µs	4.5	[119]
Orange juice	Listeria innocua	30 kV/cm, 12 µs, 54°C	6.0	[120]
Apple cider	E. coli O157:H7	90 kV/cm, 20 µs, 42°C	5.9	[121]
Orange juice	Salmonella typhimurium	90 kV/cm, 100 µs, 55°C	5.9	[122]
Apple juice	E. coli	34 kV/cm, 7.68 µs, 55°C	6.2	[123]
Grape juice	E. coli	34 kV/cm, 7.68 µs, 55°C	6.4	[123]
Orange juice	Lactobacillus brevis	35 kV/cm, 2.5–5.0 µs	5.8	[124]
Orange juice-milk mixture	L. plantarum	35 kV/cm, 2.5–5.0 µs	2.5	[125]
Apple juice	E. coli	36 kV/cm, 800 pulse s per second	6	[77]
Cherry juice	Penicillium expansum	34 kV/cm, 163 µs, 21°C	100% inactivation of spore germination	[126]
Orange juice	Salmonella enteritidis, E. coli	35 kV/cm, 4 µs, 40°C	5	[17]
Strawberry juice	E. coli O157:H7	18.6 kV/cm, 150 µs, 45°C 18.6 kV/cm, 150 µs, 50°C 18.6 kV/cm, 150 µs, 55°C	3.09 4.08 4.71	[79]

ULTRAVIOLET TECHNOLOGY (UV)

Ultraviolet technology has been utilized in the food industry to disinfect water and effectively destroy microorganisms on surfaces and packaging [80]. Ultraviolet radiation involves the use of radiation from electromagnetic spectrum from 100 to 400 nm. It is classified as UV-A (320–400 nm), UV-B (280–320 nm), and UV-C (200–280 nm) [81]. UV-C is effective against bacteria and viruses [81]. UV treatment is performed at low temperature. 254 nm wavelength of UV light is widely used in juice and beverage industry [18]. UV-C light inactivates microorganisms by damaging their DNA that absorbs UV light from 200 to 310 nm. UV creates the pyrimidine dimers which prevent microorganisms from replicating and thus rendering them inactive [81]. Novel works carried out on the effect of UV-C on the microbial load in different fruit juices have been given in Table 7.

Table 7: Effect of ultraviolet on microorganisms in fruit juices

Fruit juice	Target microorganisms	Treatment parameters mJ/cm^2	Log reduction	References
Apple juice	*Escherichia coli* K12	14.5	3-4	[127]
Orange juice	Yeasts, moulds	12.3–120	3	[128]
Apple juice	*Saccharomyces cerevisiae, Listeria innocua*, and *E. coli*	5135	1.34 4.29 5.10	[129]
Apple juice	*APC	229.5 J/L	3.5	
Guava pineapple	Yeasts	1377 J/L	3.0	
	Moulds		4.48	
Orange juice 1	APC	167 J/L		[81]
	Yeasts		1.32	
	Moulds			
Orange juice 2	APC	167 J/L		
	Yeasts		<1	
	Moulds			

APC: aerobic plate count.

ULTRASOUND

Power ultrasound has been identified as potential technology to meet US Food and Drug Administration (USFDA) requirement of 5 log reduction of Escherichia coli in fruit juices. High power ultrasound causes bubble cavitation in a liquid due to pressure changes. These resultant microbubbles collapse violently in the succeeding compression cycles of propagated ultrasonic waves resulting in localized high temperature up to 5000 K, pressures up to 50,000 kPa, and high shearing effects causing breakdown of cell walls, disruption of cell membranes, and damage of DNA of microorganisms [18, 82, 83]. Destruction of microorganisms in fruit juices by ultrasound which has been studied by various researchers is summarized in Table 8.

Table 8: Effect of ultrasound on microorganisms in fruit juices

Fruit juice(s)	Target microorganisms	Treatment parameters	Log reduction	References
Carrot juice	*Escherichia coli* K12	19.3 kHz, 700–800 W, 1 min, 60°C	2.5	[130]
Orange juice	Total mesophilic Aerobes	500 kHz, 240 W, 15 min, 60°C	3.4	[131]
Apple juice	*Alicyclobacillus acidoterrestris*	24 kHz, 300 W, 60 min	80%	[132]
Orange juice	Aerobic mesophilic count (AMC)	20 kHz, 500 W, 8 min, 10°C	1.38	[133]
	Yeast and mold counts (YMC)		0.56	

Nonthermal processes are to be used with caution in juices because they are not without harmful effects such as high hydrostatic pressure method which can alter the structure of protein and polysaccharide causing changes in the texture, physical appearance, and functionality of foods. High intensity ultrasound can denature proteins and produce free radicals that can adversely have an effect on the flavor of fruit based or high fat foods. Ultraviolet treatment is difficult to apply on fruit juices because of low UV transmittance through the juice because of high suspended and soluble solids [81]. Emerging nonthermal technology also has been so energy expensive or costly to be practical for use in food processing [84].

PRESERVATIVES

Another method to prevent microbial contamination in juice is by use of preservatives. Some chemical preservatives such as sodium benzoate and potassium sorbate are often used to prevent microbial spoilage of fruit juices [8, 18]. Consumers relate synthetic preservatives as artificial products resulting in rejection of this type of food processed [8] so demands for preservatives which have natural origin have increased drastically.

Most of these natural antimicrobials have been considered as generally recognized as safe (GRAS). Natural antimicrobials are obtained from three natural sources such as plant (herb, spices, essential oil, and vanillin), animal (lactoperoxidase, lysozyme, chitosan), and microbes (bacteriocins) [19]. The antimicrobial activity of these compounds has been studied in fruit juices by many researchers as detailed given in Table 9 and employ different mechanisms to inactivate microorganisms.

Table 9: Effect of natural antimicrobials on spoilage and pathogenic microorganisms

Natural antimicrobial	Fruit juice	Target microorganisms	Log reduction	References
Lactoperoxidase	Apple and orange juice	*Escherichia coli, Shigella* sp.	>5	[64]
Lysozyme	Orange juice	*Salmonella typhimurium*	—	[122]
Chitosan	Apple juice	Yeasts and moulds	3	[134]
		Yeasts	3–5	[135]
Cinnamon powder	Apple juice	*Listeria monocytogenes* Scott 45954	4–6	[136]
Essential oil				
	Apple, pear, melon juices	*L. innocua, S. enteritidis, E. coli*	5	[137]
Cinnamon oil	Melon, watermelon	*S. enteritidis*	3.1–3.9	[17]
		E. coli	1.4–1.9	
		L. monocytogenes	3.4–4.4	
	Apple juice	*S. hadar, E. coli* O157:H7	50%	[138]
Clove oil	Apple juice	*S. hadar, E. coli* O157:H7	50%	[138]
	Tomato juice	Native microbiota	3.9	[139]
Lemon oil	Apple juice	*S. hadar, E. coli* O157:H7	50%	[138]
Lemongrass oil	Apple, pear, melon juices	*L. innocua, S. enteritidis, E. coli*	5	[137]
	Apple juice	*S. hadar, E. coli* O157:H7	50%	[138]
Lime oil	Apple juice	*S. hadar, E. coli* O157:H7	50%	[138]
Oregano oil	Apple juice	*S. hadar, E. coli* O157:H7	50%	[138]
Carvacrol	Apple juice	*S. hadar, E. coli* O157:H7	50%	
Cinnamaldehyde	Apple juice	*S. hadar, E. coli* O157:H7	50%	[139]
Citral	Apple juice	*S. hadar, E. coli* O157:H7	50%	[138]
	Orange juice	*L. monocytogenes*	1.1–1.3	[140]
Eugenol	Apple juice	*S. hadar,* and *E. coli* O157:H7	50%	[138]

Lactoperoxidase system produces hypothiocyanite (OSCN$^-$) and hypothiocyanous acid (HOSCN), that possess antimicrobial effect by the oxidation of thiol group ($^-$SH) of cytoplasmic enzymes, damage to outer membrane, transport systems, glycolytic enzymes, and nucleic acids [85]. But lysozyme attack the bacterial peptidoglycan cell wall and are more effective against gram positive than gram negative bacteria because former contain about 90% peptidoglycan and later contain 5% to 10% [19].

Chitosan is effective against microorganisms due to its positive charge amino group at C–2 which can create polycationic structure and interact with anionic components such as lipopolysaccharide and proteins of cell surface; this binding disrupts the integrity of the outer membrane resulting in leakage of intracellular components [18]. Many herbs and plant extracts have broad spectrum activity against microorganisms [4, 86]. Essential oils are a group of terpenoids, sesquiterpenes, and possibly diterpenes with different groups of aliphatic hydrocarbons, acids, aldehydes, acyclic esters, or lactones. The antimicrobial activity of essential oil is not attributed to one specific mechanism, but there are several targets in the cell [87–89]. Hydrophobicity of essential oil enables them to partition in the lipids of bacterial cell membrane and mitochondria, disturbing the structures and rendering them more permeable [89].

Nisin possess narrow antimicrobial spectrum inhibiting only gram positive bacteria. It forms pore in cytoplasmic membrane of bacteria resulting in depletion of proton motive force and loss of cellular ions, amino acids, and ATP [19].

SYNERGISTIC EFFECT OF PHYSICAL AND NATURAL ANTIMICROBIALS

Food antimicrobials are generally biostatic and are not biocidal. Hence their effects on foods are limited. On the other hand use of combinations of antimicrobials [19] and antimicrobial along with nonthermal methods is effective against pathogenic and spoilage microorganisms. This combination improves the lethal effects of nonthermal processing and reduces the severity of nonthermal methods. So the combinations of these techniques could provide synergistic effects on prolonging the shelf life of fruit juices mentioned in Table 10and potentially as best option for traditional pasteurization methods [8, 18].

Table 10: Preservation of fruit juices by combination of different preservation methods

Fruit juice	Combination of preservation methods	Target microorganisms	Log reduction	References
Orange juice	PEF with nisin	Native microflora	6	[141]
Orange juice	PEF with nisin or lysozyme	*Escherichia coli* O157:H7	>7	[122]
Strawberry juice	PEF with cinnamon bark oil or citric acid	*E. coli* O157:H7, *Salmonella enteritidis*	>5	[17]
Apple juice	PEF with cinnamon bark oil or citric acid	*E. coli* O157:H7, *S. enteritidis*	>5	[17]
Pear juice	PEF with cinnamon bark oil or citric acid	*E. coli* O157:H7, *S. enteritidis*	>5	[17]
Carrot juice	HPH with nisin	*L. innocua*	>5	[142]
Strawberry juice	PEF with sodium benzoate and potassium sorbate	*E. coli* O157:H7, *S. enteritidis*	5.11	[79]

CONCLUSION

The demand for fruit juices has been increasing due to their health benefits. Due to change in dietary, social habits and preservation methods have led to increase in disease outbreaks linked mainly to fresh fruit juices in recent years. Pasteurization of fruit juices is very effective against a pathogenic and several spoilage microorganisms, nonetheless, sensory and nutritional properties are affected. To meet the demands for nutritious and safe foods has resulted in increased interest in nonthermal preservation techniques. The nonthermal methods described in this review have the potential to meet 5 log microbial reductions. However, only high pressure processing has been used on pilot scale. There is also need for other nonthermal methods tested on pilot scale so they become alternative of pasteurization of fruit juices.

Application of different natural antimicrobials of animal, plant, and microbial origins directly or indirectly added to fruit juices effectively reduce or inhibit pathogenic and spoilage microorganisms. Thus they also represent good alternative of thermal processing of fruit juices.

In future, the combination of nonthermal methods and natural antimicrobial compounds would be new trend of preservation of fruit juices that improve the microbiological quality while having the lowest impacts on the organoleptic properties.

REFERENCES

1.   K. R. Matthews, "Microorganisms associated with fruits and vegetables," in Microbiology of Fresh Produce, K. R. Matthews, Ed., pp. 1–19, ASM Press, Washington, DC, USA, 2006.

2.   S. Kumar, H. Thippareddi, J. Subbiah, S. Zivanovic, P. M. Davidson, and F. Harte, "Inactivation of Escherichia coli K-12 in apple juice using combination of high-pressure homogenization and chitosan,"Journal of Food Science, vol. 74, pp. M8–M14, 2009.

3.   F. Patrignani, L. Vannini, S. L. S. Kamdem, R. Lanciotti, and M. E. Guerzoni, "Potentialities of high pressure homogenization to inactivate Zygosaccharomyces bailli in fruit juices," Journal of Food Science, vol. 75, no. 2, pp. M116–M120, 2010.

4.   J. Mosqueda-Melgar, R. M. Raybaudi-Massilia, and O. Martín-Belloso, "Microbiological shelf life and sensory evaluation of fruit juices treated by high intensity electric fields and antimicrobials," Food and Bioproducts Processing, vol. 90, no. 2, pp. 205–214, 2012.

5.   R. P. Bates, J. R. Morris, and P. G. Crandall, Principals and Practices of Small and Medium Scale Fruit Juice Processing, FAO Agricultural Services Bulletin, Rome, Italy, 2001.

6.   ICMSF, "Soft drinks, fruit juices, concentrates and food preserves," in Microorganisms in Foods 6: Microbial Ecology of Food Commodity, Kluwer Academic Publisher, 2005.

7.   Bevilacqua, M. R. Corbo, D. Campaniello et al., "Shelf life prolongation of fruit juices through essential oils and homogenization: a review," in Science against Microbial Pathogens: Communicating Current Research and Technological Advances, pp. 1156–1166, 2011.

8.   A. Lima Tribst, A. De Souza Sant'ana, and P. R. De Massaguer, "Review: microbiological quality and safety of fruit juicespast, present and future perspectives Microbiology of fruit juices Tribst et al," Critical Reviews in Microbiology, vol. 35, no. 4, pp. 310–339, 2009.

9.   Vantarakis, M. Affifi, P. Kokkinos, M. Tsibouxi, and M. Papapetropoulou, "Occurrence of microorganisms of public health and spoilage significance in fruit juices sold in retail markets in Greece,"Anaerobe, vol. 17, no. 6,

pp. 288–291, 2011.

10. M. Walker and C. A. Phillips, "The growth of Propionibacterium cyclohexanicum in fruit juices and its survival following elevated temperature treatments," Food Microbiology, vol. 24, no. 4, pp. 313–318, 2007. ·

11. M. Walker and C. A. Phillips, "The effect of preservatives on Alicyclobacillus acidoterrestris andPropionibacterium cyclohexanicum in fruit juice," Food Control, vol. 19, no. 10, pp. 974–981, 2008.

12. M. Walker and C. A. Phillips, "Alicyclobacillus acidoterrestris: an increasing threat to the fruit juice industry?" International Journal of Food Science and Technology, vol. 43, no. 2, pp. 250–260, 2008.

13. E. Steyn, M. Cameron, and R. C. Witthuhn, "Occurrence of Alicyclobacillus in the fruit processing environment—a review," International Journal of Food Microbiology, vol. 147, no. 1, pp. 1–11, 2011.

14. Sospedra, J. Rubert, J. M. Soriano, and J. Mañes, "Incidence of microorganisms from fresh orange juice processed by squeezing machines," Food Control, vol. 23, no. 1, pp. 282–285, 2012.

15. J. Kuldiloke and M. N. Eshtiaghi, "Application of non thermal processing for preservation of orange juice," KMITL Science Technology Journal, vol. 8, pp. 64–74, 2008.

16. M. R. Corbo, A. Bevilacqua, D. Campaniello, C. Ciccarone, and M. Sinigaglia, "Use of high pressure homogenization as a mean to control the growth of foodborne moulds in tomato juice," Food Control, vol. 21, no. 11, pp. 1507–1511, 2010.

17. J. Mosqueda-Melgar, R. M. Raybaudi-Massilia, and O. Martín-Belloso, "Non-thermal pasteurization of fruit juices by combining high-intensity pulsed electric fields with natural antimicrobials," Innovative Food Science and Emerging Technologies, vol. 9, no. 3, pp. 328–340, 2008.

18. H. P. V. Rupasinghe and L. J. Yu, "Emerging preservation methods for fruit juices and beverages," in Food Additive, Y. El-Samragy, Ed., InTech, 2012, http://www.intechopen.com/books/food-additive/emerging-preservation-methods-3-for-fruit-juices-and-beverages.

19. R. M. Raybaudi-Massilia, J. Mosqueda-Melgar, R. Soliva-Fortuny, and O. Martín-Belloso, "Control of pathogenic and spoilage microorganisms in fresh-cut fruits and fruit juices by traditional and alternative natural antimicrobials," Comprehensive Reviews in Food Science and Food Safety, vol. 8, no. 3, pp. 157–180, 2009.

20. Rico, A. B. Martín-Diana, J. M. Barat, and C. Barry-Ryan, "Extending and

measuring the quality of fresh-cut fruit and vegetables: a review," Trends in Food Science and Technology, vol. 18, no. 7, pp. 373–386, 2007.

21. K. R. Aneja, P. Jain, and R. Aneja, A Textbook of Basic and Applied Microbiology, New Age International Publishers, New Delhi, India, 1st edition, 2008.

22. Bevilacqua, M. R. Corbo, and M. Sinigaglia, "Use of natural antimicrobials and high pressure homogenization to control the growth of Saccharomyces bayanus in apple juice," Food Control, vol. 24, no. 1-2, pp. 109–115, 2012.

23. S. Patil, V. P. Valdramidis, B. K. Tiwari, P. J. Cullen, and P. Bourke, "Quantitative assessment of the shelf life of ozonated apple juice," European Food Research and Technology, vol. 232, no. 3, pp. 469–477, 2011. ·

24. V. H. Tournas, J. Heeres, and L. Burgess, "Moulds and yeasts in fruit salads and fruit juices," Food Microbiology, vol. 23, no. 7, pp. 684–688, 2006.

25. K. A. Lawlor, J. D. Schuman, P. G. Simpson, and P. J. Taormina, "Microbiological spoilage of beverages," in Compendium of the Microbiological Spoilage of Foods and Beverages, W. H. Sperber and M. P. Doyle, Eds., Food Microbiology and Food Safety, Springer Science and Business Media, New York, NY, USA, 2009.

26. R. Arias, J. K. Burns, L. M. Friedrich, R. M. Goodrich, and M. E. Parish, "Yeast species associated with orange juice: evaluation of different identification methods," Applied and Environmental Microbiology, vol. 68, no. 4, pp. 1955–1961, 2002.

27. M. Stratford, "Food and beverage spoilage yeasts," in Yeasts in Food and Beverages Handbook, G. M. Fleet and A. Querol, Eds., Berlin, Germany, pp. 335–379, Springer, 2006.

28. L. M. Lenovich, R. L. Buchanan, N. J. Worley, and L. Restaino, "Effect of solute on sorbate resitsance inZygosaccharomyces rouxii," Journal of Food Science, vol. 53, pp. 914–916, 2006.

29. P. Wareing and R. R. Davenport, "Microbiology of soft drinks and fruit juices," in Chemistry and Technology of Soft Drinks and Fruit Juices, P. R. Ashurst, Ed., Blackwell Publishing, London, UK, 2005.

30. M. O. Moss, "Fungi, quality and safety issues in fresh fruits and vegetables," Journal of Applied Microbiology, vol. 104, no. 5, pp. 1239–1243, 2008.

31. N. Delage, A. d〉Harlingue, B. Colonna Ceccaldi, and G. Bompeix,

"Occurrence of mycotoxins in fruit juices and wine," Food Control, vol. 14, no. 4, pp. 225–227, 2003.

32. M. De Sylos and D. B. Rodriguez-Amaya, "Incidence of patulin in fruits and fruit juices marketed in Campinas, Brazil," Food Additives and Contaminants, vol. 16, no. 2, pp. 71–74, 1999.

33. V. M. Silva and P. Gibbs, "Target selection in designing pasteurization processes for shelf-stable high-acid fruit products," Critical Reviews in Food Science and Nutrition, vol. 44, no. 5, pp. 353–360, 2004. ·

34. C. M. Salomão, A. P. Slongo, and G. M. F. Aragão, "Heat resistance of Neosartorya fischeri in various juices," Food Science and Technology, vol. 40, no. 4, pp. 676–680, 2007.

35. J. A. N. Obeta and J. O. Ugwuyani, "Heat resistant fungi in Nigerian fruit juices," International Journal of Food Science and Technology, vol. 30, pp. 587–590, 2007.

36. M. Voldřich, J. Dobiáš, L. Tichá, M. Čeřovský, and J. Krátká, "Resistance of vegetative cells and ascospores of heat resistant mould Talaromyces avellaneus to the high pressure treatment in apple juice,"Journal of Food Engineering, vol. 61, no. 4, pp. 541–543, 2004.

37. R. W. Worbo and D. F. Splistosser, "Microbiology of fruit products," in Processing of Fruit Science and Technology, D. M. Barret, L. P. Somogyi, and H. S. Ramaswamy, Eds., CRC Press, London, U.K, 2004.

38. Walls and R. Chuyate, "Spoilage of fruit juices by Alicyclobacillus acidoterrestris," Food Australia, vol. 52, no. 7, pp. 286–288, 2000.

39. M. E. Parish and R. M. Goodrich, "Recovery of presumptive Alicyclobacillus strains from orange fruit surfaces," Journal of Food Protection, vol. 68, no. 10, pp. 2196–2200, 2005.

40. K. S. Bahçeci and J. Acar, "Modeling the combined effects of pH, temperature and ascorbic acid concentration on the heat resistance of Alicyclobacillus acidoterrestis," International Journal of Food Microbiology, vol. 120, no. 3, pp. 266–273, 2007.

41. W. H. Groenewald, P. A. Gouws, and R. C. Witthuhn, "Isolation and identification of species ofAlicyclobacillus from orchard soil in the Western Cape, South Africa," Extremophiles, vol. 12, no. 1, pp. 159–163, 2008.

42. W.H.Groenewald,P.A.Gouws,andR.C.Witthuhn,"Isolation,identification and typification ofAlicyclobacillus acidoterrestris and Alicyclobacillus acidocaldarius strains from orchard soil and the fruit processing environment in South Africa," Food Microbiology, vol. 26, no. 1, pp. 71–76, 2009.

43. J. D. Wisotzkey, P. Jurtshuk Jr., G. E. Fox, G. Deinhard, and K. Poralla, "Comparative sequence analyses on the 16S rRNA (rDNA) of Bacillus acidocaldarius, Bacillus acidoterrestris, and Bacillus cycloheptanicusand proposal for creation of a new genus, Alicyclobacillus gen. nov.," International Journal of Systematic Bacteriology, vol. 42, no. 2, pp. 263–269, 1992.

44. Y. Smit, M. Cameron, P. Venter, and R. C. Witthuhn, "Alicyclobacillus spoilage and isolation: a review,"Food Microbiology, vol. 28, no. 3, pp. 331–349, 2011.

45. K. Goto, K. Mochida, Y. Kato et al., "Proposal of six species of moderately thermophilic, acidophilic, endospore-forming bacteria: Alicyclobacillus contaminans sp. nov.,Alicyclobacillus fastidiosus sp. nov.,Alicyclobacillus kakegawensis sp. nov., Alicyclobacillus macrosporangiidus sp. nov., Alicyclobacillus saccharisp. nov. and Alicyclobacillus shizuokensis sp. nov," International Journal of Systematic and Evolutionary Microbiology, vol. 57, no. 6, pp. 1276–1285, 2007.

46. R. C. Witthuhn, W. Duvenage, and P. A. Gouws, "Evaluation of different growth media for the recovery of the species of Alicyclobacillus," Letters in Applied Microbiology, vol. 45, no. 2, pp. 224–229, 2007.

47. S.-S. Chang and D.-H. Kang, "Alicyclobacillus spp. in the fruit juice industry: history, characteristics, and current isolation/detection procedures," Critical Reviews in Microbiology, vol. 30, no. 2, pp. 55–74, 2004. ·

48. J. M. Jay, M. J. Loessner, and D. A. Golden, "Extrinsic and intrinsic parameters of food that affect microbial growth," in Modern Food Microbiology, Springer Science; Business Media Incorporation, New York, NY, USA, 2005.

49. M. Z. Durak, J. J. Churey, M. D. Danyluk, and R. W. Worobo, "Identification and haplotype distribution of Alicyclobacillus spp. from different juices and beverages," International Journal of Food Microbiology, vol. 142, no. 3, pp. 286–291, 2010.

50. M. D. Danyluk, L. M. Friedrich, C. Jouquand, R. Goodrich-Schneider, M. E. Parish, and R. Rouseff, "Prevalence, concentration, spoilage, and mitigation of Alicyclobacillus spp. in tropical and subtropical fruit juice concentrates," Food Microbiology, vol. 28, no. 3, pp. 472–477, 2011.

51. R. C. Witthuhn, Y. Smit, M. Cameron, and P. Venter, "Guaiacol production by Alicyclobacillus and comparison of two guaiacol detection methods," Food Control, vol. 30, no. 2, pp. 700–704, 2013.

52. K. Kusano, H. Yamada, M. Niwa, and K. Yamasato, "Propionibacterium

cyclohexanicum sp. nov., a new acid-tolerant ω-cyclohexyl fatty acid-containing Propionibacterium isolated from spoiled orange juice,"International Journal of Systematic Bacteriology, vol. 47, no. 3, pp. 825–831, 1997.

53. B. Siegmund and B. Pöllinger-Zierler, "Growth behavior of off-flavor-forming microorganisms in apple juice," Journal of Agricultural and Food Chemistry, vol. 55, no. 16, pp. 6692–6699, 2007. ·

54. M. Stratford, P. D. Holman, and M. B. Cole, "Fruit juices, fruit drinks and soft drinks," in Microbiological Safety and Quality of Food, B. lund, A. C. Baird Parker, and W. G. Grahame, Eds., ASPEN, Gaithersburg, Md, USA, 2000.

55. H. Daryaei and V. M. Balasubramaniam, "Kinetics of Bacillus coagulans spore inactivation in tomato juice by combined pressure-heat treatment," Food Control, vol. 30, no. 1, pp. 168–175, 2013.

56. FDA (U.S. Food and Drug Administration), "Federal register proposed rules—63 FR 20449 April 24, (1998). HACCP; procedures for the safe and sanitary processingand importing of juice; food labeling : warning notice statements; labeling of juice products," Federal Register, vol. 63, pp. 20449–20486, 1998,http://www.fda.gov/Food/FoodSafety/HazardAnalysisCriticalControlPointsHACCP/JuiceHACCP/ucm082031.htm.

57. Van Opstal, C. F. Bagamboula, T. Theys, S. C. M. Vanmuysen, and C. W. Michiels, "Inactivation ofEscherichia coli and Shigella in acidic fruit and vegetable juices by peroxidase systems," Journal of Applied Microbiology, vol. 101, no. 1, pp. 242–250, 2006.

58. J. E. Lewis, P. Thompson, B. Rao, C. Kalavati, and B. Rajanna, "Human Bacteria in street vended fruit juices: a case study of Vishakhapatnam city, India," Internet Journal of Food Safety, vol. 8, pp. 35–38, 2006.

59. P. Mahale, R. G. Khade, and K. V. Vaidya, "Microbiological Analysis of Street vended Fruit juices From Mumbai city, India," Internet Journal of Food Safety, vol. 10, pp. 31–34, 2008.

60. T. Ketema, T. Gadissa, and K. Bacha, "Microbiological safety of fruit juices served in cafes/restaurants, Jimma town, South west, Ethopia," Ethiopian Journal of Health Science, vol. 18, pp. 95–100, 2008.

61. Titarmare, P. Dabholkar, and S. Godbole, "Bacteriological analysis of street vended fresh fruit and vegetable juices in Nagpur city, India," Internet Journal of Food Safety, vol. 11, pp. 1–3, 2009.

62. H. Tambeker, V. J. Jaiswal, D. V. Dhanorker, P. B. Gulhane, and M. N. Dudhane, "Microbial quality and safety of street vended fruit juices: a

case study of Amravati city," Internet Journal of Food Safety, vol. 10, pp. 72–76, 2009.

63. CDC, Annual Listing of Food Borne Disease Outbreaks, United States, 1990–2004, CDC, Atlanta, Ga, USA, 2007.

64. van Opstal, C. F. Bagamboula, T. Theys, S. C. M. Vanmuysen, and C. W. Michiels, "Inactivation ofEscherichia coli and Shigella in acidic fruit and vegetable juices by peroxidase systems," Journal of Applied Microbiology, vol. 101, no. 1, pp. 242–250, 2006.

65. CDC, "Outbreaks of Salmonella serotype Muenchen infections associated with unpasteurized orange juice—United State and Canada," Morbidity and Mortality Weekly Report, vol. 48, pp. 582–585, 1999.

66. Krause, R. Terzagian, and R. Hammond, "Outbreak of Salmonella serotype anatum infection associated with unpasteurized orange juice," Southern Medical Journal, vol. 94, no. 12, pp. 1168–1172, 2001.

67. CDC, "Foodborne Outbreak Online Database (FOOD)," 2011,http://wwwn.cdc.gov/foodborneoutbreaks/default.aspx.

68. CDC, "Outbreaks of Escherichia coli O157:H7 infection and cryptosporidiosis associated with drinking unpasteurized apple cider—connecticut and New York, October 1996," Morbidity and Mortality Weekly Report, vol. 46, pp. 4–8, 1997.

69. S. H. Cody, M. K. Glynn, J. A. Farrar et al., "An outbreak of Escherichia coli O157:H7 infection from unpasteurized commercial apple juice," Annals of Internal Medicine, vol. 130, no. 3, pp. 202–209, 1999. ·

70. FDA, "Federal register final rule- 66Fr 6137, January 19, 2001: hazard analysis and critical control point (HACCP) ; procedures for the safe and sanitary processing and importing of juices," Federal Register, vol. 66, no. 13, pp. 6137–6202, 2001.

71. K. R. Aneja, Experiments in Microbiology, Plant Pathology and Biotechnology, New Age International Publishers, New Delhi, India, 4th edition, 2003.

72. R. A. Samson, E. S. Hoekstra, and J. S. Frisvad, Introduction to Food and Airborne Fungi, Ponson and Looyen, Wageningen, The Netherlands, 7th edition, 2004.

73. W. B. Whitman, M. Goodfellow, P. Kämpfer et al., Bergey›s Manual of Systematic Bacteriology, vol. 5, 2012.

74. S. Kutama, I. Yusuf, and M. Hayatu, "Detection of heat resistant molds in some canned fruit juices sold in Kano, Nigeria," Bioscience Research Communications, vol. 22, pp. 221–225, 2009.

75. M. McKay, M. Linton, J. Stirling, A. Mackle, and M. F. Patterson, "A comparative study of changes in the microbiota of apple juice treated by high hydrostatic pressure (HHP) or high pressure homogenisation (HPH)," Food Microbiology, vol. 28, no. 8, pp. 1426–1431, 2011.

76. Vercammen, B. Vivijs, I. Lurquin, and C. W. Michiels, "Germination and inactivation of Bacillus coagulans and Alicyclobacillus acidoterrestris spores by high hydrostatic pressure treatment in buffer and tomato sauce," International Journal of Food Microbiology, vol. 152, no. 3, pp. 162–167, 2012.

77. V. Charles-Rodríguez, G. V. Nevárez-Moorillón, Q. H. Zhang, and E. Ortega-Rivas, "Comparison of thermal processing and pulsed electric fields treatment in pasteurization of apple juice," Food and Bioproducts Processing, vol. 85, no. 2, pp. 93–97, 2007.

78. Z. Cserhalmi, Á. Sass-Kiss, M. Tóth-Markus, and N. Lechner, "Study of pulsed electric field treated citrus juices," Innovative Food Science and Emerging Technologies, vol. 7, no. 1-2, pp. 49–54, 2006.

79. J. B. Gurtler, R. B. Bailey, D. J. Geveke, and H. Q. Zhang, "Pulsed electric field inactivation of E. coliO157:H7 and non-pathogenic surrogate E. coli in strawberry juice as influenced by sodium benzoate, potassium sorbate, and citric acid," Food Control, vol. 22, no. 10, pp. 1689–1694, 2011.

80. S. L. Chia, S. Rosnah, M. A. Noranizan, and W. D. Wan Ramli, "The effect of storage on the quality attributes of ultraviolet-irradiated and thermally pasteurised pineapple juices," International Food Research Journal, vol. 19, no. 3, pp. 1001–1010, 2012.

81. M. Keyser, I. A. Muller, F. P. Cilliers, W. Nel, and P. A. Gouws, "Ultraviolet radiation as a non-thermal treatment for the inactivation of microorganisms in fruit juice," Innovative Food Science and Emerging Technologies, vol. 9, no. 3, pp. 348–354, 2008.

82. K. Tiwari, C. P. O›Donnell, and P. J. Cullen, "Effect of non thermal processing technologies on the anthocyanin content of fruit juices," Trends in Food Science and Technology, vol. 20, no. 3-4, pp. 137–145, 2009.

83. D. Char, E. Mitilinaki, S. N. Guerrero, and S. M. Alzamora, "Use of high-intensity ultrasound and UV-C light to inactivate some microorganisms in fruit juices," Food and Bioprocess Technology, vol. 3, no. 6, pp. 797–803, 2010.

84. I. V. Ross, M. W. Griffiths, G. S. Mittal, and H. C. Deeth, "Combining nonthermal technologies to control foodborne microorganisms," International Journal of Food Microbiology, vol. 89,

no. 2-3, pp. 125–138, 2003.

85. V. Touch, S. Hayakawa, S. Yamada, and S. Kaneko, "Effects of a lactoperoxidase-thiocyanate-hydrogen peroxide system on Salmonella enteritidis in animal or vegetable foods," International Journal of Food Microbiology, vol. 93, no. 2, pp. 175–183, 2004.

86. M. M. Tajkarimi, S. A. Ibrahim, and D. O. Cliver, "Antimicrobial herb and spice compounds in food,"Food Control, vol. 21, no. 9, pp. 1199–1218, 2010.

87. P. Skandamis, K. Koutsoumanis, K. Fasseas, and G.-J. E. Nychas, "Inhibition of oregano essential oil and edta on Escherichia coli O157:H7," Italian Journal of Food Science, vol. 13, no. 1, pp. 65–75, 2001.

88. F. Carson, B. J. Mee, and T. V. Riley, "Mechanism of action of Melaleuca alternifolia (tea tree) oil onStaphylococcus aureus determined by time-kill, lysis, leakage, and salt tolerance assays and electron microscopy," Antimicrobial Agents and Chemotherapy, vol. 46, no. 6, pp. 1914–1920, 2002.

89. S. Burt, "Essential oils: their antibacterial properties and potential applications in foods—a review,"International Journal of Food Microbiology, vol. 94, no. 3, pp. 223–253, 2004.

90. Piecková and R. A. Samson, "Heat resistance of Paecilomyces variotii in sauce and juice," Journal of Industrial Microbiology and Biotechnology, vol. 24, no. 4, pp. 227–230, 2000.

91. R. E. Besser, S. M. Lett, J. T. Weber et al., "An outbreak of diarrhea and hemolytic uremic syndrome fromEscherichia coli O157:H7 in fresh-pressed apple cider," Journal of the American Medical Association, vol. 269, no. 17, pp. 2217–2220, 1993.

92. R. Singh, S. B. Kulshreshtha, and K. N. Kapoor, "Orange juice-borne diarrhoeal outbreak due to Enterotoxigenic E.coli," Journal of Food Science and Technology, vol. 32, no. 6, pp. 504–506, 1995.

93. P. S. Millard, K. F. Gensheimer, D. G. Addiss et al., "An outbreak of cryptosporidiosis from fresh-pressed apple cider," Journal of the American Medical Association, vol. 272, no. 20, pp. 1592–1596, 1994.

94. J. C. Buzby and S. R. Crutchfield, "New Juice Regulations Underway. Food Review," 1999,http://www.fda.gov/Food/FoodSafety/HazardAnalysisCriticalControlPointsHACCP/JuiceHACCP/ucm082031.htm.

95. CDC, "Outbreak of Salmonella Hartford infections among travelers to Orlando, Florida," EPI-AID Trip Report, pp. 95–162, 1995.

96. K. A. Cook, T. E. Dobbs, W. G. Hlady et al., "Outbreak of Salmonella serotype Hartford infections associated with unpasteurized orange juice," Journal of the American Medical Association, vol. 280, no. 17, pp. 1504–1509, 1998.

97. M. Parish, "Relevancy of Salmonella and pathogenic E.coli to fruit juices," in Proceedings of the IFU-Workshop "Microbiology", Fruit Processing, vol. 10, pp. 246–250, 2000.

98. Thurston, J. Stuart, B. McDonnell, S. Nicholas, and T. Cheasty, "Fresh orange juice implicated in an outbreak of Shigella flexneri among visitors to a South African game reserve," Journal of Infection, vol. 36, no. 3, p. 350, 1998.

99. CDC, "Outbreak of Escherichia coli O157:H7 infections associated with drinking unpasteurized commercial apple juice—British Columbia, California, Colorado and Washington. October 1996,"Morbidity and Mortality Weekly Report, vol. 45, p. 975, 1996.

100. S. Tamblyn, J. deGrosbois, D. Taylor, and J. Stratton, "An outbreak of Escherichia coli O157:H7 infection associated with unpasteurized non-commercial, custom-pressed apple cider—ontario, 1998," Canada Communicable Disease Report, vol. 25, no. 13, pp. 113–120, 1999.

101. M. E. Butler, "Salmonella outbreak leads to juice recall in Western states," Food Chemical News, 2000.

102. J. D. Vojdani, L. R. Beuchat, and R. V. Tauxe, "Juice-associated outbreaks of human illness in the United States, 1995 through 2005," Journal of Food Protection, vol. 71, no. 2, pp. 356–364, 2008.

103. S. Jain, S. A. Bidol, J. L. Austin et al., "Multistate outbreak of Salmonella Typhimurium and Saintpaul infections associated with unpasteurized orange juice-United States, 2005," Clinical Infectious Diseases, vol. 48, no. 8, pp. 1065–1071, 2009.

104. K. S. Pereira, F. L. Schmidt, A. M. A. Guaraldo, R. M. B. Franco, V. L. Dias, and L. A. C. Passos, "Chagas› disease as a foodborne illness," Journal of Food Protection, vol. 72, no. 2, pp. 441–446, 2009.

105. CDC, "Botulism associated with commercial carrot juice—Georgia and Florida," Morbidity and Mortality Weekly Report, vol. 55, pp. 1098–1099, 2006.

106. FDA, "DHMH Issues Consumer Alert Regarding Recall of Baugher's Apple Cider,"http://www.fda.gov/Safety/Recalls/ucm232878.htm.

107. M. Linton, J. M. J. McClements, and M. F. Patterson, "Inactivation of Escherichia coli O157:H7 in orange juice using a combination of high

pressure and mild heat," Journal of Food Protection, vol. 62, no. 3, pp. 277–279, 1999.

108. S. Ramaswamy, E. Riahi, and E. Idziak, "High-pressure destruction kinetics of E. coli (29055) in apple juice," Journal of Food Science, vol. 68, no. 5, pp. 1750–1756, 2003.

109. Bayindirli, H. Alpas, F. Bozoglu, and M. Hızal, "Efficiency of high pressure treatment on inactivation of pathogenic microorganisms and enzymes in apple, orange, apricot and sour cherry juices," Food Control, vol. 17, no. 1, pp. 52–58, 2006.

110. Avure Technologies, http://www.avure.com.

111. A. Guerrero-Beltran, G. V. Barbosa-Canovas, and J. Welti-Chanes, "High hydrostatic pressure effect on natural microflora, Saccharomyces cerevisiae, Escherichia coli, and Listeria Innocua in navel orange juice,"International Journal of Food Engineering, vol. 7, no. 1, article 14, 2011.

112. W. J. Briñez, A. X. Roig-Sagués, M. M. H. Herrero, and B. G. López, "Inactivation by ultrahigh-pressure homogenization of Escherichia coli strains inoculated into orange juice," Journal of Food Protection, vol. 69, no. 5, pp. 984–989, 2006.

113. P. Campos and M. Cristianini, "Inactivation of Saccharomyces cerevisiae and Lactobacillus plantarum in orange juice using ultra high-pressure homogenisation," Innovative Food Science and Emerging Technologies, vol. 8, no. 2, pp. 226–229, 2007.

114. M. McKay, "Inactivation of fungal spores in apple juice by high pressure homogenization," Journal of Food Protection, vol. 72, no. 12, pp. 2561–2564, 2009.

115. Patrignani, L. Vannini, S. L. S. Kamdem, R. Lanciotti, and M. E. Guerzoni, "Effect of high pressure homogenization on Saccharomyces cerevisiae inactivation and physico-chemical features in apricot and carrot juices," International Journal of Food Microbiology, vol. 136, no. 1, pp. 26–31, 2009.

116. Patrignani, L. Vannini, S. L. S. Kamdem, R. Lanciotti, and M. E. Guerzoni, "Potentialities of high pressure homogenization to inactivate Zygosaccharomyces bailli in fruit juices," Journal of Food Science, vol. 75, no. 2, pp. M116–M120, 2010.

117. A. Evrendilek, Q. H. Zhang, and E. R. Richter, "Inactivation of Escherichia coli O157:H7 andEscherichia coli 8739 in apple juice by pulsed electric fields," Journal of Food Protection, vol. 62, no. 7, pp. 793–796, 1999.

118. Z. T. Jin and Q. H. Zhang, "Pulsed electric field inactivation of microorganisms and preservation of quality of cranberry juice," Journal of Food Processing and Preservation, vol. 23, no. 6, pp. 481–497, 1999. ·

119. A. Evrendilek, Z. T. Jin, K. T. Ruhlman, X. Qiu, Q. H. Zhang, and E. R. Richter, "Microbial safety and shelf-life of apple juice and cider processed by bench and pilot scale PEF systems," Innovative Food Science and Emerging Technologies, vol. 1, no. 1, pp. 77–86, 2000.

120. J. McDonald, S. W. Lloyd, M. A. Vitale, K. Petersson, and F. Innings, "Effects of pulsed electric fields on microorganisms in orange juice using electric field strengths of 30 and 50 kV/cm," Journal of Food Science, vol. 65, no. 6, pp. 984–989, 2000.

121. Iu, G. S. Mittal, and M. W. Griffiths, "Reduction in levels of Escherichia coli O157:H7 in apple cider by pulsed electric fields," Journal of Food Protection, vol. 64, no. 7, pp. 964–969, 2001.

122. Z. Liang, G. S. Mittal, and M. W. Griffiths, "Inactivation of Salmonella typhimurium in orange juice containing antimicrobial agents by pulsed electric field," Journal of Food Protection, vol. 65, no. 7, pp. 1081–1087, 2002.

123. V. Heinz, S. Toepfl, and D. Knorr, "Impact of temperature on lethality and energy efficiency of apple juice pasteurization by pulsed electric fields treatment," Innovative Food Science and Emerging Technologies, vol. 4, no. 2, pp. 167–175, 2003.

124. P. Elez-Martínez, J. Escolà-Hernández, R. C. Soliva-Fortuny, and O. Martín-Belloso, "Inactivation of Lactobacillus brevis in orange juice by high-intensity pulsed electric fields," Food Microbiology, vol. 22, no. 4, pp. 311–319, 2005.

125. F. Sampedro, A. Rivas, D. Rodrigo, A. Martínez, and M. Rodrigo, "Pulsed electric fields inactivation of Lactobacillus plantarum in an orange juice-milk based beverage: effect of process parameters," Journal of Food Engineering, vol. 80, no. 3, pp. 931–938, 2007.

126. A. Evrendilek, F. M. Tok, E. M. Soylu, and S. Soylu, "Inactivation of Penicillum expansum in sour cherry juice, peach and apricot nectars by pulsed electric fields," Food Microbiology, vol. 25, no. 5, pp. 662–667, 2008.

127. T. Koutchma, S. Keller, S. Chirtel, and B. Parisi, "Ultraviolet disinfection of juice products in laminar and turbulent flow reactors," Innovative Food Science and Emerging Technologies, vol. 5, no. 2, pp. 179–189, 2004.

128. M. T. Tran and M. Farid, "Ultraviolet treatment of orange juice," Innovative Food Science & Emerging Technologies, vol. 5, no. 4, pp. 495–502, 2004.

129. A. Guerrero-Beltrán and G. V. Barbosa-Cánovas, "Reduction of Saccharomyces cerevisiae, Escherichia coli and Listeria innocua in apple juice by ultraviolet light," Journal of Food Process Engineering, vol. 28, no. 5, pp. 437–452, 2005.

130. M. Zenker, V. Heinz, and D. Knorr, "Application of ultrasound-assisted thermal processing for preservation and quality retention of liquid foods," Journal of Food Protection, vol. 66, no. 9, pp. 1642–1649, 2003.

131. M. Valero, N. Recrosio, D. Saura, N. Muñoz, N. Martí, and V. Lizama, "Effects of ultrasonic treatments in orange juice processing," Journal of Food Engineering, vol. 80, no. 2, pp. 509–516, 2007. ·

132. Y. Yuan, Y. Hu, T. Yue, T. Chen, and Y. M. Lo, "Effect of ultrasonic treatments on thermoacidophilicAlicyclobacillus acidoterrestris in apple juice," Journal of Food Processing and Preservation, vol. 33, no. 3, pp. 370–383, 2009.

133. V. M. Gómez-López, L. Orsolani, A. Martínez-Yépez, and M. S. Tapia, "Microbiological and sensory quality of sonicated calcium-added orange juice," LWT: Food Science and Technology, vol. 43, no. 5, pp. 808–813, 2010.

134. S. Roller and N. Covill, "The antifungal properties of chitosan in laboratory media and apple juice,"International Journal of Food Microbiology, vol. 47, no. 1-2, pp. 67–77, 1999.

135. G. Kisko, R. Sharp, and S. Roller, "Chitosan inactivates spoilage yeasts but enhances survival ofEscherichia coli O157:H7 in apple juice," Journal of Applied Microbiology, vol. 98, no. 4, pp. 872–880, 2005.

136. Yuste and D. Y. C. Fung, "Inactivation of Listeria monocytogenes Scott A 49594 in apple juice supplemented with cinnamon," Journal of Food Protection, vol. 65, no. 10, pp. 1663–1666, 2002.

137. R. M. Raybaudi-Massilia, J. Mosqueda-Melgar, and O. Martín-Belloso, "Antimicrobial activity of essential oils on Salmonella enteritidis, Escherichia coli, and Listeria innocua in fruit juices," Journal of Food Protection, vol. 69, no. 7, pp. 1579–1586, 2006.

138. M. Friedman, P. R. Henika, C. E. Levin, and R. E. Mandrell, "Antibacterial activities of plant essential oils and their components against Escherichia coli O157:H7 and Salmonella enterica in apple juice," Journal of Agricultural and Food Chemistry, vol. 52, no. 19, pp. 6042–6048, 2004.

139. P. Nguyen and G. S. Mittal, "Inactivation of naturally occurring microorganisms in tomato juice using pulsed electric field (PEF) with and without antimicrobials," Chemical Engineering and Processing: Process Intensification, vol. 46, no. 4, pp. 360–365, 2007.

140. S. Ferrante, S. Guerrero, and S. M. Alzamora, "Combined use of ultrasound and natural antimicrobials to inactivate Listeria monocytogenes in orange juice," Journal of Food Protection, vol. 70, no. 8, pp. 1850–1856, 2007.

141. M. Hodgins, G. S. Mittal, and M. W. Griffiths, "Pasteurization of fresh orange juice using low-energy pulsed electrical field," Journal of Food Science, vol. 67, no. 6, pp. 2294–2299, 2002.

142. P. Pathanibul, T. M. Taylor, P. M. Davidson, and F. Harte, "Inactivation of Escherichia coli and Listeria innocua in apple and carrot juices using high pressure homogenization and nisin," International Journal of Food Microbiology, vol. 129, no. 3, pp. 316–320, 2009.

Chapter 8

ROLE OF BIOPRESERVATION IN IMPROVING FOOD SAFETY AND STORAGE

Swarnadyuti Nath, S. Chowdhury, Prof. K.C. Dora and S. Sarkar

Faculty of Fishery Sciences, West Bengal University of Animal and Fishery Sciences, Kolkata, India

ABSTRACT

Biopreservation refers to the use of antagonistic microorganisms or their metabolic products to inhibit or destroy undesired microorganisms in foods to enhance food safety and extend shelf life. In order to achieve improved food safety and to harmonize consumer demands with the necessary safety standards, traditional means of controlling microbial spoilage and safety hazards in foods are being replaced by combinations of innovative technologies that include biological antimicrobial systems such as lactic acid bacteria (LAB) and/or their metabolites. Bacillus spp. have an antimicrobial action against Gram positive and Gram negative bacteria, as well as fungi, can therefore be used as a potential biopreservative in food processing due to its wide antimicrobial spectra. Bacteriocins are peptides or complex proteins biologically active with antimicrobial action against other bacteria, principally closely related species. Bacteriocins produced by lactic acid bacteria (LAB) have received particular attention in recent years due to their potential application in food industry as natural preservatives. Bacteriocin production in Bacillus spp. has been studied over the past few decades which include Subtilin from B. subtilis, Megacin from B. megaterium and Thermacin from B. stearothermophilus. Biopreservation may be effectively used in combination with other preservative factors (called hurdles) to inhibit microbial growth and achieve food safety. Using an adequate mix of hurdles is not only economically attractive; it also serves to improve microbial stability and safety, as well as the sensory and nutritional qualities of a food.

INTRODUCTION

Modern technologies in food processing and microbiological food safety standards have reduced but not eliminated the likelihood of food-related illness and product spoilage in industrialized countries. Food spoilage refers to the damage of the original nutritional value, texture, flavour of the food that eventually render food harmful to people and unsuitable to eat. The increasing consumption of precooked food especially seafood, prone to temperature abuse, and the import of raw seafood from developing countries results in outbreak of food borne illness [1]. One of the concerns in food industry is the contamination by pathogens, which are frequent cause of food borne diseases. In the USA, acute gastroenteritis affects 250 to 350 million people with more than 500 human deaths annually and approximately 22 to 30% of these cases are thought to be food borne diseases with the main foods implicated including meat, poultry, eggs, seafood and dairy products [2]. Several bacterial pathogens including Salmonella, Campylobacter jejuni, Escherichia coli 0157:H7, Listeria monocytogenes, Staphylococcus aureus and Clostridium botulinum are found associated with such outbreaks [1]. In order to achieve improved food safety against such pathogens, food industry makes use of chemical preservatives or physical treatments (e.g. high temperatures). These preservation techniques have many drawbacks which includes the proven toxicity of the chemical preservatives (e.g. nitrites), the alteration of the organoleptic and nutritional properties of foods, and especially recent consumer demands for safe but minimally processed products without additives. To harmonize consumer demands with the necessary safety standards, traditional means of controlling microbial spoilage and safety hazards in foods are being replaced by combinations of innovative technologies that include biological antimicrobial systems such as lactic acid bacteria (LAB) and/or their metabolites [1]. The increasing demand for safe food has increased the interest in replacing chemical additives by natural products, without injuring the host or the environment. Biotechnology in the food-processing sector targets the selection, production and improvement of useful microorganisms and their products, as well as their technical application in food quality. The use of non-pathogenic microorganisms and/or their metabolites to improve microbiological safety and extend the shelf life of foods is defined as biopreservation [3]. Antagonistic properties of LAB allied to their safe history of use in traditional fermented food products make them very attractive to be used as biopreservatives [4].

BIO-PRESERVATION

The use of non-pathogenic microorganisms and/or their metabolites to improve microbiological safety and extend the shelf life of foods is defined as biopreservation [1,3]. Bio-preservation refers to extended storage life and enhanced safety of foods using the natural microflora and (or) their antibacterial products. It can be defined as the extension of shelf life and food safety by the use of natural or controlled microbiota and/or their antimicrobial compounds [1,5]. One of the most common forms of food biopreservation is fermentation, a process based on the growth of microorganisms in foods, whether natural or added. It employs the breakdown of complex compounds, production of acids and alcohols, synthesis of Vitamin-B12, riboflavin and Vitamin-C precursor, ensures antifungal activity and improvement of organoleptic qualities such as, production of flavor and aroma compounds. In fish processing, biopreservation is achieved by adding antimicrobials or by increasing the acidity of the fish muscle. Efforts have concentrated on identification and development of protective bacterial cultures with antimicrobial effects against known pathogens and spoilage organisms. Following compounds such as organic acids, bacteriocins, diacetyl and acetaldehyde, enzymes, CO_2, hydrogen peroxide etc. are contributing to antimicrobial activity by Microbiota.

BACTERIOCIN

Bacteriocins are peptides or complex proteins biologically active with antimicrobial action against other bacteria, principally closely related species. They are produced by bacteria and are normally not termed antibiotics in order to avoid confusion and concern with therapeutic antibiotics, which can potentially illicit allergic reactions in humans and other medical problems [6]. Bacteriocins differ from most therapeutic antibiotics in being proteinaceous agents that are rapidly digested by proteases in the human digestive tract. Since, bacteriocins are ribosomally synthesized; there exists a possibility of improving their characteristics to enhance their intensity and spectra of action [1,7]. Colicine was the first bacteriocin, discovered in 1925 by André Gratia and his workgroup [8]. Bacteriocin production could be considered as an advantage for food and feed producers since, in sufficient amounts, these peptides can kill or inhibit pathogenic bacteria that compete for the same ecological niche or nutrient pool. This role is supported by the fact that many bacteriocins have a narrow host range, and is likely to be most effective against related bacteria with nutritive demands for the same scarce resources [9].

LAB BACTERIOCINS

A large number of new bacteriocins in lactic acid bacteria (LAB) have been characterized in recent years. Most of the new bacteriocins belong to the class II bacteriocins which are small (30–100 amino acids) heat-stable and commonly not posttranslationally modified. While most bacteriocin producers synthesize only one bacteriocin, it has been shown that several LAB produce multiple bacteriocins (2–3 bacteriocins). The production of some class II bacteriocins (plantaricins of Lactobacillus plantarum C11 and sakacin P of Lactobacillus sake) have been shown to be transcriptionally regulated through a signal transduction system which consists of three components: an induction factor (IF), histidine protein kinase (HK) and a response regulator (RR). Some bacteriocin-producing strains can be applied as protective cultures in a variety of food products and LAB bacteriocins possess many attractive characteristics that make them suitable candidates for use as food preservatives, such as:

- Protein nature, inactivation by proteolytic enzymes of gastrointestinal tract
- Non-toxic to laboratory animals tested and generally non-immunogenic
- Inactive against eukaryotic cells
- Generally thermo-resistant (can maintain antimicrobial activity after pasteurization and sterilization)
- Broad bactericidal activity affecting most of the Gram-positive bacteria and some, damaged, Gram-negative bacteria including various pathogens such as Listeria monocytogenes, Bacillus cereus, Staphylococcus aureus and Salmonella.

Genetic determinants generally located in plasmid, which facilitates genetic manipulation to increase the variety of natural peptide analogues with desirable characteristics [1]. In general, the following features should be considered when selecting bacteriocin-producing strains for food applications:

- strains for food applications:
- The producing strain should preferably have GRAS (generally regarded as safe) status. • Depending on the application, the bacteriocin should have a broad spectrum of inhibition that includes pathogens or else high specific activity.
- Thermostability.
- Beneficial effects and improved safety.
- No adverse effect on quality and flavour.

BIOPRESERVATION WITH BACTERIOCIN: FIELD OF APPLICATION

Before bacteriocin can be applied in foods their cytolytic abilities should be assessed in detail. This is a very important issue since recently a cytolysin produced by E. faecalis was described that possesses both haemolytic and bacteriocin activities [10]. Recombinant DNA technology is currently applied, to enhance production, to transfer bacteriocin genes to other species and for mutation and selection of bacteriocin variants with increased and/or broad activity spectra [11]. Continued study of the physical and chemical properties, mode of action and structure-function relationships of bacteriocins is necessary if their potential in food preservation is to be exploited. Further research into the synergistic reactions of these compounds and other natural preservatives, in combination with advanced technologies could result in replacement of chemical preservatives, or could allow less severe processing (eg. heat) treatments, while still maintaining adequate microbiological safety and quality in foods.

CLASSIFICATION

On the basis of structure and mode of action bacteriocins are divided in 4 major groups [12].

Class I

- It is termed lantibiotics, constitue a group of small peptides that are characterized by their content of several unusual amino acids [13].

- They are small peptides that are differentiated from other bacteriocins by their content in dehydroamino acids and thioether amino acids. They include nisin, discovered in 1928 [14], lacticin 481 of L. lactis [15], citolysin of E. faecalis [10], and lacticin 3147 of L. lactis [16], among others.

Class II

- These are small, nonmodified, heat stable peptides [17].

- In general, they have an amphiphilic helical structure, which allows them to insert into the membrane of the target cell, leading to depolarisation and death.

- They comprises the (<10 kDa) thermostable non-lantibiotic linear peptides. They are divided into three subclasses on the basis of either a distinctive N-terminal sequence, the pediocinlike bacteriocins (class II.1) (e.g. pediocin PA- 1/AcH produced by Pediococcus [18], the lack of leader peptide (class II.2) (e.g. enterocin EJ97 by E. faecalis [19], or neither of the above traits (class II.3)(e.g. enterocin L50A by E. faecalis [20].

Class III

- It is formerly termed bacteriolysins, large (> 30 KDa), heat-labile protein bacteriocins [21], such as helveticin J of L. helveticus [22] and bacteriocin Bc-48 of E. faecalis [23].

- It can function directly on the cell wall of Grampositive targets, leading to death and lysis of the target cell.

Class IV

- It is presently reserved for cyclic bacteriocins composed not only from protein (also lipid or cidrate) [24].

    Factors inhibiting bacteriocin production [25] include:

- inadequate physical conditions and chemical composition of food (pH, temperature, nutrients, etc.);

- spontaneous loss in production capacity;

- inactivation by phage of the producing strain;

- and antagonism effect of other microorganisms in foods.

    The effectiveness of bacteriocin activity in food is negatively affected by:

- resistance development of pathogens to the bacteriocin;

- inadequate environmental conditions for the biological activity;

- higher retention of the bacteriocin molecules by food system components (e.g. fat);

- inactivation by other additives; slower diffusion and solubility and/or irregular distribution of bacteriocin molecules in the meat matrix [26].

Several factors, such as the presence of salts, other food ingredients, poor solubility and the uneven distribution of the bacteriocin, have all shown to effect the efficacy of bacteriocins in food [27].

BACTERIOCINS OF VARIOUS GRAM POSITIVE BACTERIA

Although mast research has focused on antimicrobial agents produced by lactic acid bacteria, many bacteriocins of other Gram positive bacteria have been isolated and documented [28,29]. One such area of interest is the use of these bacteriocins to control the growth of undesirable microorganisms, particularly those of public health concern, e.g., C. botulinum and L. monocytogenes. Bacteriocin production has been documented for a variety of Gram positive bacteria, including Staphylococcus, Clostridium, and

Bacillus spp. Although the use of these bacteriocins may be precluded from foods because the producer strain may be a pathogen, recent developments in genetic engineering techniques have made the transfer of genes encoding for bacteriocin production from bath Gram positive and Gram negative bacteria to food grade microorganisms possible [30].

BACILLUS AS BIOPRESERVATIVE

Bacteriocin production in Bacillus spp. has been studied over the past few decades and several reports describe the production, isolation and characterization of bacteriocins from these species [31,32,33], which include Subtilin from B. subtilis, Megacin from B. megaterium and Thermacin from B. stearothermophilus [34,35,36]. Subtilin, a cationic peptide produced by B. subtilis, having molecular mass of 3317 Da.; is a member of the group of bacteriocins known as the lantibiotics. The structure of subtilin was determined by Gross et al. (1973) [34]. Bacteriocin production by different strains of B. stearothermophilus, the main spoilage microorganisms of low acid canned products, was first described by Shafia (1966) [36]. He found that 12 out of 22 strains of B. stearothermophilus produced an inhibitory substance ta species of the same genus. The bacteriocin produced by these species was named as Thermacin. More recently, Naclerio et al. (1993) [33] isolated the bacteriocin Gerein from B. cereus. The bacteriocin is relatively heat stable (75°C far 15 min), sensitive to proteolytic enzymes and has an apparent molecular mass of 9KDa. Bacteriocins produced by Bacillus spp. could be alternatives to those produced by lactic acid bacteria for several reasons: i. Bacillus spp., like lactic acid bacteria, have been used for hundreds of years in making food and various enzymes from Bacillus have been used intensively in food processing worldwide. No adverse effects have been demonstrated in humans from consuming foods made from Bacillus spp., and/or their products. Bacteriocins from these microorganisms would be safe for humans and would be no more of a risk than lactic acid bacteria. ii. Bacillus spp. have an antimicrobial action against Gram positive and Gram negative bacteria, as well as fungi, and therefore have a greater antimicrobial spectra than lactic acid bacteria and their bacteriocins. iii. The metabolic diversities of Bacillus spp. may result in bacteriocins with various properties such as inhibitory activity at alkaline, acidic condition or after thermal processing and would be suitable for food processing. iv. The physiology/genetics of Bacillus are well understood, second only to those of Escherichia coli. Molecular biological techniques would provide safe/reliable tools for producing bacteriocins for the food industry.

BIOPRESERVATION OF SEAFOOD PRODUCTS

Although bacteriocins are produced by many Gram-positive and Gram-negative species, those produced by LAB and now a days, Bacillus sp. are of particular interest to the food industry, since these bacteria have generally been regarded as safe. Among the lactic acid bacteria, a high diversity of bacteriocins is produced and several have been patented for their applications in foods. To date, the only commercially produced bacteriocins are the group of nisins produced by Lactoccocus lactis, and pediocin PA-1, produced by Pediococcus acidilactici [1,37]. Bacteriocins produced by lactic acid bacteria (LAB) have received particular attention in recent years due to their potential application in food industry as natural preservatives [38]. Biopreservation refers to the use of antagonistic microorganisms or their metabolic products to inhibit or destroy undesired microorganisms in foods to enhance food safety and extend shelf life [26]. Three approaches are commonly used in the application of bacteriocins for biopreservation of foods [1,26]: 1) Inoculation of food with LAB that produce bacteriocin in the products. The ability of the LAB to grow and produce bacteriocin in the products is crucial for its successful use. 2) Addition of purified or semi-purified bacteriocins as food preservatives. 3) Use of a product previously fermented with a bacteriocin producing strain as an ingredient in food processing. The effectiveness of bacteriocins and protective cultures to control growth of L. monocytogenes in vacuum-packed cold smoked salmon has been demonstrated by several researchers. Katla et al, (2001) [39] examined the inhibitory effect of sakacin P and/or L. sake cultures (sakacin P producer) against L. monocytogenes in cold-smoked salmon. Nilsson et al, (1997) [40] showed that a nonbacteriocin-producing strain of C. piscicola was as effective as a bacteriocin-producing strain of C. piscicola in the inhibition of L. monocytogenes in vacuum-packed cold-smoked salmon. They suggested that the growth inhibition of L. monocytogenes was probably due to the competitive growth of C. piscicola that resulted in depletion of essential nutrients. The inhibitory effect of nisin in combination with carbon dioxide and low temperature on the survival of L. monocytogenes in cold-smoked salmon has also been investigated [40]. The effectiveness of nisin Z, carnocin UI49, and a preparation of crude bavaricin A on shelflife extension of brined shrimp was evaluated by Einarsson and Lauzon (1995) [41]. In a study using vacuum-packed cold-smoked rainbow trout, Niskänen et al, (2000) [42] examined the inhibition of L. monocytogenes and mesophilic aerobic bacteria by nisin, sodium lactate, or their combination.

CONCLUSION

Bio-preservation offers the potential to extend the storage life and food safety using the natural microflora and (or) their antibacterial products. Shelf life of sea foods can be extended and safety ensured by the use of natural or controlled microbiota and/or their antimicrobial compounds. Bio-preservation may be effectively used in combination with other preservative factors (called hurdles) to inhibit microbial growth and achieve food safety [43]. Using an adequate mix of hurdles is not only economically attractive; it also serves to improve microbial stability and safety, as well as the sensory and nutritional qualities of a food. The principle hurdles employed in food safety are temperature (higher or lower), water activity (aw), pH, redox potential (Eh), chemical preservatives, vacuum packaging, modified atmosphere, HHP, UV and competitive flora (LAB producing antimicrobial compounds)

ACKNOWLEDGEMENTS

The corresponding author is grateful to Department of Science & Technology, Ministry of Science of Technology, Govt. of India for present financial assistance.

REFERENCES

1. S. Nath, S. Chowdhury, S. Sarkar, and K.C. Dora, Lactic Acid Bacteria – A Potential Biopreservative In Sea Food Industry, International Journal of Advanced Research, 1(6), 2013, 471-475

2. P. S. Mead, L. Slutsker, V. Dietz, L. F. McCaig, J. S. Bresee, C. Shapiro, P. M. Griffin, and R.V. Tauxe, Emerging Infectious Diseases, 5, 1999, 607-625.

3. De Martinis, C. P. Elaine. and D. G. M. Bernadette, B.D.G.M. Franco, Inhibition of Listeria monocytogenes in a pork product by a Lactobacillus sake strain, International Journal of Food Microbiology, 42(1-2), 2001, 119-126.

4. E. Caplice, and G.F. Fitzgerald, Food fermentations: role of microorganisms in food production and preservation, International Journal of Food Microbiology, 50 (1-2),1999, 131-149.

5. M. E. Stiles, Biopreservation by lactic acid bacteria. Antonie van Leuwenhoek,70,1996, 331.

6. S.F. Deraz, E.N. Karlsson, M. Hedström, M.M. Andersson, and B. Mattiasson, Purification and characterisation of acidocin D20079, a

bacteriocin produced by Lactobacillus acidophilus DSM 20079, Journal of Biotechnology, 117(4), 2005, 343-354.

7.   J.M. Saavedra, A. Abi-Hanna, N. Moore, and R.H. Yolken, Long-term consumption of infant formulas containing live probiotic bacteria: tolerance and safety, American Journal of Clinical Nutrition, 79(2), 2004, 261-267.

8.   F. Jacob, A. Lwoff, A. Siminovitch, and E. Wollman, Definition de quelques termes relatifs a la Iysogenie, Am. Inst. Pasteur, 84, 1953, 222-224.

9.   L.H. Deegan, P.D. Cotter, C. C. Hill, and P. Ross, Bacteriocins: Biological tools for biopreservation and shelf-life extension, International Dairy Journal, 16(9), 2006, 1058-1071.

10. M. S. Gillmore, R. A. Segarra, M. C. Booth, C. P. Bogie, L. R. Hall, and D. B. Clewell, Genetic structure of the Enterococcus faecalis plasmid pAD1 encoded cytolytic toxin system and its relationship to lantibiotic determinants, Journal of Bacteriology, 176, 1990, 1335.

11. O. Osmanagaoglu, and Y. Beyatli, The Use of Bacteriocin Produced by Lactic Acid bacteria in Food preservation, Turk Mikrobiyol Cem Derg, 32, 2001, 295-306.

12. P.D. Cotter, C. Hill, R. P. Ross, Bacteriocins: developing innate immunity for food, Nature Reviews, Microbiology, 3(10), 2006, 777–788.

13. A. Guder, I. Wiedeman, and H.G. Sahl, Post translationally modified bacteriocins, the lantibiotics. Bioploymers 55, 2000, 62-73.

14. L. A.Rogers, and E.O. Whittier, Limiting factors in the lactic fermentation, Journal of Bacteriology, 16(4), 1928, 211–229.

15. J.C. Piard, P.M. Muriana, M.J. Oesmazeaud, and T.R. Klaenhammer, Purification and partial characterization of Lactiein 481, a lanthionine-containing bacterioein produced by Lactocoecus lactis subsp. lactis CNRZ 481, Applied Environmental Microbiology, 58, 1992, 279-284.

16. M. P. Ryan, R. P. Ross, M. Galvin, O. McAuliffe, S. M. Morgan, D. P. Twomey, W. J. Meaney, and C. Hill, Developing applications for lactococcal bacteriocins, Lactic Acid Bacteria: Genetics, Metabolism and Applications, 1999, 337-346.

17. I. F. Nes and H. Holo, Class II antimicrobial peptides from lactic acid bacteria, Peptide Science,55(1), 2000, 50–61.

18. J.T. Henderson, A. L. Chops, and P. D.V. Wassenar, Purification and primary structure of Pediocin PA-1 produced by Pediococcus acidilactici PAC 1.0, Archive of Biochemistry and Biophysics, 295(1), 1992, 5-12.

19. M. Maqueda, M. Sánchez-Hidalgo, M. Fernández, M. Montalbán-López, E. Valdivia, M.Martínez-Bueno, Genetic features of circular bacteriocins produced by Gram-positive bacteria, FEMS Microbiology Reviews, 32(1), 2008, 2–22.

20. L. M. Cintas, M. P. Casaus, C.C. Herranz, I. F. Nes, and P. E. Hernández, Characterization of Garvicin ML, a Novel Circular Bacteriocin Produced by Lactococcus garvieae DCC43, Isolated from Mallard Ducks (Anas platyrhynchos), Applied Environmental Microbiology, 77, 2000, 369-373.

21. R. Bauer, M. L. Chikindas, and L.M.T. Dicks, Purification, partial amino acid sequence and mode of action of pediocin PD-1, a bacteriocin produced by Pediococcus damnosus, International Journal of Food Microbiology, 101(1), 2005, 17–27.

22. M.C. Joerger, and T.R. Klaenhammer, Characterization and purification of Helveticin J and evidence for a chromosomally determined bacteriocin produced by Lactobacil/us helveticus 481, Journal of Bacteriology, 167,1986,439-446.

23. I. López-Lara, A. Gálvez, M. MartínezBueno, M. Maqueda, and E. Valdivia, Purification, characterization, and biological effects of a second bacteriocin from Enterococcus faecalis ssp. liquefaciens S-48 and its mutant strain B-48-28, Canadian Journal of Microbiology, 37(10), 1991, 769- 774.

24. T.R. Klaenhammer, Genetics of bacteriocin produced by Lactic acid bacteria. FEMS Microbiology Reviews, 12(1-3), 1993, 39- 85.

25. J. Cleveland, T. J. Montville, I. F. Nes, M. L. Chikindas, Bacteriocins: safe, natural antimicrobials for food preservation, International Journal of Food Microbiology, 71(1), 2001, 1–20.

26. U. Schillinger, R. Geisen, and W. H. Holzapfel, Potential of antagonistic microorganisms and bacteriocins for the biological preservation of foods, Trends in Food Science & Technology, (5),1996, 158- 164.

27. L. DeVuyst, Production and application of bacteriocins from lactic acid bacteria, Bioactive peptides for future food preservation, Cerevisia 21,1996, 71-74.

28. J. R. Tagg, A. S. Dajani, and L.W. Wannamaker, Bacteriocins of Gram positive bacteria, Bacterielogy Reviews, 40(3), 1976, 722-756.

29. R.W. Jack, J.R. Tagg, and B. Ray, Bacteriocins of Gram-positive bacteria. Microbiology Reviews, 59(2), 1995, 171- 200.

30. J.K. McCormick, T.R. Klaenhammer, and M.E. Stiles, Colicin V can be

produced by lactic acid bacteria, Letters Applied Microbiology, 29, 1999, 37-41.

31. S. Banerjee, and J. Hansen, Structure and expression of a gene encoding the precursor of Subtilin: a small protein antibiotic, The Journal of Biological Chemistry, 263(19), 1988, 9508-9514.

32. Y.J. Chung, M. Steen, and J. Hansen, The Subtilin gene of Bacillus subtilis ATCC 6633 is encoded in an operon that contains a homology of the hemolysin B transport protein, Journal of Bacteriology, 174, 1992, 1417-1422.

33. G. Naclerio, E. Ricca, M. Sacco, and M. Defelice, Antimicrobial activity of a newly identified bacteriocin of Bacillus cereus. Applied Environmental Microbiology, 59, 1993, 4313-4316.

34. E. Gross, H.H. Kiltz, and E. Nebelin, Subtilin VI. Die Struktur des Subtilins. Hoppe Seylers Z Physiol Chem, 354(7), 1973, 810-812.

35. G. Ivanovic, and L. Alfoldi, A new antimicrobial principle: Megacin, Nature,174, 1954, 465.

36. F. Shafia, Thermocins of Bacillus stearothermophilus, Journal of Bacteriology, 92, 1966, 524-525.

37. R. Schöbitz, T. Zaror, O. León, M. Costa, A bacteriocin from Carnobacterium piscicola for the control of Listeria monocytogenes in vacuum-packaged meat, Food Microbiology, 16, 1999, 249-55.

38. E. Rodriguez, J. Calzada, J. L. Arques, J. M. Rodrigues, M. Nunez, and M. Medina, Antimicrobial activity of Pediocinproducing Lactococcus lactis on Listeria monocytogenes, Staphylococus aureus and Escherichia coli 0157:H7 in cheese. International Dairy Journal, 15,2002, 51.

39. T. Katla, T. Møretrø, I.M. Aasen, A. Holck, L. Axelsson, and K. Naterstad, Inhibition of Listeria monocytogenes in cold smoked salmon by addition of sakacin P and/or live Lactobacillus sakei cultures. Food Microbiology, 18, 2001, 431-9.

40. L. Nilsson, H. H. Huss, and L. L. Gram, Inhibition of Listeria monocytogenes on cold-smoked salmon by nisin and carbon dioxide atmosphere, International Journal of Food Microbiology, 38(2-3), 1997, 217- 227.

41. H. Einarsson and H.L. Lauzon, Biopreservation of Brined Shrimp (Pandalus borealis) by Bacteriocins from Lactic Acid Bacteria, Applied Environmental Microbiology, 61(2), 1995, 669-676.

42. A. Niskanen, and E. Nurmi, Effect of starter culture on staphylococcal enterotoxin and thermonuclease production in dry sausage, Applied Environmental Microbiology, 31, 2000, 11–20.

43. S. Ananou, M. Maqueda, M. MartinezBueno, and E. Valdivia, Biopreservation, an ecological approach to the safety and shelflife of foods, Communicating Current Research and Educational Topics and Trends in Applied Microbiology. A. MendezVilas (Ed.), 2007, 475.

Chapter 9

BIOPRESERVATION OF FRESH ORANGE JUICE USING ANTILISTERIAL BACTERIOCINS101 AND ANTILISTERIAL BACTERIOCIN103 PURIFIED FROM LEUCONOSTOC MESENTEROIDES

Backialakshmi S, Meenakshi RN, Saranya A, Jebil MS, Krishna AR, Krishna JS and Suganthi Ramasamy

Department of Biotechnology, Dr. G R Damodaran College of Science, Coimbatore-14, Tamil Nadu, India

INTRODUCTION

The increasing demand for fresh-tasting, healthy, nutritious and ready-to-eat foods has stimulated the expansion of minimally processed fruit and vegetable markets worldwide [1-3]. Processing of the products resulting from natural fruits and vegetables have been observed to increase certain reactions leading to susceptibility to microbes [4]. From a consumer perspective, increasing scientific evidence for consumption of fresh fruits for prevention of biological problems, demand for low-calorie diet and increasing microbiological and pesticide content in processed food has increased the consumption of ready-to-eat vegetables and fruits [5]. This health based option for customers has been short lived due many inappropriate or manipulative storage conditions that again lead to microbiological spoilage and diseases [6].

Listeria monocytogenes, a major contaminant of food products, is a Gram positive, facultative, intrecellular bacterium that multiplies mainly on dairy, milk and meat products including industrial food products that are distributed or procured under refrigerated conditions [7,8]. L. monocytogenes, is considered a serious food-borne pathogen in immune-compromised individuals such as cancer subjects, HIV patients, elderly and pregnant women. Ingestion of L. monocytogenes can cause listeriosis in human and animals [9,10] which lead to gastroenteritis, septicaemia, perinatal infections, stillbirth, abortion, meningitis and meningoencephalitis in immunocompromised individuals

[11]. Listeriosis affects a wide variety of mammals including monogastric, ruminants (mostly sheep) and human with morality rate of 20-30% [12].

Traditionally, the shelf-life stability of juices has been achieved by thermal processing. Low temperature long time (LTLT) and high temperature short time (HTST) treatments are the most commonly used techniques for juice pasteurization. However, thermal pasteurization tends to reduce the product quality and freshness. Apart from thermal pasteurization, some chemical preservatives are also widely used for the extension of the shelf-life of fruit juices and beverages. Two of the most commonly used preservatives are potassium sorbate and sodium benzoate. However, consumer demand for natural origin, safe and environmental friendly food preservatives has been increasing since 1990s. Apart from thermal pasteurization, some chemical preservatives are also widely used for the extension of the shelf-life of fruit juices and beverages. Two of the most commonly used preservatives are potassium sorbate and sodium benzoate [13]. Nitrate and nitrite as sodium or potassium salts has also been used as food additives to improve the microbiological safety of food and to extend their safe shelf life [14].

Use of microorganisms in food fermentation is one of the oldest methods for producing and preserving food [15]. Lactic acid bacteria (LAB) have an important role in preserving foods and preventing poisoning [16]. This inhibition is due to the production of various compounds as organic acids, diacetyl, hydrogen peroxide and bacteriocins [17]. Bacteriocins as biopreservatives against food spoilage especially led by microbiological contamination are generally regarded as safe (GRAS) and include proteins with varying molecular weight, biochemical properties and mode of activity [18].

The keen interest towards bacteriocin of lactic acid bacteria worldwide is due to their essential role in majority of food fermentation, flavour development and preservation of food products along with proving safer for health. In the present study, antilisterial bacteriocin isolated from lactic acid bacteria has been targeted for its biopreservative potential.

MATERIALS AND METHODS

Maintenance of Listeria Monocytogens MTCC 657

In the present study Listeria monocytogenes was used as indicator microorganism [19,20]. The indicator microorganism Listeria monocytogens MTCC 657 was obtained from IMTECH, Chandigarh and revived using Brain Heart Infusion broth (Himedia, India) at 30°C for 2 days. Listeria monocytogenes MTCC 657 was subcultured and maintained in Tryptone Glucose Extract agar (Casein enzymic hydrolysate - 5 g/l, Yeast extract-3 g/l, Glucose -1 g/l, Agar - 15 g/l and Final pH at 25°C-7.0 ± 0.2).

Isolation of lactic acid bacteria

Toddy was selected as a source for isolating Bacteriocin-producing Lactic Acid Bacteria. Toddy is one of the traditional, social and local drinks extracted from either a coconut tree or a palm tree flower and called as 'kallu'. Toddy samples were collected from Kerala and Tamil Nadu, serially diluted and spread plated on Man Rogosa Sharpe (MRS) agar (Himedia, India; Final pH at 25°C, 6.5 ± 0.2). The plates were incubated overnight at 37°C and further screening of the isolates was done by inoculating the colonies into MRS medium.

Screening Of the Isolates for Antilisterial Bacteriocin Production

The individual colonies that showed growth in MRS agar (BPB GRD 101 and Leuconostoc mesenteroides BPB GRD 103) were selected and screened for the production of bacteriocin by spot-on-lawn method. 4 ml of TGE soft agar (0.7%) containing 30 µl of overnight culture of the indicator strain Listeria monocytogenes MTCC 657 was overlaid on petriplates pre poured with TGE hard agar (1.8%). Each of the individual colony isolates were spotted on the lawn and the plates were incubated for 24 h at 30°C.

Production and Purification of Antilisterial Bacteriocins Using Amberlite Xad-16

A single colony of the isolated Leuconostoc mesenteroides BPB GRD 101 and Leuconostoc mesenteroides BPB GRD 103 strains were inoculated into 10 ml of production media (Tryptone-2%, Glucose-1%, Yeast extract- 2%, MnSO4-0.005%, MgSO4- 0.005%, Tween 80-0.2%, pH-6.8) and incubated overnight at 30°C without shaking. This 10 ml culture was used as the inoculum for 250 ml of production media. The culture was grown at 30°C without shaking for 16-18 hours and centrifuged at 10,000 rpm for 10 minutes. The supernatant was heated at 100°C for 10 minutes to inactivate proteases before further purification. The heat inactivated cell-free supernatant was used for further purification steps.

Five grams of Amberlite XAD-16 was soaked in 50% isopropanol and stored at 4°C. Traces of isopropanol was removed by washing repeatedly with deionized distilled water before use. Five grams of Amberlite XAD-16 was added to 250 ml of heat inactivated cell-free supernatant and incubated with shaking at room temperature for four hours. The mixture was transferred to a chromatographic column and the matrix was washed with 20 ml of deionized distilled water and 20 ml of 40% (v/v) ethanol. The ALB101 and ALB103 were eluted with 20 ml of 70% (v/v) isopropanol followed by washing with 20 ml of absolute isopropanol. The eluted fraction was evaporated to half of its volume and the pH was adjusted to 5.7 with 5N NaOH. This fraction termed as Amberlite fraction was determined for Antilisterial Bacteriocin activity by agar well diffusion method. The protein concentration of this fraction was estimated by measuring the absorbance at 280 nm.

Antilisterial Bacteriocin101 and Antilisterial Bacteriocin103 Activity Assay

The activity of Antilisterial Bacteriocin101 (ALB101) and Antilisterial Bacteriocin103 (ALB103) was assayed based on the critical dilution of antagonistic activity caused by the Antilisterial Bacteriocin producing strain. The crude ALB101 and ALB103 samples were prepared and assayed against the indicator organism Listeria monocytogenes MTCC 657 by agar well diffusion method. Wells were punched on TGE agar (1.8%) seeded with the indicator strain (overnight culture of Listeria monocytogenes MTCC 657). Dilutions were aliquoted into the appropriately labelled wells and incubated at 4°C for 4 h and incubated at 30°C overnight. The Antilisterial Bacteriocin activity was defined as the reciprocal of the highest two fold dilution showing complete inhibition of the indicator lawn and was expressed in activity units (AU) per ml of culture media. The temperature based activity of the isolated bacteriocins were undertaken by subjecting them to varied temperatures and the activity was assayed.

Biopreservative potential of antilisterial bacteriocin101 and antilisterial bacteriocin103 in unpasteurized and pasteurized fresh orange juice at (4°C)

The effect of Antilisterial Bacteriocins (ALBs) as biopreservative was checked on fresh orange juice. The juice samples were procured from local juice corner (Coimbatore). The experiment was carried in two sets of four containers (A, B, C and D) each containing 10 ml of juice samples inoculated with indicator strain i.e. L. monocytogenes MTCC 657 (6.75 log CFU/ml). First bottle (A) labelled as 'Control' (without any preservative), second and third bottle (B and C with purified ALBs) labelled as 'Test' and fourth bottle (C) having sodium nitrite (chemical preservative). Purified ALBs and sodium nitrite were added (40 ppm and 12 ppm) respectively into each pathogen treated food sample to observe biopreservative effect of different preservatives against test indicator and the assay was performed in triplicates. The storage studies for stability were done for 12 days at 24 h interval and log CFU/ml was noted. The procured juice samples were also taken in sterilized glass containers and pasteurized by keeping it in hot water at 72°C for 2 min and subjected to the above mentioned analysis.

RESULTS AND DISCUSSION

Isolation of Lactic Acid Bacteria

Lactic acid bacteria isolated from toddy on MRS agar plate were subjected to morphological, biochemical identification and assayed for its bacteriocin activity. Two strains designated as BPB GRD 101 and BPB GRD 103 were subjected to 16S rRNA sequencing and phylogenetic analysis. Based on the microbiological, physiological tests and 16S rRNA phylogeny, the strains BPB GRD 101 and BPB GRD 103 showed 99% similarity with Leuconostoc mesenteroides. The sequences BPB GRD 101 and BPB GRD 103 defined asLeuconostoc mesenteroides BPB GRD 101 and Leuconostoc mesenteroides GRD BPB 103 was deposited in NCBI under the GenBank accession number JX174179 and JX174180.

Screening Of Antilisterial Bacteriocin Activity

The Spot-On-Lawn antimicrobial activity was the conventional method of monitoring bacteriocin production which is fast and reliable. Various authors had used Spot-On-Lawn method for screening of bacteriocin activity [17,18,21]. The Antilisterial Bacteriocins (ALB101 and ALB103) were tested for antimicrobial activity against L. monocytogenes MTCC 657 by using the spot-on-lawn method with modifications. ALB101 and ALB103 showed a largest clear zone of inhibition against L.monocytogenes MTCC 657 (Figure 1). The Antilisterial Bacteriocins producing LAB strains were further characterized for bacteriocin activity. The crude and partially purified Antilisterial Bacteriocins were tested for their antimicrobial activity against L. monocytogenes MTCC 657 by two fold dilution method and well diffusion assay (WDA). The crude and partially purified ALB101 and ALB103 showed the anntilisterial activity against L. monocytogenes MTCC 657. Activity was measured as the reciprocal of the highest dilution showing antimicrobial activity. When the bacteriocin was subjected to various temperatures ranging from -80°C to -121°C, the stability of the bacteriocin was observed as the activity remained stable.

Production and Purification of Antilisterial Bacteriocins

Antilisterial bacteriocin production was carried out using the TGE + Tween 80 (pH 6.8) medium, which supported maximum bacteriocin production. Bacteriocin activity was detected at early exponential phase and maximized at stationary phase. Tween 80 is a non-ionic detergent, it might help in releasing bacteriocin molecules from the producer cell wall into the medium [22].

Numerous purification strategies have been reported for bacteriocins all with varying degrees of success, which may be attributable to the extremely heterogeneous nature of bacteriocin [23]. The purification methods commonly employed include ammonium sulphate precipitation, ion-exchange chromatography, hydrophobic chromatography coupled with cation-exchange column chromatography, reverse-phase high-performance liquid chromatography (HPLC), Amberlite XAD- 2, Sephadex G-25 gel filtration, ultrafiltration and gel permeation chromatography and ethanol precipitation. Each purification method has its own drawbacks, which may include issues with low yield and purity, cost, and the requirement for a skilled operator [24]. Among the methods tested, the Amberlite XAD-16 method seemed the most appropriate in partially purifying the sample, especially since it is used in the previously reported purification scheme of the NKR-5-3 enterocins [25]. Among the methods tested, the Amberlite XAD-16 method seemed the most appropriate in partially purifying the sample, especially since it is used in the previously reported purification scheme of the NKR-5-3 enterocins.

Biopreservative potential of antilisterial bacteriocin101 and antilisterial bacteriocin103 on total viable counts of L. monocytogenes mtcc 657 in unpasteurized fresh orange juice

Microbial counting was carried out using pour plate method [26] which is a standard method for viable counting. Bacteriocins used in this study, when applied to fresh orange juice had inhibited the multiplication of L. monocytogenes MTCC 657 when compared to the control and showed maximum reduction on bacterial population (Figures 2 and 3). The results further revealed that microbial count drastically decreased in both the fresh and pasteurized sample.

The initial density of L. monocytogenes MTCC 657 on introduction to unpasteurized fresh orange juice was 6.75 log CFU ml-1 on day 0. On day 12, L. monocytogenes MTCC 657 was increased to 7.14 log CFU ml-1 in control. Addition of ALB101, decreased the viable count of microbial population slowly from 6.75 to 6.56 log CFU ml-1 till 4th day after which an increase in proliferation was observed. For ALB103, the reduction potential was observed as from 6.75 to 6.49 log CFU ml-1 during 4th day of storage which followed by proliferation of viable counts. Reduction of viable count was observed from 6.75 to 6.7 log CFU ml-1 at 2nd day on addition of sodium nitrite (Figure 2).

Biopreservative potential of antilisterial bacteriocin101 and antilisterial bacteriocin103 on total viable counts of L. monocytogenes mtcc 657 in pasteurized fresh orange juice

The initial density of L. monocytogenes MTCC 657 on introduction to pasteurized fresh orange juice was 6.75 log CFU ml-1 on day 0. ALB101 showed an obvious gradual decrease in viable extent from 6.75 log CFU ml-1 as 6.54 log CFU ml-1 and 6.66 log CFU ml-1 for 4thand 5th days of storage, respectively, followed by new viable cells. Potential of microbial inhibition of ALB103 was obvious in reduction of viable cells from three to six days as 6.58, 6.51, 6.61 and 6.69 log CFU ml-1. In case of chemical preservative (Sodium nitrate), reduction of microbial reduction at 4th day was observed as 6.62 log CFU ml-1 which then increased following incubation period (Figures 3 and 4).

These results suggest that ALB103 shows increased reduction of microbial population comparing with ALB101 and chemical preservative. However, both ALB101 and ALB103 were effective in reduction of microbial population than chemical preservative. The ability of foodborne pathogens to contaminate fruits and vegetables has led to impose hazard analysis and critical control point requirements on juice processors.

A recent study demonstrated the effectiveness of Nisin as nonthermal technology for the inactivation of microorganisms as E. coli and L. innocua, in fruit and vegetable juices and the synergism of Nisin with the high pressure homogenization in apple and carrot juices [27]. The primary action site of Nisin against vegetative cells is considered to be the cytoplasmic membrane where Nisin acting as voltage dependent polarizer. Analogous to our findings, the clear effectiveness of Nisin in small concentration was lost by increasing the storage time was reported [28]. Analogous to our findings, reports indicate the inhibitory effect of nisin at low concentration was lost by increasing the storage time . These findings are consistent with the work of several other investigators who studied the antimicrobial activity of Nisin. The increased concentration of bacteriocins isolated from CA44 showed increased biopreservative effect in fruit juice. The biopreservative potential of purified bacteriocin from isolate CA44 against B. cereus which increased the preservative effect in fruit juice with the increase in the concentration of bacteriocin which is accordance to our study was also observed [29]. Reports on effective control of contaminant bacteria on ready-to-use vegetables have been revealed on onion and smoked salmon respectively [30-34].

Although the treatment of the juice by pasteurization, in this study, did not affect the growth of L. monocytogenes MTCC 657, an obvious effect was observed on the surviving bacteria after exposure to ALB101 and ALB 103. Similar results explaining the heat synergistic effect of bacteriocins have been studied [35,36]. However, instances of crossresistance to nisin after mild treatment due to antagonism of heat shock proteins have also been revealed [37,38]. Further studies that reveal the mechanism of inhibition and effectiveness of LAB produced bacterions on other food products are required.

CONCLUSION

LAB are known for the safety and shelf life extension of the food products and also serve as alternatives to chemical preservatives/ additives in food preservation. Bacteriocin-producing strains of LAB are of great interest as they are generally recognised as safe (GRAS) organisms and their antimicrobial products as biopreservatives. To meet the demand for very potent natural biopreservative, bacteriocin producing lactic acid bacteria was isolated from toddy. Isolated bacteriocins were thermostable and active against the foodborne pathogenL. monocytogenes MTCC 657. As a conclusion of spoton- lawn and well diffusion assay, the Antilisterial Bacteriocins (ALB101 and ALB103) isolated from Leuconostoc mesenteroides stains showed strong antagonism against indicator microorganism. The biopreservative potential of ALB101 and ALB103 along with chemical preservative (Sodium nitrite) was carried out using fresh orange juice (unpasteurized and pasteurized) and the results showed that low concentrations (40 ppm) of antilisterial Bacteriocins were potentially reduced the viability of L. monocytogenes MTCC 657 in unpasteurized as well as pasteurized orange juice than chemical preservative. However, ALB103 was more potent than ALB101. So it would be recommended to use these Antilisterial Bacteriocins as biopreservatives during the food preservation process.

REFERENCES

1. Abadias M, Usall J, Oliveira M, Alegre I, Viñas I (2008) Efficacy of neutral electrolyzed water (NEW) for reducing microbial contamination on minimally-processed vegetables. Int J Food Microbiol 123: 151-158.

2. Oms-Oliu G, Rojas-Graü M, González LA, Varela P, Soliva-Fortuny R, et al. (2010) Recent approaches using chemical treatments to preserve quality of fresh-cut fruit: A review. Postharvest Bio Techno l57: 139-148.

3. Xiao Z, Luo Y, LuoY,Wang Q (2011) Combined effects of sodium chlorite dip treatment and chitosan coatings on the quality of fresh-cut d'Anjou pears. Postharvest Bio Tech 62: 319-333.

4. Botelho MC, Leme SC, Lima LCO, Abrahao SA, Silqueira HH et al. (2010) Qualidade de palmitopupunhaminimamenteprocessado: aplicação de antioxidants. Ciência e Agrotecnologia34: 1312-1319.

5. Rico D, Martın-Diana AB, Barat JM, Barry-Ryan C (2007) Extending and measuring the quality of fresh-cut fruit and vegetables: a review. Trends Food SciTechnol 18: 373-386.

6. Dfaz-Cinco ME, Acedo-Felix E,Garcia-Galaz A (2005) Principales microorganisms' pat´ogenosy de deterioro, In: Gonz´alez-Aguilar GA, Gardea AA, Cuamea-Navarro F, (eds). Nuevastecnolog´ıas de conservaci ´on de productosvegetales frescos cortados.

7. Portnoy DA, AuerbuchV, Ian J Glomski IJ. (2002) The Cell Biology of Listeria monocytogenes Infection: the Intersection of Bacterial Pathogenesis and Cell-Mediated Immuntiy. The J Cell Biol 158: 409-414.

8. Vazquez-Boland, Jose A, Kuhn M, Berche P,Chakraborty T et al. (2001) Listeria Pathogenesis and Molecular Virulence Determinants. ClinMicrobiol Rev 14: 584-640.

9. Chen BY, Pyla R, Kim TJ, Silva JL, Jung YS. (2010) Prevalence and contamination patterns of Listeria monocytogenesin catfish processing environment and fresh fillets. Food Microbiol 27: 645-652.

10. Pagadala S, Parveen S, Rippen T, Luchansky JB, Call JE et al. (2012) Prevalence, characterization and sources of Listeria monocytogenesin blue crab (Callinectussapidus) meat and blue crab processing plants. Food Microbiol 31: 263-270.

11. Barbuddhe SB, Malik SVS, Kumar JA, Kalorey DR, Chakraborty T (2012) Epidemiology and risk management of listeriosis in India. Int J Food Microbiol 154: 113-118.

12. Chaturongkasumrit Y, Takahashi H, Keeratipibul S, Kuda T, Kimura B (2011) The effect of polyesterurethane belt surface roughness on Listeria monocytogenes biofilm formation and its cleaning efficiency. Food Cont 22: 1893-1899.

13. Walker C, Philips CA (2008) The effect of preservatives on Alicyclobacillus acidoterrestris and Propionibacterium cyclohexanicum in fruit juice. Food Cont 19: 974-981.

14. Davidson MP, Juneja VK, Branen JK (2002) Antimicrobial agents.In: Food Additives Second Edition Revised and Expanded, Marcel Dekkar, Inc.

15. Mukherjee A, Speh D, Dyck E, Diez-Gonzalez F (2004) Preharvest evaluation of coliforms, Escherichia coli, Salmonella andEscherichia coli O157:H7 in organic and conventional produce grown by Minnesota farmers. J Food Prot 67: 894-900.

16. Salminen S, Von-wright A, Ouwehand A (2004) Lactic acid bacteria. Microbiological and Functional Aspects, Marcel Dekker.

17. Oyetayo VO, Adetuyi FC, Akinyosoye FA (2003) Safety and protective effect of Lactobacillus acidophilus and Lactobacillus caseiused as probiotic agent invivo. Afr J Biotechnol 2: 442-452.

18. Jack RW, Tagg JR, Ray B (1995) Bacteriocins of Gram- positive bacteria. Microbil Rev 59: 171-200.

19. Char C, Guerrero S, Alzamora SM (2009) Survival of Listeria innocua in thermally processed oranje juice as affected by vanillin addition. Food Cont 20: 67-74.

20. Guerrero-Belton, Jose A, Gustavo V, Barbosa-Canovas, Weltichanes J (2011) High hydrostatic pressure effect on natural microflora,Saccharomyces cerevisiae, Escherichia coli and Listeria innocua in navel orange juice. Int J Food Eng

21. Shin MS, Han SK, Ryu JS, Kim KS, Lee WK (2008) Isolation and partial characterization of a bacteriocin produced byPediococcuspentosaceus K 23-2 isolated from Kimchi. J Appl Microbiol 105: 331-339.

22. Perez RH, Himeno K, Ishibashi N, Masuda Y, Zendo T et al. (2012) Monitoring of the multiple bacteriocin production by Enterococcus faecium NKR-5-3 through a developed liquid chromatography and mass spectrometry-based quantification system. J Biosci Bioengin 114: 490-496.

23. Maldonado-Barragán A, Cárdenas N, Martínez B, Ruiz-Barba JL, Fernández-Garayzábal JF et al. (2013) A Novel Class IIdBacteriocin from Lactococcusgarvieaethat inhibits septum formation in L. garvieaeStrains. Appl Environ Microbiol 79: 4336-4346.

24. Keren T, Yarmus M, Halevy G, Shapira R (2004) Immunodetection of the Bacteriocin Lacticin RM: Analysis of the Influence of Temperature and Tween 80 on Its Expression and Activity. Appl Environ Microbiol 70: 2098-2104.

25. Klaenhammer T R (1993) Genetics of bacteriocins produced by lactic acid bacteria. FEMS Microbiol Rev 12: 39-85.

26. Pratush A, Gupta A, Kumar A, Vyas G (2012) Application of purified bacteriocin produced by Lactococcuslactis AP2 as food biopreservative in acidic foods. Ann Food Sci Tech. 13: 82-87.

27. Abriouel H, Valdivia E, Martinez-Bueno M, Maqueda M, Galvez AN (2003) A simple method for semi-preparative-scale production and recovery of enterocin AS-48 derived from Enterococcus faecalis subsp.

Liquefaciens A-48-32. J Microbiol Methods 55: 599-605.

28. Ishibashi N, Himeno K, Fujita K, Masuda Y, Perez R H et al. (2012) Purification and characterization of multiple bacteriocins and an inducing peptide produced by Enterococcus faecium NKR-5-3 from Thai fermented fish. Biosci Biotechnol Biochem 76: 947-953.

29. Pathanibul P, Taylor TM, Davidson PM, Harte F (2009) Inactivation of Escherichia coli and Listeria innocua in apple and carrot juices using high pressure homogenization and nisin. Int J Food Microbial129: 316-320

30. Assous MTM, Khalaf-Allah AM, Sobhy HM, Amani MIH (2012) Inhibition of Bacillus cereus in Fresh Guava-Nectar by Plantaricin and Nisin. World J Dairy Food Sci7: 93-100.

31. Joshi VK, Sharma S, Rana NS (2006) Production, Purification, Stability and Efficacy of Bacteriocin from Isolates of Natural Lactic Acid Fermentation of Vegetables. Food Technol Biotechnol 44: 435-439.

32. Vescovo M, Orsi C, Scolari G, Torriani S (1995) Inhibitory effect of selected lactic acid bacteria on microflora associated with ready-to-use vegetables. LettApplMicrobiol 21: 121-125.

33. Vescovo M, Scolari G, Zacconi C (2006) Inhibition of Listeria innocua growth by antimicrobial-producing lactic acid cultures in vacuum-packed cold-smoked salmon. Food Microbiol 23: 689-693.

34. Ghalfi H, Allaoui A, Destain J, Benkerroum N, Thonart P (2006) Bacteriocin activity by Lactobacillus curvatus CWBI-B28 to inactivateListeria monocytogenes in cold-smoked salmon during 4°C storage. J Food Protect 69: 1066-1071.

35. Ueckert JE, TerSteeg PF, Coote PJ (1998) Synergistic antibacterial action of heat in combination with nisin and magainin II amide. J Appl Microbial 85: 487-494.

36. Beuchat LR, Clavero MR, Jaquette CB (1997) Effects of nisin and temperature on survival, growth, and enterotoxin production characteristics of psychrotrophicBacillus cereus in beef gravy. Appl Environ Microbial 63: 1953-1958.

37. Periago PM, Palop A, Fernandez PS (2001) Combined effect of nisin, carvacrol and thymol on the viability of Bacillus cereus heat-treated vegetative cells. Food Sci Tech Int 7: 487-492.

38. Periago PM, Van Schaik W, Abee T, Wouters JA (2002) Identification of proteins involved in the heat stress response of Bacillus cereus ATCC 14579. Appl Environ Microbiol 68: 3486-3495.

Chapter 10

THE USE OF PLANT ANTIMICROBIAL COMPOUNDS FOR FOOD PRESERVATION

Tana Hintz,[1] Karl K. Matthews,[1] and Rong Di[2]

[1]Department of Food Science, Rutgers, The State University of New Jersey, New Brunswick, NJ 08901, USA

[2]Department of Plant Biology, Rutgers, The State University of New Jersey, New Brunswick, NJ 08901, USA

ABSTRACT

Foodborne disease is a global issue with significant impact on human health. With the growing consumer demand for natural preservatives to replace chemical compounds, plant antimicrobial compounds must be thoroughly investigated for their potential to serve as biopreservatives. This review paper will focus on the plant-derived products as antimicrobial agents for use in food preservation and to control foodborne pathogens in foods. Structure, modes of action, stability, and resistance to these plant compounds will be discussed as well as their application in food industries and possible technologies by which they can be delivered. Benefits as well as challenges, such as the need for further research for implementation and governmental regulation, will be highlighted.

INTRODUCTION

The use of plants for healing dates to prehistory. As early as 60,000 years ago, the Neanderthals, in present-day Iraq, used plants including hollyhock for healing. These plants are still used globally [1]. Hippocrates wrote about several hundred medicinal plants in the late fifth century B.C. and the Bible mentions healing plants, such as frankincense and myrrh, which have antiseptic properties [1]. Plant oils and plant extracts have been utilized for thousands of years, serving many purposes, such as food preservatives and medical therapeutic agents [2]. The compounds that are found in some spices and produced by herbs act as self-defense mechanisms to protect the plant against infectious organisms [3] They are also used by many cultures as flavoring agents and as natural preservatives in food. For example, in foods of India and in traditional Indian medicine, many spices, including garlic, black pepper, cumin, clove, ginger, and caraway, are used [4].

The majority of western plant pharmaceutical information was destroyed during the fall of ancient civilizations, but the Renaissance saw a revival in the use of medicinal plants in the western world [1]. In North America, indigenous cultures have used medicinal plants since prehistory and Americans of European origin began using botanicals in the 19th century to counter the toxic medical practices of that time, such as the use of mercury baths to treat syphilis [1]. Asian culture focuses on the use of herbs to treat diseases and illnesses. Throughout China's history, extensive research was conducted to learn the curative powers of plants. The Imperial Grace Formulary that was compiled in 985 C.E. contains 16,834 herbal entries. Indeed, Chinese medicine reflects traditions developed over 3,000 years and is a holistic approach that takes into account a condition in relation to the whole body in contrast to Western medicine that focuses on a specific cause and attempts to control it. Although Chinese and Western medicine are based on very different philosophies, Chinese herbal medicine has become well-known in the US and in Britain over the last few decades [5].

APPROACHES TO CONTROL BACTERIA: HUMAN HEALTH AND FOOD

Food safety is a global issue with significant implications for human health. The World Health Organization reports that, annually, unsafe food results in the illnesses of at least 2 billion people worldwide and can be deadly. Some countries have made great progress in controlling foodborne diseases, but the number of those affected by foodborne diseases is growing globally (WHO, 2004). In the United States, the Centers for Disease Control and Prevention (CDC) estimate that each year about 1 in 6 Americans becomes ill and thousands die of foodborne diseases (http://www.cdc.gov/foodborneburden/2011-foodborne-estimates.html).

Thermal processing is a common method of destroying vegetative microorganisms to ensure food safety, but this technique may cause undesirable nutritional and quality effects [3]. Preservatives are commonly used to reduce the risk of foodborne illnesses. Increasing regulatory restrictions and consumer negative response to chemical compounds and to the use of antibiotics in agriculture have contributed to the pressure for the development of alternative compounds for use as antimicrobial agents [6]. Antimicrobial agents have been predominantly isolated from bacteria and fungi and either produced through fermentation or produced chemically [1]. In the United States, one-quarter to one-half of pharmaceuticals are derived from plants, but very few are used as antimicrobials. Worldwide, spending on anti-infective agents has increased in recent years due to the limited effective lifespan of antibiotics as new resistant microbes emerge [1]. New sources, including plants, must be thoroughly investigated for identification of novel antimicrobial compounds. For example, it is known that some spices and herbs confer antimicrobial activity. Although there are conflicting reports in the literature about the absolute efficacy of various spices and herbs, Holley and Patel [7] state that the spices, cinnamon, mustard, vanillin, clove, and allspice, and some herbs, specifically oregano, rosemary, thyme, sage, and basil, all confer strong antimicrobial activity. They continue by stating that there are many others that show limited or moderate antimicrobial activity as well. However, Nychas [8] suggests that Gram-positive bacteria are generally more sensitive than Gram-negative bacteria to the antimicrobial compounds of spices.

Alternatives to traditional antimicrobial compounds include bacteriophages, antimicrobial phytochemicals, and antimicrobial peptides. There is less extensive research on the application of antimicrobial compounds from sources other than bacteria. This review paper will focus on the use of plant products as antimicrobial agents, specifically for use in food preservation and safety. Given the consumer demand for natural preservatives and the rapid rate of plant species extinction, it is imperative that more research is focused on the application of plant antimicrobials to food safety.

PLANT ANTIMICROBIAL PEPTIDES (PAMPS)

First discovered in 1942, antimicrobial peptides are produced by bacteria, animals, and plants to serve as natural defense compounds against pathogens. Although generally accepted to be small molecules, there is debate amongst researchers about their exact size. According to Joerger [6], they have a molecular mass between 1 and 5 kDa, but, according to Choon Koo et al. [9] and Garcia-Olmedo et al. [10], they range in size between 2 and 9 kDa. Furthermore, they are predominantly positively charged and are amphiphilic [11]. Although there are many sources of AMPs, this section will focus on plant-derived antimicrobial peptides and their modes of action.

Research clearly demonstrates that antimicrobial peptides are a key part of a plant's defense against pathogens, serving roles both in the defense response upon infection and as part of the preexisting defense barrier. Research has demonstrated that a peptide serves a defense role based on whether peptide-sensitive mutants of pathogens show decreased virulence in plant tissues containing the respective peptide and/or whether overexpression of the peptide results in enhanced tolerance of the plant to the pathogen [10]. Some peptides show specificity towards Gram-negative or Gram-positive bacteria, but most are able to inhibit the activity of both [12]. Plant AMPs are predominantly cysteine-rich compounds that have been isolated from different plant species and from different tissues. A recent review has listed more than twenty different pAMPs [13]. Although there is debate amongst researchers about the number of families, they can be divided into distinct protein families based on structure and amino acid sequence characteristics [9]. These plant antimicrobial peptide families are shown below.

Thionins

The first discovered pAMPs, thionins, are toxic to yeast, fungi, and Gram-positive and Gram-negative bacteria [14]. The example of wheat purothionin-α1 [15] is shown in Table 1 with 45 amino acids.

Table 1: Molecular structures of selected pAMPs.

pAMP	Amino acid (AA) sequence	Plant source	Reference
Thionin-α1	KSCCRSTLGRNCYNLCRARGAQ KLCAGVCRCKISSGLSCPKGFPK	*Triticum aestivum* (wheat)	[15]
IbAMP1	QWGRRCCGWGPGRRYCVRWC	*Impatiens balsamina*	[22]
Lipid transfer protein 2	AITCGQVSSALGPCAAYAKG SGTSPSAGCCSGVKRLAGLA RSTADKQATCRCLKSVAGA YNAGRAAGIPSRCGVSVPY TISASVDCSKIH	*Hordeum vulgare* (barley)	[10]
MBP1	153 AA repeats with 13 AA motif: SGKGTDSGSST(K/Q)D 8 AA motif: GSQGGQGG	*Arabidopsis thaliana*	[17, 25]
Hevein	EQCGRQAGGKLCPNNLCCSQWG WCGSTDEYCSPDHNCQSNCKD	*Hevea brasiliensis*	[26]
Snakin1	GSNFCDSKCKLRCSKAGLADR CLKYCGVCCEECKCVPSGTYG NKHECPCYRDKKNSKGKSKCP	*Solanum tuberosum* (potato)	[27]
Kalata B1	CGETCVGGTCNTPGCTCSWPV CTRNGLPV	*Oldenlandia affinis*	[28]

Table 2: MIC values of antimicrobial agents.

AMP agent	Bacterial target	MIC	Recorded condition	Reference
Potato defensin	L. monocytogenes	>25 µg/mL	24 h at 37°C	[14]
Thionins	L. monocytogenes	2 µg/mL	24 h at 37°C	[14]
Snakin	L. monocytogenes	10 µg/mL	24 h at 37°C	[14]
Lipid transfer protein	L. monocytogenes	no effect	24 h at 37°C	[14]
Potato defensin	L. ivanovii	>25 µg/mL	24 h at 37°C	[14]
Thionins	L. ivanovii	5 µg/mL	24 h at 37°C	[14]
Snakin	L. ivanovii	10 µg/mL	24 h at 37°C	[14]
Lipid transfer protein	L. ivanovii	>25 µg/mL	24 h at 37°C	[14]
Carvacrol	Salm. Typhimurium	1 m/mol	16 h at 37°C	[34]
Cinnamic acid	Salm. Typhimurium	7.5 m/mol	16 h at 37°C	[34]
Diacetyl	Salm. Typhimurium	12.5 m/mol	16 h at 37°C	[34]
Eugenol	Salm. Typhimurium	3.0 m/mol	16 h at 37°C	[34]
Thymol	Salm. Typhimurium	1.0 m/mol	16 h at 37°C	[34]
Carvacrol	E. coli	1.5 m/mol	16 h at 37°C	[34]
Cinnamic acid	E. coli	5.0 m/mol	16 h at 37°C	[34]
Diacetyl	E. coli	7.5 m/mol	16 h at 37°C	[34]
Eugenol	E. coli	2.5 m/mol	16 h at 37°C	[34]
Thymol	E. coli	1.2 m/mol	16 h at 37°C	[34]
Thymol	E. coli	250 ppm	48 h at 37°C	[35]
Carvacrol	E. coli	375 ppm	48 h at 37°C	[35]
Cyclotide CyO2	E. coli	2.2 µM	37°C	[20]
Cyclotide Kalata B1	E. coli	≥100 µM	37°C	[20]
Cyclotide Kalata B2	E. coli	≥35 µM	37°C	[20]
Cyclotide Vaby A	E. coli	32.5 µM	37°C	[20]
Cyclotide Vaby D	E. coli	50 µM	37°C	[20]
Cyclotide CyO2	S. aureus	>50 µM	37°C	[20]
Cyclotide Kalata B1	S. aureus	>100 µM	37°C	[20]
Cyclotide Kalata B2	S. aureus	35 µM	37°C	[20]
Cyclotide Vaby A	S. aureus	>90 µM	37°C	[20]
Cyclotide Vaby D	S. aureus	>90 µM	37°C	[20]
Thymol	S. aureus	250 ppm	48 h at 37°C	[35]
Carvacrol	S. aureus	225 ppm	48 h at 37°C	[35]
Cyclotide CyO2	S. enterica	8.75 µM	37°C	[20]
Cyclotide Kalata B1	S. enterica	>100 µM	37°C	[20]
Cyclotide Kalata B2	S. enterica	>35 µM	37°C	[20]
Cyclotide Vaby A	S. enterica	90 µM	37°C	[20]
Cyclotide Vaby D	S. enterica	>90 µM	37°C	[20]

They are able to induce leakage of intracellular material in bacteria and yeast. Regarding mode of action, it has been shown that they cause cell permeability to isoaminobutyric acid and affect electrical currents in artificial membranes. Purified genetic variants of thionins exhibited differences in activity and some differences in specificity [10]. López-Solanilla et al. [16] used an in vitro method to show that thionins purified from wheat flour have a strong inhibitory effect on multiple strains of Listeria monocytogenes and on a strain of Listeria ivanovii, which is another pathogenic species of the genus Listeria. The MICs (minimum inhibitory concentrations) are listed in Table 2. The authors also studied the effect of temperature on listerial susceptibility to AMPs and determined that a shift from environmental temperature (20°C) to mammalian host temperature (37°C) made L. monocytogenes more sensitive to thionin, but the opposite was shown for L. ivanovii.

Plant Defensins

Structurally related to insect and mammalian defensins [14], plant defensins are able to inhibit bacteria and fungi [10]. The high antifungal activity of plant defensins underscores the significance of fungal pathogens in the plant world, which differs from the high antibacterial activity of animal defensins [14]. Defensins have been identified in locations of first contact and entry by plant pathogens, including peripheral cell layers, xylem, and stomatal cells and in cells lining the substomatal cavity. They have been isolated from tubers, leaves, pods, seeds, and flowers [10]. Defensin gene expression can be developmentally regulated or influenced by external stimuli. Pea, tobacco, radish, and Arabidopsis have defensin genes that are expressed upon pathogen infection.

There are four defensin groups, which are classified by structural properties. Group I inhibits Gram-positive bacteria and fungi, group II inhibits fungi, group III inhibits Gram-positive bacteria and Gram-negative bacteria, and group IV inhibits Gram-positive bacteria, Gram-negative bacteria, and fungi. Importantly, there has been no reported toxicity of plant defensins to animal or plant cells [10] which is very significant from a food safety standpoint should these antimicrobials be leveraged as biopreservatives. The mode of action of antifungal defensins is potentially dependent on electrostatic interactions between hyphal membranes and peptides that cause a rapid Ca^{2+} influx and K^+ efflux [17].

IbAMP1 from the seeds of Impatiens balsamina represents one of the smallest pAMPs with only 20 residues (Table 1) and two disulfide bonds. IbAMP1 has been shown to be active against fungi, Gram-positive bacteria, and Gram-negative bacteria at micromolar levels [18–21]. Wu et al. [22] have demonstrated the concentration-dependent effect of Ib-AMP1 on the cell membrane of Shiga toxin-producing E. coli O157:H7. They showed that Ib-AMP1 exerted its bactericidal activity by interfering with outer and inner membrane integrity permitting efflux of ATP and interfering with intracellular biosynthesis of DNA, RNA, and protein [22].

López-Solanilla et al. [16] used an in vitro method that showed that potato defensin was only weakly inhibitory to L. monocytogenes and L. ivanovii at 37°C (MICs are listed in Table 2). However, at 20°C, the two species were resistant. Potentially, this shows an adaptive technique by L. monocytogenes and L. ivanovii to improve survival when in their primary natural habitat of decaying plant matter-filled soil, which is about 20°C.

Lipid Transfer Proteins (LTPs)

These peptides were once thought to be involved in the transfer of lipids between organelles but have been shown to be involved in plant defense. They seem to have an important role in pathogen defense as well as during low temperature and salt stress. LTPs and defensins can synergistically inhibit fungal and bacterial growth in plants [17]. They are expressed in many areas of a plant, especially in exposed surfaces and in vascular tissues. LTPs have been isolated from barley (Table 1) [10], maize, spinach, Arabidopsis, broccoli, and radish and have demonstrated some specificity [10].

There is limited research on the mode of action of plant LTPs. In vitro research suggests plant LTPs function in plant defense against pathogens based on their ability to inhibit microbial growth. However, there is little direct evidence of the basis of their antimicrobial activity and, unlike other AMPs, they are thought to have many other roles in vivo. Ha-AP10, a LTP, completely inhibits the germination of spores at a concentration of 40 μg/mL. Regente et al. [23] demonstrated that Ha-AP10 acts as a fungicidal compound by not only inducing liposome leakage but also modifying the permeability in Fusarium solani spores. Although other factors may also contribute to the fungicidal activity, the membrane permeabilization mechanism is common to other pAMPs. In addition, the authors demonstrated that Ha-AP10 was able to permeabilize fungal cells in media containing 1 mM $CaCl_2$. This is significant because it more closely represents environmental conditions since the physiological concentration of free Ca^{2+} is 0.1–1 mM in the apoplast, where the fungal-plant contact is likely to occur. Furthermore, the authors demonstrated the selective toxicity of Ha-AP10 for fungal cells over plant (potato host cells). By conducting a follow-up experiment that used model membranes with encapsulated fluorescent probes, the authors hypothesized that this differentiation was due to the composition of phospholipids in the plant and fungal membranes. The activity of Ha-AP10, a cationic and hydrophobic peptide, may therefore be mediated by its electrostatic interaction with anionic membrane phospholipids [23].

Myrosinase-Binding Proteins (MBPs)

Myrosinase (EC 3.2.3.1) is a glucosinolate-degrading enzyme mainly found in the Brassicaceae special idioblasts, myrosin cells [24]. MBPs are involved in plant development and defense activities, primarily against pathogens and insects. The amino acid features of the MBP1 from Arabidopsis are shown in Table 1 [17, 25]. It has been discovered that the potential MBP mode of action is to act as ionophores over microbial membranes [17]. The mechanism of action has been proposed by Capella et al. [25] to be the hydrolysis of glucosinolates by myrosinase enzymes producing molecules with diverse modes of action against fungi, bacteria, and insects.

Hevein- and Knottin-Like Peptides

Both of these peptides inhibit fungi and Gram-positive bacteria in vitro. They have been primarily isolated from seeds, but hevein-like peptides (HTPs) have been found in other tissues as well [10]. Hevein is a small chitin-binding peptide that was initially isolated from rubber latex (Table 1) [26]. This peptide can be used as a fungicide but it may be allergenic, presenting a significant barrier from a safety and labeling standpoint should this peptide be considered as a food preservative. The mechanism of action of HTPs is hyphal penetration that leads to cell burst [17].

Snakins

Snakins are antimicrobial peptides that have been isolated from potatoes (Table 1) [27]. The snakin-1 peptide is active against fungi and Gram-positive and Gram-negative bacteria at concentrations less than $10\,\mu M$ [10]. The peptide is able to aggregate bacteria in vitro but does not mediate leakage or aggregation of artificial liposomes at low- or high-salt concentrations and does not destroy lipid membranes [17]. The mechanism of action of this peptide has yet to be elucidated.

Using an in vitro method, López-Solanilla et al. [16] showed that a snakin peptide had a strong inhibitory effect on L. monocytogenes and L. ivanovii (MICs are listed in Table 2). The Listeria species exhibited differential susceptibility to various pAMPs, which could be potentially linked to the differential fate of Listeria in different areas of the plant.

Cyclotides

Plant cyclotides do not have N- or C-termini because this peptide has a cyclic structure, which serves an important role in the peptide's activity and stability. The structure is a head-to-tail backbone with six conserved cysteine residues, forming a knot motif. They can be classified into two subfamilies: Mobius and bracelet. Mobius cyclotides have a twist formation in the backbone whereas bracelet cyclotides do not have a twist. A third group in the cyclotide family consists of proteinase inhibitors that were isolated from Momordica cochinchinensis, and another cyclotide structure, kalata B8, isolated from Oldenlandia affinis, seems to be a hybrid of the Mobius and bracelet subfamilies [28]. Kalata B1 is listed in Table 1 [28].

These peptides serve various plant defensive roles, including cytotoxicity to plant tumor cells, antiviral and insecticidal activities, and proteinase inhibition. Some also exhibit antibacterial activity, with peptide kalata B1 and circulin A active against Gram-positive bacteria, such as Staphylococcus aureus, and circulin B and cyclopsychotride active against both Gram-positive and Gram-negative bacteria [28]. Pränting et al. [29] conducted research to determine the antibacterial activity of various cyclotides against several Gram-positive and Gram-negative bacterial strains (MIC values are listed in Table 1). From the five evaluated cyclotides, cycloviolacin O2 (cyO2) was determined to be the most potent antibacterial cyclotide, showing high efficacy against Gram-negative bacterial species, including E. coli. This activity is significant given the evidence that other antimicrobial agents, such as nisin, are generally less active against Gram-negative species. However, the mode(s) of action for cyclotides have not yet been elucidated. More research is needed to determine mechanism(s) of action and further biological functions of cyclotides, but these compounds have great potential to serve as novel antibiotics and antiviral therapies to control infectious diseases [28].

Peptides from Hydrolysates

Plant protein hydrolysates can be a source of bioactive peptides [13]. Hydrolysis is either done enzymatically or by acids. Hydrolysates from leguminous plants are particularly favored as they are parts of the food ingredients for many countries in the world [13]. As summarized by Salas et al. [13], the enzymatic hydrolysates from common bean varieties of Phaseolus demonstrated antimicrobial activities against S. aureus and Shigella flexneri with MIC values in the range of 0.1 to 0.99 mg/mL. Another report has shown that the alcalase hydrolysates of rapeseed (Brassica napus) protein inhibited the protease activity of human immunodeficiency virus (HIV) that was expressed in E. coli cells [30].

Some industry by-products represent another source of bioactive peptides that possess antimicrobial properties. For example, the palm kernel expeller (PKE) is produced after palm kernel oil production [31]. Tan et al. tested the efficacy of the purified alcalase- and tryptic-hydrolysates of PKE, PAH, and PTH on Bacillus cereus [31, 32]. It was shown that both PAH and PTH disrupted the membrane integrity of B. cereus, allowing efflux of K^+, depleted the ATP molecules, and inhibited the intracellular macromolecule metabolism especially the RNA of the bacterium.

STRUCTURE OF PAMPS RELATED TO THEIR MODES OF ACTION

Plant AMPs have similar physical properties but diverse primary amino acid sequences. In addition, pAMPs have a range of secondary structures, but there are at least four major themes: loop structures, amphiphilic peptides with two to four β-strands, amphipathic α-helices, and extended structures [11]. However, there are peptides that do not fit into this structure classification, such as many bacterially produced peptides that have two domains, one of which is α-helical and the other of which has a β structure. In addition, there is little scientific literature describing the tertiary structures of pAMPs. However, in silico analyses have shown that pAMPs have similarities in their three-dimensional structures [12].

The antibacterial mode of action for most pAMPs involves cell membranes of targeted organisms and is driven by net positive charge, flexibility, and hydrophobicity to enable interaction with bacterial membranes [11]. Although it was originally thought that the sole mode of action was permeabilization of the bacterial cell membrane, research suggests there may be alternative modes of action or that the pAMPs act upon multiple cell targets. However, interaction with the bacterial cell membrane is critical. There are several models in the literature that illustrate interaction at the cell membrane, each of which uses a different intermediate that leads to either formation of a transient channel, translocation across the membrane, or micellization or dissolution of the membrane. The modes of action are therefore either membrane acting (permeabilizing) or nonmembrane acting (nonpermeabilizing) since translocation does not cause membrane disruption but allows entrance to target essential intracellular processes. In addition, a peptide may target both the cell membrane and intracellular components [11].

The antifungal mode of action was first thought to only involve cell lysis or interference with the synthesis of the fungal cell wall. However, research indicates additional modes of action, including permeabilization, binding to ergosterol/cholesterol in the membrane, depolymerization of the actin cytoskeleton, and targeting intracellular organelles, such as mitochondria [11]. The mode of action of some antifungal peptides is still debated amongst researchers. Plant antimicrobial peptides with predominantly antifungal efficacy tend to be rich in polar and neutral amino acids. The mode of action by plant defensins against fungi has recently been reviewed by Vriens et al. [33].

Antiviral activity is often related to a direct effect on the viral envelope or related to the viral adsorption and entry process [11]. Some antiprotozoal modes of action are similar to antibacterial, antifungal, and antiviral mode of action, such as cell membrane disruption via pore formation or direct interaction with the lipid bilayer. However, there are conflicting reports found in the literature which indicate that antiprotozoal activity may be dependent on peptides that are different from viral, fungal, and bacterial activities [11].

RESISTANCE TO PAMPS

Microbial resistance to antimicrobial agents used in food preservation and sanitation is a major concern. Antibiotic resistance is generally caused by transfer of resistance genes between bacterial cells [34]. Laboratory and clinical studies have determined that resistance to AMPs is less likely than resistance to conventional antibiotics [35]. This is likely due to their membrane-targeting mechanism of action that is more difficult to develop resistance to antibiotics, which generally target macromolecular synthesis (DNA, RNA, and protein) [36]. However, it has been demonstrated that specific genes can confer resistance to pAMPs. For example, thepagP gene increases resistance to the bactericidal effects of some antimicrobial peptides in Salmonella [37]. Changes to the targeted organism's cell membrane may also lead to resistance. It has been difficult to develop resistant strains from previously sensitive strains to particular antimicrobial peptides [6]. This is an ongoing area of research.

It is imperative to note that commensal bacteria, which are beneficial to the host organism, such asLactobacillus in the intestines of humans, are relatively resistant to the action of endogenous antimicrobial peptides [35]. This type of resistance and AMP specificity suggests that pAMPs could potentially be used in food application after further toxicology studies are conducted.

PHYTOCHEMICALS

Phytochemicals are nonnutritive plant components that confer organoleptic properties and serve as antimicrobial agents. The concentration, composition, structure, and functional groups serve an important role in determining antimicrobial activity. Phenolic compounds are generally the most effective [7]. Based on their chemical structures, they may be divided into different categories including simple phenolic compounds, flavonoids, quinones, tannins, and coumarins. The most important phytochemicals used as food preservatives are essential oils, which have been used by humans across the continents since ancient times. Some alkaloids from plants have also been used as antimicrobials in food. Recently, many different phytochemicals have been listed by Negi [38] and their antibacterial activities have been summarized. The antifungal and antifungal toxin activities from various plant extracts including phenolic compounds and essential oils have also been recently reviewed [39]. Polyphenolic compounds from fruits such as cranberry, pomegranate, blueberry, raspberry, and grape were also summarized in 2014 for their antiviral activities against human enteric viruses [40]. Here we briefly review these phytochemicals for their antimicrobial activities in food applications with some examples.

Simple Phenolic Compounds

Simple bioactive phytochemicals are comprised of a single substituted phenolic ring [41]. There is some evidence that the sites and number of hydroxyl groups on the phenolic ring are related to the degree of toxicity to microorganisms, with increased hydroxylation resulting in increased toxicity. It has also been suggested that higher oxidation confers greater inhibition [1]. The mode of action is enzyme inhibition by the oxidized compounds. Phenolic compounds are known to alter microbial cellular permeability, resulting in loss of macromolecules, and interact with membrane proteins, causing structural changes [39]. A simple phenol example is caffeic acid, which is found in thyme and tarragon and is active against fungi, viruses, and bacteria. Eugenol is a phenolic compound found in clove oil that is active against bacteria and fungi [1].

Flavones, Flavonols, and Flavonoids

Flavones are phenolic compounds with one carbonyl. Flavonols are phenolic compounds with a carbonyl and a 3-hydroxyl group. Flavonoids are hydroxylated phenolic structures with a C3–C6 aromatic ring linkage. They are effective against many microorganisms because of their ability to bind to and inactivate proteins and to complex with bacterial cell walls. Catechins provide the antimicrobial activity in oolong teas. The green tea polyphenol, epigallocatechin-3-gallate, was shown to be antiviral against hepatitis B virus replication in vitro[42]. Unlike simple phenolic compounds, the degree of hydroxylation does not predict the level of toxicity to microorganisms [1].

Quinones

Quinones are aromatic rings with two carbonyls, providing a stable source of free radicals. In addition to serving as antioxidants, they are potent antimicrobial compounds. Similar to flavones, flavonols, and flavonoids, the antimicrobial mode of action is to bind to and inactivate proteins. In addition, they may make substrates unavailable to the microorganism [1]. The 6-(4, 7-dihydroxy-heptyl) quinone isolated from the leaves of Pergularia daemia (Forsk.), a traditional medicinal plant, was shown to be effective against several food-contaminating pathogens including Bacillus subtilis, Staphylococcus aureus, and Escherichia coli [43].

Tannins and Coumarins

Tannins are polymeric phenolic substances that are divided into hydrolysable and condensed tannins (also known as proanthocyanidins). The latter are based on flavonoid monomers and hydrolysable tannins are based on gallic acid. Tannins may be formed by polymerization of quinones or by condensation of flavan derivatives. Their antimicrobial mode of action is similar to that of quinones and they have been shown to be toxic to bacteria, yeasts, and some fungi [1]. Tannins naturally occur in many fruits, nuts, and seeds. A recent review by Lipińska et al. [44] shows that the hydrolysable ellagitannins found in pomegranate, strawberry, blackberry, raspberry, walnuts, almonds, and seeds exhibit antimicrobial activity against fungi, viruses, and, importantly, bacteria, including antibiotic-resistant strains such as methicillin-resistant Staphylococcus aureus (MRSA). Additionally, a recent article has comprehensively reviewed the antimicrobial activities of bioactive components from berries including flavonoids (anthocyanins, flavonols, and catechins), phenolic acids, stilbenes, and tannins [45].

Coumarins are phenolic structures comprised of a fused benzene and alpha-pyrone ring. Although toxic to some animals, they have been shown to have species-dependent metabolism, with toxic coumarin derivatives excreted in human urine without adverse health effects. A recent review summarizes not only the anti-inflammatory, anticoagulant, anticancer, antihypertensive, antitubercular, anticonvulsant, antiadipogenic, antihyperglycemic, antioxidant, and neuroprotective properties of coumarins but also their antibacterial, antifungal, and antiviral activities [46].

Essential Oils

Essential oils, or terpenes, are secondary metabolites and are based on an isoprene structure. They are volatile compounds that provide the fragrance of plants and are mainly responsible for the flavor and aroma of spices. Terpenoids contain additional elements, such as oxygen, and confer activity against bacteria, fungi, protozoa, and viruses [1]. Research has shown that essential oils have anti-inflammatory, bactericidal, antiviral, and anticancer effects and possess antioxidant activity [47]. For example, Delaquis et al. [48] determined that the essential oil of cilantro was particularly effective against Listeria monocytogenes, potentially because of long chain alcohols and aldehydes since the antimicrobial properties of alcohols are known to increase with molecular weight. A review by Seow et al. [49] has included 47 different essential oils as antimicrobials. The antibacterial and antifungal activities of these essential oils have been listed. As essential oils contain highly diverse groups of phytochemicals, their antimicrobial modes of action have been suggested to involve multiple targets. The unique hydrophobicity features of essential oils render their abilities to react with lipids on the bacterial cell membranes, increasing the membrane permeability and disturbing the original cell structure [50,51]. The clover essential oil has been shown to cause an extensive lesion of fungal cell membrane [52]. Essential oil has also been shown to inhibit viral protein synthesis at multiple stages of viral infection and replication [53].

Alkaloids

Alkaloids are heterocyclic nitrogen compounds and have demonstrated limited microbicidal activity as well as possessing an antidiarrheal activity. An example of an alkaloid is berberine [1]. Berberine is the main antibacterial substance of rhizoma Coptidis (Coptis chinensis Franch) and cortex Phellodendri (Phellodendron amurense Ruprecht) [54]. The MICs of berberine against methicillin-resistant Staphylococcus aureus (MRSA) bacteria ranged from 32 to 128 µg/mL. Ninety percent inhibition of MRSA was obtained with 64 µg/mL or less of berberine.

STABILITY OF PLANT ANTIMICROBIAL COMPOUNDS

The effect of food processing on plant-derived antimicrobial compounds must be evaluated. Plant-derived foods are often exposed to acidic or alkali conditions or high heat during processing to destroy microorganisms to enhance microbial food safety. The aforementioned conditions are used to peel fruits and vegetables and to recover proteins from cereals and legumes [55]. However, these conditions may destroy natural plant antimicrobial compounds, which serve as defense mechanisms to the plant and have potential to serve as natural antimicrobial compounds against human pathogens.

Research by Friedman and Jürgens [55] indicated that the chemical structure of phenolic compounds has a significant effect on their susceptibility to destruction at alkali conditions. By using ultraviolet spectroscopy, they determined that gallic acid, caffeic acid, and chlorogenic acid were not stable in high pH and that the spectral transformations were not reversible when the pH was reduced back to neutral conditions. The phenolic OH groups were hypothesized to be primarily responsible for the spectral changes since ferulic acid (with one OH group) was more stable in high pH versus caffeic acid (with two OH groups) and gallic acid (with three OH groups). Furthermore, the ionized and resonance forms of multiring structures conferred more resistance to high pH versus monoring structures. Therefore, multiring catechin, epigallocatechin, and rutin had less spectral transformations at high pH conditions versus monoring gallic acid, caffeic acid, and chlorogenic acid. This research provides a foundation that suggests structural elements that may confer stability at alkali conditions. It is critical to determine whether natural antimicrobial compounds are stable in food processing conditions in order to determine their feasibility as alternative antimicrobial compounds to foodborne pathogens in food systems.

In addition to this foundational research under artificial conditions, it is critical for natural antimicrobials to be tested in food systems. For example, solubility and food constituents, such as proteins and lipids, could potentially impact efficacy and stability of plant-derived antimicrobial compounds. Research conducted by Aureli et al. [56] indicated that thyme essential oil reduced viable Listeria monocytogenes L28 cells in a meat matrix but noted that there was a decreased efficacy in the food system compared to in vitro testing. Pränting et al. [29] reported that the antibacterial efficacy of the pAMP cyclotide cyO2 was reduced in media containing salt, which suggests that this peptide would be less efficacious as a preservative in a high-salt food. The authors do note, however, that even though some research has indicated similar influence of media composition on antibacterial efficacy of AMPs, the AMPs were still active against bacteria in biological systems. It is essential that the antimicrobial compounds effectively reduce pathogenic bacteria to allowable limits or completely inactivate pathogens.

Because pasteurization can affect organoleptic and nutritional properties and increase processing costs and postpasteurization contamination can occur, Friedman and Jürgens [55] conducted a study to assess the stability of a naturally occurring antimicrobial, chlorogenic acid, in apple juice. The authors determined that the phenolic compound, chlorogenic acid, was stable in the low pH, heat treatment, and storage of apple juice, offering a promising candidate to combat contamination by E. coli O157:H7 and Salmonella Typhimurium of nonfermented apple juice products. Payne et al. [57] determined that the phenolic compounds propyl paraben (propyl ester of p-hydroxybenzoic acid) and tertiary butylhydroquinone (TBHQ) were significantly more effective than potassium sorbate, a commonly used antimicrobial, against Listeria monocytogenes in a model milk system containing 10% nonfat milk solids at

35°C. However, TBHQ was inconsistent in its activity. Although the authors suggest that this study indicates that inhibition would be achieved at refrigeration temperatures, it seems that there may have been an error in conversion of Celsius to Fahrenheit, since 35°C is much greater than refrigeration temperatures. The authors do raise a valid point because it is important to conduct studies at the proper storage temperature of the food product in order to mimic normal food shelf conditions and because temperature greatly affects the survival rate of foodborne pathogens.

APPLICATIONS OF PLANT ANTIMICROBIAL COMPOUNDS IN FOOD INDUSTRY

For the food industry, plant antimicrobial compounds have potential use as biopreservatives and bioinsecticides, with potential use for development of genetically modified crop plants with increased disease resistance as well as use against foodborne pathogens. In fact, research has already shown the effectiveness of plant compounds against virulent foodborne pathogens. For example, essential oils have been used to help control Listeria monocytogenes. Aureli et al. [56] used a paper disc diffusion method to demonstrate that 12 out of 32 essential oils were active against Listeria monocytogenes, with clove, cinnamon, pimento, origanum, and thyme showing the greatest inhibition. However, the application of plant antimicrobial compounds for controlling growth of foodborne pathogens must incorporate the range of activity against the microorganisms of concern associated with a particular product. For example, essential oils are typically more effective against Gram-positive than Gram-negative bacteria. But some, such as clove and cinnamon, have been shown to be effective against both [3]. Olasupo et al. [58] demonstrated that 5 natural organic compounds were effective against the Gram-negative bacteria E. coli and Salmonella Typhimurium with efficacy in the following order: thymol > carvacrol > eugenol > cinnamic acid > diacetyl. Table 1 shows the MIC values of various antimicrobial agents against pathogenic bacterial targets at various conditions. This table illustrates that the efficacy of natural antimicrobials is impacted by environmental conditions and antimicrobial agents have a wide range of activity against bacterial targets. This information must be fully verified for a particular food product when considering the use of plant-derived antimicrobials in the food product. Although this review focused on efficacy against pathogenic organisms, it should be noted that the common spoilage bacteria,Pseudomonas, are generally resistant to plant antimicrobials due to the production of exopolysaccharide layers that offer protection and delay penetration of antimicrobials [3].

Another potential use of antimicrobial peptides in the food industry is replacement of antibiotics used in animal production to increase feed efficiency. Although there are some conflicting reports in the literature, Jin et al. [59] determined that pigs fed increasing levels (0 to 600 ppm) of refined potato protein (RPP) from Solanum tuberosum L. cv. Gogu Valley demonstrated linear improvements in performance and a linear decrease in fecal and intestinal bacteria, suggesting that RPP was effective at higher levels. Since the potato tubers of Gogu Valley are known to contain the AMP potamin-1 (PT-1), the authors suggest that the AMP may have caused the decrease in microbe numbers, contributing towards increased performance. Given the linear effects demonstrated with increasing levels of RPP in the pigs' diets, future studies that evaluate higher levels of RPP in pigs (as well as other production animals) should be completed to more fully elucidate the potential of RPP as an alternative to antibiotics. Furthermore, other AMPs should be evaluated as potential performance enhancers and chemical modification or encapsulation methods should be evaluated to prevent degradation by proteolytic enzymes within the digestive system of production animals [6]. However, it is known that cyclotides are stable against proteases, such as pepsin and trypsin, and their cyclic structure confers protection against exopeptidases [29].

TECHNOLOGIES BY WHICH PLANT ANTIMICROBIALS CAN BE DELIVERED

There are various methods by which plant antimicrobials could be delivered. The most suitable methods must incorporate cost-effectiveness, stability, and efficacy of the compound under processing, transportation, and storage conditions. A recent technology incorporates antimicrobials into packaging materials instead of the food itself. This technique offers the advantage of concentrating the antimicrobial at the surface of the food product, which is where potential pathogens grow, and reduces obstruction from food particles [7].

Microencapsulated antimicrobial agents incorporated in food packaging have been demonstrated to successfully destroy a range of microorganisms, offering a controlled-release preservation technique. This application is a type of active packaging (AP), in which the conditions of the packaged food are changed to better preserve the sensory attributes, safety, and shelf-life of the product. Since microbial contact primarily occurs on the surface of a packaged food, antimicrobial activity should be focused on solid or semisolid surfaces by either indirect contact using antimicrobial volatiles or via direct contact between the antimicrobial package and the food surface. Research has successfully incorporated antimicrobial peptides, such as bacteriocins, as well as phytochemicals, such as essential oils, into packaging materials [60].

Guarda et al. [60] demonstrated the antimicrobial activity of microencapsulated carvacrol and its isomer thymol, which are phenolic compounds that are major components of essential oils with known antimicrobial activity. By creating emulsions with the compounds and applying them to a polymer film, the authors demonstrated their antimicrobial activity by the agar plate method. The minimal inhibitory concentration (MIC) of thymol was 125–250 ppm and the MIC of carvacrol was 75–375 ppm against both pathogenic and nonpathogenic microorganisms: Escherichia coli O157:H7, Staphylococcus aureus, Listeria innocua, Saccharomyces cerevisiae, and Aspergillus niger. The highest synergism was determined to be 50% of each compound and various concentrations were studied to determine the required concentration for the most resistant microorganism, E. coli, based on zones of inhibition. The authors found similar findings to others in the literature that Gram-positive bacteria were more sensitive than Gram-negative bacteria, such as E. coli, potentially due to the outer membrane that surrounds the lipopolysaccharide wall inhibiting the penetration of the hydrophobic phenolic compounds. The required concentration of the antimicrobial agent to confer antimicrobial activity therefore depends on the type of targeted organism in a given food product. Furthermore, it is important to note that loss of antimicrobial activity may occur during preparation of the packaging. Lower losses were observed in the coating process utilized [60] than high-temperature processes, such as extrusion blow-molding. The authors noted that the release rate of the microcapsules was lower and more controlled as compared to films with antimicrobial agents directly incorporated into the matrix, potentially due to the affinity of the carrier, gum arabic, to a polar compound and the good film forming capability of gum arabic. This research provides a strong foundation for the use of these GRAS approved materials in active packaging, but more research is required on the use of this packaging on food systems.

Chitosan is a hydrophilic polymer that is obtained commercially by N-deacetylation of chitin. Chitosan exhibits some antimicrobial activity against fungi, algae, and some bacteria. Antimicrobial efficacy is influenced by the type of chitosan, molecular weight, and environmental conditions. Chitosan is limited by its insolubility in water, high viscosity, tendency to coagulate with proteins, and poor solubility at high pH. However, water-soluble salts can be formed by neutralization with acids. The exact antimicrobial action of chitosan is debated based on review of the literature, but it has been suggested that chitosan interacts with negatively charged microbial cell membranes, leading to cellular leakage. The polymer also acts as a chelating agent that binds trace metals, inhibiting toxin production and microbial growth. Research is also being conducted to determine its ability to elicit natural plant defenses when applied to plant tissues or cultured plant cells [61].

Chitosan is a biopolymer that is safe for human consumption and has several effective delivery methods, including the use as a seed treatment and as an edible antifungal coating material for postharvest produce. Chitosan films are semipermeable, durable, long-lasting, natural, inexpensive, and nontoxic and have been successfully used to delay decay of various fruits and vegetables potentially due to decreased rates of respiration, delay of ripening from the reduction of ethylene and carbon dioxide production, and fungal inhibition. Additional antimicrobial agents can be applied to chitosan films so that they serve as an active type of packaging, releasing the biopreservatives in a controlled manner onto the food surface to inhibit microbial growth. N,O-Carboxymethyl chitin films have been approved for use in fruits in both USA and Canada. Further research is needed to determine novel derivatives of chitosan with increased antimicrobial activity [61].

Plant antimicrobials can be delivered via plant extracts or consumed whole. The literature generally cites that spice extracts are less antimicrobial than the whole spice. For example, Lachowicz et al. [62] used an agar well diffusion method to demonstrate that essential oils extracted from five different varieties of Ocimum basilicumL. plants showed equivalent or better antimicrobial activity against a range of Gram-positive and Gram-negative bacteria, yeasts, and molds compared to the purified components linalool and methyl chavicol either separately or together. In contrast, Delaquis et al. [48] determined that fractions of dill and cilantro oil had greater antimicrobial activity than the whole oil. The latter authors aptly suggest that essential oils, which are naturally variable in composition, could more reliably be used as preservative agents if the concentrations of antimicrobial components could be adjusted to consistent levels for the needed spectrum and strength of microbial inhibition. More research is required in this area.

REGULATION OF PLANT-DERIVED ANTIMICROBIALS FOR FOOD APPLICATION

The exact number of plant-derived antimicrobials approved for food application globally is difficult to discern due to the limited amount of information available. Food additives are closely controlled by legislation worldwide, but there is little agreement between countries regarding food additives that are safe, permitted concentrations, and specific permitted food uses [63]. In the US, the Food and Drug Administration (FDA) evaluates the safety of unapproved food additives to determine whether they should be approved. The evaluation includes the amount of the substance that would normally be consumed as well as short- and long-term health effects and other safety considerations. When a food additive is approved, the FDA issues a regulation that may include the types of foods in which it can be used, the maximum amount allowed, and its proper identification on food labels. These regulations are published in the Title 21 of the Code of Federal Regulations. According to the FDA's Guidance for Industry: Antimicrobial Food Additives, the FDA has regulatory authority over food additives. However, the Environmental Protection Agency (EPA) sets tolerances for pesticide chemicals and pesticide chemical residues in or on foods, which are enforced by the FDA. It should be noted that antimicrobials applied to, or included in, food packaging materials are excluded from the definition of "pesticide chemical" and thus are regulated as food additives by FDA.

Although regulation varies worldwide, other countries have similar departments or agencies in place to evaluate the safety and provide guidance and regulation on food additives. In Europe, there are relatively few compounds that are allowed as food preservatives, and these are primarily organic acids [63]. There is a strict protocol in place in order for food additives to be approved for human consumption. When applying for authorization of a new food additive, an applicant submits a formal request to the European Commission, which is the executive body of the European Union, and includes information on the substance, including scientific data concerning safety. Upon acceptance of the application, the Commission requests the European Food Safety Authority (EFSA) to issue an opinion on the safety of the substance for its intended uses. In addition to carrying out safety evaluations of new food additives before they can be authorized for use in the European Union (EU), EFSA reviews certain food additives that have new scientific information and/or changing conditions (http://www.efsa.europa.eu/en/topics/topic/additives.htm). Similarly, the Chinese government employs the China Food Additives Association (CFAA) as the only registered nationwide food additive and ingredients industry organization to make evaluations and assist the government in making regulations in the food and food additives industry (http://www.cfaa.cn/english.htm).

Given the strict regulation and the relatively limited in vivo research on pAMPs, a literature review and a search of the web pages of the governmental agencies listed above yielded no known approved plant AMPs in those countries for food application. The agencies are constantly evaluating new potential food additives, including biopreservatives, so regulation of plant-derived antimicrobials can be expected to be seen in the future.

CONCLUSION

Food processing and some preservation techniques, such as heating, may alter food's nutritional or organoleptic properties. Microbial resistance to current antimicrobial compounds has increased in recent years worldwide; therefore, alternative compounds must be investigated and developed. This review has cited many of the benefits and lists some of the current hurdles of implementing plant-derived antimicrobials in food application. Most significantly, despite extensive in vitro research of plant-derived antimicrobials, there are limited in vivo studies, yielding knowledge about the toxicology of the extensive repertoire of compounds. Future toxicology research will aid the governmental food safety agencies in their evaluation and regulation of these compounds for food application.

Because of variation in stability and efficacy to various food processing parameters and food systems, it is critical that plant-derived antimicrobials be selected and delivered so that they are active against potential pathogens in particular food and are stable throughout the food's shelf life. Effects of these compounds in combination with other compounds or techniques must be more thoroughly investigated. For example, hevein- and knottin-like peptides are active against Gram-positive bacteria and fungi. However, research should be conducted to verify if they would also be active against Gram-negative bacteria if chelators that perturb the outer membrane of Gram-negative bacteria were also present. In addition to these scientific evaluations, sensory studies must be conducted to ensure that the organoleptic properties of a food are not impacted by the natural antimicrobial peptides.

Given the consumer demand for more natural products and the growing need for alternative preservatives to ensure food safety, it is imperative that plant-derived antimicrobial compounds be fully assessed for their feasibility for food application. This new field of research has great potential for more evaluation to meet regulatory requirements and to fully elucidate the possibility of employing antimicrobials from the extensive source of plants worldwide.

REFERENCES

1. M. M. Cowan, "Plant products as antimicrobial agents," Clinical Microbiology Reviews, vol. 12, no. 4, pp. 564–582, 1999. ·

2. K. A. Hammer, C. F. Carson, and T. V. Riley, "Antimicrobial activity of essential oils and other plant extracts," Journal of Applied Microbiology, vol. 86, no. 6, pp. 985–990, 1999.

3. B. K. Tiwari, V. P. Valdramidis, C. P. O›Donnell, K. Muthukumarappan, P. Bourke, and P. J. Cullen, "Application of natural antimicrobials for

food preservation," Journal of Agricultural and Food Chemistry, vol. 57, no. 14, pp. 5987–6000, 2009.

4. D. S. Arora and J. Kaur, "Antimicrobial activity of spices," International Journal of Antimicrobial Agents, vol. 12, no. 3, pp. 257–262, 1999.

5. E. R. Rogans, Chinese Herbal Medicine: A Step-by-Step Guide, Element Books, Rockport, Mass, USA, 1997.

6. R. D. Joerger, "Alternatives to antibiotics: bacteriocins, antimicrobial peptides and bacteriophages,"Poultry Science, vol. 82, no. 4, pp. 640–647, 2003.

7. R. A. Holley and D. Patel, "Improvement in shelf-life and safety of perishable foods by plant essential oils and smoke antimicrobials," Food Microbiology, vol. 22, no. 4, pp. 273–292, 2005.

8. G. J. E. Nychas, Natural Antimicrobials from Plants, Blackie Academic and Professional, Glasgow, UK, 1995.

9. J. Choon Koo, H. Jin Chun, H. Cheol Park et al., "Over-expression of a seed specific hevein-like antimicrobial peptide from Pharbitis nil enhances resistance to a fungal pathogen in transgenic tobacco plants," Plant Molecular Biology, vol. 50, no. 3, pp. 441–452, 2002.

10. F. Garcia-Olmedo, A. Molina, J. M. Alamillo, and P. Rodriguez-Palenzuela, "Plant defense peptides,"Biopolymers, vol. 47, no. 6, pp. 479–491, 1998. ·

11. H. Jenssen, P. Hamill, and R. E. W. Hancock, "Peptide antimicrobial agents," Clinical Microbiology Reviews, vol. 19, no. 3, pp. 491–511, 2006.

12. P. B. Pelegrini, R. P. Del Sarto, O. N. Silva, O. L. Franco, and M. F. Grossi-de-Sa, "Antibacterial peptides from plants: what they are and how they probably work," Biochemistry Research International, vol. 2011, Article ID 250349, 9 pages, 2011.

13. C. E. Salas, J. A. Badillo-Corona, G. Ramírez-Sotelo, and C. Oliver-Salvador, "Biologically active and antimicrobial peptides from plants," BioMed Research International, vol. 2015, Article ID 102129, 11 pages, 2015. ·

14. R. E. W. Hancock and D. S. Chapple, "Peptide antibiotics," Antimicrobial Agents and Chemotherapy, vol. 43, no. 6, pp. 1317–1323, 1999. ·

15. F. J. Colilla, A. Rocher, and E. Mendez, "γ-Purothionins: amino acid sequence of two polypeptides of a new family of thionins from wheat endosperm," FEBS Letters, vol. 270, no. 1-2, pp. 191–194, 1990.

16. E. López-Solanilla, B. González-Zorn, S. Novella, J. A. Vázquez-Boland,

and P. Rodríguez-Palenzuela, "Susceptibility of Listeria monocytogenes to antimicrobial peptides," FEMS Microbiology Letters, vol. 226, no. 1, pp. 101–105, 2003.

17. L. S. Tavares, M. D. O. Santos, L. F. Viccini, J. S. Moreira, R. N. G. Miller, and O. L. Franco, "Biotechnological potential of antimicrobial peptides from flowers," Peptides, vol. 29, no. 10, pp. 1842–1851, 2008.

18. R. H. Tailor, D. P. Acland, S. Attenborough et al., "A novel family of small cysteine-rich antimicrobial peptides from seed of Impatiens balsamina is derived from a single precursor protein," The Journal of Biological Chemistry, vol. 272, no. 39, pp. 24480–24487, 1997. ··

19. D. G. Lee, S. Y. Shin, D.-H. Kim et al., "Antifungal mechanism of a cysteine-rich antimicrobial peptide, Ib-AMP1, from Impatiens balsamina against Candida albicans," Biotechnology Letters, vol. 21, no. 12, pp. 1047–1050, 1999.

20. K. Thevissen, I. E. J. A. François, L. Sijtsma et al., "Antifungal activity of synthetic peptides derived fromImpatiens balsamina antimicrobial peptides Ib-AMP1 and Ib-AMP4," Peptides, vol. 26, no. 7, pp. 1113–1119, 2005.

21. K. Matsuzaki, "Control of cell selectivity of antimicrobial peptides," Biochimica et Biophysica Acta—Biomembranes, vol. 1788, no. 8, pp. 1687–1692, 2009.

22. W.-H. Wu, R. Di, and K. R. Matthews, "Antibacterial mode of action of Ib-AMP1 against Escherichia coliO157:H7," Probiotics and Antimicrobial Proteins, vol. 5, no. 2, pp. 131–141, 2013. ··

23. M. C. Regente, A. M. Giudici, J. Villalaín, and L. de la Canal, "The cytotoxic properties of a plant lipid transfer protein involve membrane permeabilization of target cells," Letters in Applied Microbiology, vol. 40, no. 3, pp. 183–189, 2005.

24. E. Andréasson, L. B. Jørgensen, A.-S. Höglund, L. Rask, and J. Meijer, "Different myrosinase and idioblast distribution in Arabidopsis and Brassica napus," Plant Physiology, vol. 127, no. 4, pp. 1750–1763, 2001.

25. Capella, M. Menossi, P. Arruda, and C. Benedetti, "COI1 affects myrosinase activity and controls the expression of two flower-specific myrosinase-binding protein homologues in Arabidopsis," Planta, vol. 213, no. 5, pp. 691–699, 2001.

26. Broekaert, H.-I. Lee, A. Kush, N.-H. Chua, and N. Raikhel, "Wound-induced accumulation of mRNA containing a hevein sequence in laticifers of rubber tree (Hevea brasiliensis)," Proceedings of the National

Academy of Sciences of the United States of America, vol. 87, no. 19, pp. 7633–7637, 1990.

27. Segura, M. Moreno, F. Madueño, A. Molina, and F. García-Olmedo, "Snakin-1, a peptide from potato that is active against plant pathogens," Molecular Plant-Microbe Interactions, vol. 12, no. 1, pp. 16–23, 1999.

28. P. B. Pelegrini, B. F. Quirino, and O. L. Franco, "Plant cyclotides: an unusual class of defense compounds,"Peptides, vol. 28, no. 7, pp. 1475–1481, 2007.

29. M. Pränting, C. Lööv, R. Burman, U. Göransson, and D. I. Andersson, "The cyclotide cycloviolacin O_2 from Viola odorata has potent bactericidal activity against Gram-negative bacteria," Journal of Antimicrobial Chemotherapy, vol. 65, no. 9, Article ID dkq220, pp. 1964–1971, 2010. ··

30. M. D. M. Yust, J. Pedroche, C. Megías et al., "Rapeseed protein hydrolysates: a source of HIV protease peptide inhibitors," Food Chemistry, vol. 87, no. 3, pp. 387–392, 2004.

31. Y. N. Tan, K. R. Matthews, R. Di, and M. K. Ayob, "Bacteriostatic mode of action of trypsin-hydrolyzed palm kernel expeller peptide against Bacillus cereus," Probiotics and Antimicrobial Proteins, vol. 4, no. 1, pp. 59–65, 2012.

32. Y. N. Tan, K. R. Matthews, R. Di, and M. K. Ayob, "Comparative antibacterial mode of action of purified alcalase- and tryptic-hydrolyzed palm kernel cake proteins on the food-borne pathogen Bacillus cereus,"Food Control, vol. 31, no. 1, pp. 53–58, 2013.

33. K. Vriens, B. P. A. Cammue, and K. Thevissen, "Antifungal plant defensins: mechanisms of action and production," Molecules, vol. 19, no. 8, pp. 12280–12303, 2014. ··

34. J. Cleveland, T. J. Montville, I. F. Nes, and M. L. Chikindas, "Bacteriocins: safe, natural antimicrobials for food preservation," International Journal of Food Microbiology, vol. 71, no. 1, pp. 1–20, 2001.

35. M. Zasloff, "Antimicrobial peptides of multicellular organisms," Nature, vol. 415, no. 6870, pp. 389–395, 2002.

36. Y. Sang and F. Blecha, "Antimicrobial peptides and bacteriocins: alternatives to traditional antibiotics,"Animal Health Research Reviews, vol. 9, no. 2, pp. 227–235, 2008.

37. M. Robey, W. O›Connell, and N. P. Cianciotto, "Identification of Legionella pneumophila rcp, a pagP-like gene that confers resistance to cationic antimicrobial peptides and promotes intracellular infection,"Infection

and Immunity, vol. 69, no. 7, pp. 4276–4286, 2001. · ·

38. P. S. Negi, "Plant extracts for the control of bacterial growth: efficacy, stability and safety issues for food application," International Journal of Food Microbiology, vol. 156, no. 1, pp. 7–17, 2012.

39. L. D. C. Cabral, V. F. Pinto, and A. Patriarca, "Application of plant derived compounds to control fungal spoilage and mycotoxin production in foods," International Journal of Food Microbiology, vol. 166, no. 1, pp. 1–14, 2013.

40. D. H. D›Souza, "Phytocompounds for the control of human enteric viruses," Current Opinion in Virology, vol. 4, pp. 44–49, 2014.

41. M. Galal, "Natural product-based phenolic and nonphenolic antimicrobial food preservatives and 1,2,3,4-tetrahydroxybenzene as a highly effective representative: a review of patent literature 2000–2005,"Recent patents on anti-infective drug discovery, vol. 1, no. 2, pp. 231–239, 2006.

42. J.-Y. Pang, K.-J. Zhao, J.-B. Wang, Z.-J. Ma, and X.-H. Xiao, "Green tea polyphenol, epigallocatechin-3-gallate, possesses the antiviral activity necessary to fight against the hepatitis B virus replication in vitro,"Journal of Zhejiang University: Science B, vol. 15, no. 6, pp. 533–539, 2014.

43. S. Ignacimuthu, M. Pavunraj, V. Duraipandiyan, N. Raja, and C. Muthu, "Antibaccterial activity of a novel quinone from the leaves of Pergularia daemia (Forsk.), a traditional medicinal plant," Asian Journal of Traditional Medicines, vol. 4, no. 1, pp. 36–40, 2009.

44. L. Lipińska, E. Klewicka, and M. Sójka, "Structure, occurrence and biological activity of ellagitannins: a general review," Acta Scientiarum Polonorum, Technologia Alimentaria, vol. 13, no. 3, pp. 289–299, 2014.

45. S. H. Nile and S. W. Park, "Edible berries: bioactive components and their effect on human health,"Nutrition, vol. 30, no. 2, pp. 134–144, 2014.

46. K. N. Venugopala, V. Rashmi, and B. Odhav, "Review on natural coumarin lead compounds for their pharmacological activity," BioMed Research International, vol. 2013, Article ID 963248, 14 pages, 2013.

47. T. Aburjai and F. M. Natsheh, "Plants used in cosmetics," Phytotherapy Research, vol. 17, no. 9, pp. 987–1000, 2003.

48. P. J. Delaquis, K. Stanich, B. Girard, and G. Mazza, "Antimicrobial activity of individual and mixed fractions of dill, cilantro, coriander and eucalyptus essential oils," International Journal of Food Microbiology, vol. 74, no. 1-2, pp. 101–109, 2002.

49. Y. X. Seow, C. R. Yeo, H. L. Chung, and H.-G. Yuk, "Plant essential oils as active antimicrobial agents,"Critical Reviews in Food Science and Nutrition, vol. 54, no. 5, pp. 625–644, 2014.

50. K. Knobloch, N. Weis, and H. Weigand, "Mechanism of antimicrobial activity of essential oils," Planta Medica, vol. 52, no. 6, pp. 556–556, 1986. ·

51. J. Sikkema, J. A. M. de Bont, and B. Poolman, "Interactions of cyclic hydrocarbons with biological membranes," The Journal of Biological Chemistry, vol. 269, no. 11, pp. 8022–8028, 1994. ·

52. E. Pinto, L. Vale-Silva, C. Cavaleiro, and L. Salgueiro, "Antifungal activity of the clove essential oil fromSyzygium aromaticum on Candida, Aspergillus and dermatophyte species," Journal of Medical Microbiology, vol. 58, no. 11, pp. 1454–1462, 2009.

53. S. Wu, K. B. Patel, L. J. Booth, J. P. Metcalf, H.-K. Lin, and W. Wu, "Protective essential oil attenuates influenza virus infection: an in vitro study in MDCK cells," BMC Complementary and Alternative Medicine, vol. 10, article 69, 2010.

54. H.-H. Yu, K.-J. Kim, J.-D. Cha et al., "Antimicrobial activity of berberine alone and in combination with ampicillin or oxacillin against methicillin-resistant Staphylococcus aureus," Journal of Medicinal Food, vol. 8, no. 4, pp. 454–461, 2005.

55. M. Friedman and H. S. Jürgens, "Effect of pH on the stability of plant phenolic compounds," Journal of Agricultural and Food Chemistry, vol. 48, no. 6, pp. 2101–2110, 2000.

56. P. Aureli, A. Costantine, and S. Zolea, "Antimicrobial activity of some essential oils against Listeria monocytogenes," Journal of Food Protection, vol. 55, no. 5, pp. 344–348, 1992.

57. K. D. Payne, E. Rico-Munoz, and P. M. Davidson, "The antimicrobial activity of phenolic compounds against Listeria monocytogenes and their effectiveness in a model milk system," Journal of Food Protection, vol. 52, no. 3, pp. 151–153, 1989.

58. N. A. Olasupo, D. J. Fitzgerald, M. J. Gasson, and A. Narbad, "Activity of natural antimicrobial compounds against Escherichia coli and Salmonella enterica serovar Typhimurium," Letters in Applied Microbiology, vol. 37, no. 6, pp. 448–451, 2003.

59. Z. Jin, P. L. Shinde, Y. X. Yang et al., "Use of refined potato (Solanum tuberosum L. cv. Gogu valley) protein as an alternative to antibiotics in weanling pigs," Livestock Science, vol. 124, no. 1–3, pp. 26–32, 2009.

60. Guarda, J. F. Rubilar, J. Miltz, and M. J. Galotto, "The antimicrobial activity of microencapsulated thymol and carvacrol," International Journal of Food Microbiology, vol. 146, no. 2, pp. 144–150, 2011.

61. E. I. Rabea, M. E.-T. Badawy, C. V. Stevens, G. Smagghe, and W. Steurbaut, "Chitosan as antimicrobial agent: applications and mode of action," Biomacromolecules, vol. 4, no. 6, pp. 1457–1465, 2003.

62. K. J. Lachowicz, G. P. Jones, D. R. Briggs et al., "The synergistic preservative effects of the essential oils of sweet basil (Ocimum basilicum L.) against acid-tolerant food microflora," Letters in Applied Microbiology, vol. 26, no. 3, pp. 209–214, 1998.

63. M. Stratford and T. Eklund, Organic Acids and Esters, Kluwer Academic/ Plenum Publishers, New York, NY, USA, 2003.

Chapter 11

ANTIMICROBIAL EDIBLE FILMS AND COATINGS FOR MEAT AND MEAT PRODUCTS PRESERVATION

Irais Sánchez-Ortega,[1,2] Blanca E. García-Almendárez,[1] Eva María Santos-López,[2] Aldo Amaro-Reyes,[1] J. Eleazar Barboza-Corona,[3] and Carlos Regalado[1]

[1]DIPA, PROPAC, Facultad de Química, Universidad Autónoma de Querétaro, 76010 Querétaro, QRO, Mexico

[2]Área Académica de Química, Instituto de Ciencias Básicas e Ingeniería, Universidad Autónoma del Estado de Hidalgo, Ciudad del Conocimiento, Carr. Pachuca-Tulancingo Km 4.5 Col Carboneras, 42184 Mineral de la Reforma, HGO, Mexico

[3]División Ciencias de la Vida, Universidad de Guanajuato, Campus Irapuato-Salamanca, 36500 Irapuato, GTO, Mexico

ABSTRACT

Animal origin foods are widely distributed and consumed around the world due to their high nutrients availability but may also provide a suitable environment for growth of pathogenic and spoilage microorganisms. Nowadays consumers demand high quality food with an extended shelf life without chemical additives. Edible films and coatings (EFC) added with natural antimicrobials are a promising preservation technology for raw and processed meats because they provide good barrier against spoilage and pathogenic microorganisms. This review gathers updated research reported over the last ten years related to antimicrobial EFC applied to meat and meat products. In addition, the films gas barrier properties contribute to extended shelf life because physicochemical changes, such as color, texture, and moisture, may be significantly minimized. The effectiveness showed by different types of antimicrobial EFC depends on meat source, polymer used, film barrier properties, target microorganism, antimicrobial substance properties, and storage conditions. The perspective of this technology includes tailoring of coating procedures to meet industry requirements and shelf life increase of meat and meat products to ensure quality and safety without changes in sensory characteristics.

INTRODUCTION

Animal origin foods (AOF) constitute a good nutrients source for human diet, where their protein provides high biological value and essential amino acids which complement the quality of cereals and other vegetable proteins [1]. However, AOF are susceptible to chemical deterioration and microbiological spoilage and therefore represent a high risk for consumer health, in addition to producer economic losses. According to the Centers for Disease Control (CDC), every year foodborne illnesses account for about 48 million cases, 3,000 deaths, and 128,000 hospitalizations, reaching US $77.7 billion economic burden in the United States. In addition, reduced consumer confidence, recall losses, or litigation costs should be met by the food industry, whereas public health agencies pay the cost of responding to illnesses and outbreaks [2]. Losses can be greater in countries where less stringent regulation system and sanitary control is practiced. Outbreaks involving AOF comprise 40% of total US reported cases [3]. The presence of foodborne pathogens in a country food supply not only affects the health of local population but also represents a potential for pathogens spread by tourists and consumers where these food products are exported [4].

Edible coatings are food grade suspensions which may be delivered by spraying, spreading, or dipping, which upon drying form a clear thin layer over the food surface. Coatings are a particular form of films directly applied to the surface of materials and are regarded as part of the final product [5]. On the other hand, edible films are obtained from food grade filmogenic suspensions that are usually cast over an inert surface, which after drying can be placed in contact with food surfaces. Films can form pouches, wraps, capsules, bags, or casings through further processing and one of the main differences between films and coatings is their thickness.

The use of films in foods dates back to the 12th century in China where waxes were used to coat citric fruits to retard water loss, whereas the first edible film used for food preservation was made in the 15th century from soymilk (Yuba) in Japan. In England lard or fats were used as coating to prolong shelf life of meat products in the 16th century and in Europe; this process was known as "larding" [6, 7]. In the nineteenth century, a US patent was issued in relation to preservation of meat products by gelatin coatings [7, 8].

Edible films and coatings (EFC) are an alternative to extend the shelf life of AOF by acting as barriers to water vapor, oxygen, and carbon dioxide and as carriers of substances to inhibit pathogenic and spoilage microorganisms. Natural antimicrobial agents may be incorporated into the corresponding suspensions, adding functionality to edible films and coatings, leading to the antimicrobial edible films and coatings (AEFC) obtaining.

There is increased interest in development and use of AEFC to preserve meat quality for longer shelf life periods while maintaining food safety, which is based on consumers demand for natural and safe products. Industry is concerned about these issues, while keeping competitive production costs [9]. Other key issues are sustainability through the use of biodegradable packaging materials and applications of by-products from the food industry that can generate added value [5].

Due to similar properties of edible films and coatings this review discusses characteristics of both types of coverings applied to meat products. This work focuses on a critical discussion of issues raised by recent research findings on the effectiveness of antimicrobial films and coatings and their potential application to enhance safety and quality of meat products.

MEAT PRODUCTS

Meat Importance and Consumption

Meat (including poultry and fish) is the first-choice source of animal protein for many people all over the world [10]. According to the Codex Alimentarius [11] meat is defined as "all parts of an animal that are intended for, or have been judged as safe and suitable for human consumption." Worldwide meat production is expected to be >250 million tons in 2014, with pork as the main product (108.9 million tons), whereas poultry production is expected around 87 million tons. Fish and seafood is also an important market, since world production in 2008 reached 142,287 tons [12]. Meat industry represents a significant share of national economies and therefore production and marketing systems should follow meat sanitation practices and additionally emerging preservation technologies such as the AEFC to extend shelf life and to avoid economic losses. Nutritionally, meat importance is derived from its high quality protein containing all essential amino acids and its highly bioavailable minerals and vitamins [13]. In 2010, the average annual red meat consumption per capita in developing countries was 32.4 kg, whereas in industrialized countries was 79.2 kg, increasing to 124 kg/capita in the USA. Global annual poultry consumption rose from 11.1 to 13.6 kg/capita between 2000 and 2012. In 2009, fish accounted for 16.6% of world population intake of AOF and 6.5% of all protein consumed [13].

Meat Spoilage

Meat quality is highly dependent on preslaughter handling of livestock and postslaughter handling of meat [10]. Among the main factors affecting meat quality is pH, which is determined by the glycogen content of the muscle and varies from 5.4 to 5.7 in postrigor muscle; another important factor is temperature, which must be quickly decreased from 37°C to refrigeration temperatures (4–8°C) [14].

There are three mechanisms involved in meat and meat products deterioration during processing and storage: microbial spoilage, lipid oxidation, and enzymatic autolysis. Microbial population may arrive from native microflora of the intestinal tract and skin of the animals or through environmental, human, handling, and storage conditions associated to the production chain [15]. Microbial growth in meat can result in slime formation, structural components degradation, decrease in water holding capacity, off odors, and texture and appearance changes [10]. Lipid oxidation depends on fatty acids composition, vitamin E concentration, and prooxidants such as free iron in muscles. Oxidation products, such as hydroperoxides, aldehydes, and ketones, can cause loss of color and nutritive value due to degradation of lipids, pigments, proteins, carbohydrates, and vitamins [10, 16]. Enzymatic autolysis of carbohydrates, fats, and proteins of the tissues results in softening and greenish discoloration of meat and may lead to microbial decomposition. Proteolytic enzymes are active even at low temperatures (5°C) leading to microbial growth, loss of water holding capacity, and biogenic amines production [17].

Meat Related Outbreaks

Meat products outbreaks are often due to inadequate cooking or cross-contamination from other foods. However, contamination may occur while meat is processed, cut, packaged, transported, sold, or handled. Pathogenic microorganisms do not survive thorough meat cooking, but several of their toxins and spores do [10].

Red meat is frequently involved in outbreaks, mainly due to the presence of Salmonella spp., Listeria spp.,Clostridium spp., and Staphylococcus spp. [3]. Most outbreaks reported in the EU in 2010 were due to meat and meat products consumption in which Salmonella was the main pathogen involved [18]. Listeria infection is often considered as the most lethal; for instance, in 1998 hot dogs consumption caused 21 deaths and >100 illnesses [3]. Recently, an outbreak of Salmonella typhimurium was linked to the consumption of ground beef which caused hospitalization of seven people [19].

In the case of poultry Salmonella and Campylobacter account for most of the cases of food poisoning associated with chicken [3, 20]. In 2010 turkey contaminated with Clostridium perfringens caused 135 illnesses in Kansas (USA), whereas in 2011 ground turkey contaminated with Salmonella Heidelberg infected 136 people in 34 USA states [3].

Most outbreaks caused by fish and fish products are caused by natural toxins (scombrotoxin and ciguatoxin), rather than by bacteria or viruses. However, outbreaks caused mainly by Vibrio parahaemolyticus and V. cholerae in raw oysters have been reported; additionally Clostridium botulinum, Staphylococcus aureus,Salmonella enterica, and Escherichia coli were also involved in illnesses due to fish consumption [3, 21].

EDIBLE FILMS AND COATING TYPES

EFC act as barrier between food and the surrounding environment to enhance the quality of food products protecting them from physical, chemical, and biological deterioration. Design and application of EFC on meat products arises from the search of new preservation methods, the need to add value to by-products from renewable sources, the desire to give food products a more natural or ecological image, and reduction of environmental impact of using oil-derived plastic packaging materials [22]. Additionally, they may provide moisture loss reduction during storage of fresh or frozen meats, prevention of juice dripping, and decrease in myoglobin oxidation of red meats. There are two commercially available edible films, New Gem™, which contains spices and bilayer protein films that are used to enhance ham glaze and Coffi™, that is made from collagen nettings used to wrap boneless meat products [23]. Antimicrobials or antioxidant compounds incorporated into the polymer matrix may prevent growth of spoilage and pathogenic microorganisms, delay of meat fat rancidity, discoloration prevention, and even improvement of the nutritional quality of coated foods [24, 25].

Composition and Properties of Lipid-Based Films and Coatings

A wide range of hydrophobic compounds has been used to produce EFC, including animal and vegetable oils and fats (peanut, coconut, palm, cocoa, lard, butter, fatty acids, and mono-, di-, and triglycerides), waxes (candelilla, carnauba, beeswax, jojoba, and paraffin), natural resins (chicle, guarana, and olibanum), essential oils and extracts (camphor, mint, and citrus fruits essential oils), and emulsifiers and surface active agents (lecithin, fatty alcohols, and fatty acids) [26]. In meat products, emulsifiers and surface active agents are sometimes used as gas and moisture barriers. However, pure lipids can be combined with hydrocolloids such as protein, starch, cellulose, and their derivatives providing a multicomponent system able to be applied as meat coatings [27]. In fresh and processed meats, lipid incorporation into EFC can improve hydrophobicity, cohesiveness, and flexibility, making excellent moisture barriers, leading to prolongation of freshness, color, aroma, tenderness, and microbiological stability [24].

Palmitoylated alginate is the only lipid-containing material of AEFC recently reported to wrap beef muscle and ground beef [28] (Table 1). However, essential oil extracts have been widely used to promote antimicrobial activity of AEFC (column 3, Tables 1–3).

Table 1: Use of antimicrobial films and coatings in meat and meat products

Product	Coating material	Antimicrobial compound	Target microorganism	Inoculation technique	Conditions	Results	Reference
Sliced bologna and summer sausage	Whey protein isolate (WPI, pH 5.2) films	0.5 to 1.0% p-aminobenzoic acid (PABA) and/or sorbic acid (SA)	L. monocytogenes, E. coli O157:H7, and S. enterica typhimurium (10^6 CFU/g)	0.1 mL of inoculum spread onto both surfaces	4°C, 21 days (d) Slices placed in plates covered with edible film Stored in samples	WPI films with SA or PABA reduced L. monocytogenes, E. coli, and S. typhimurium populations by 3.4–4.1, 3.3–3.6, and 3.1–4.1 log CFU/g, respectively, on both products	[29]
Hot dogs (beef 60%, pork 40%)	Whey protein isolate films (casings)	p-aminobenzoic acid (PABA) 1%	L. monocytogenes (10^3 CFU/g)	Immersion in L. monocytogenes culture for 1 min and dried in a safety cabinet	4°C, 42 d Vacuum-packaged samples	Growth inhibition for 42 d in refrigeration but no population reduction. Controls increased around 2.5 CFU/g	[30]
Beef muscle slices	Milk protein films	Oregano essential oil (OR) 1.0% (w/v), Pimento essential oil (PI) 1.0% (w/v), or 1% OR-PI (1:1)	Escherichia coli O157:H7 or Pseudomonas spp. (10^3 CFU/cm²)	Spreading over meat surface Samples placed in plates, covered on either side with the corresponding film	4°C, 7 d Meat sterilized by radiation and then inoculated Samples in plates were hermetically sealed	Film with OR was the most effective against both bacteria Reduction of 0.95 log of Pseudomonas spp. and 1.12 log reduction of E. coli O157:H7	[31]
Sterile beef muscle slices or ground beef	Palmitoylated alginate films Activated alginate beads	Covalently immobilized nisin (N) to activated alginate beads (AAB) (0–1000 IU/mL), or ground beef mixed with 0–1000 IU/mL of N	Staphylococcus aureus (10^3 CFU/g)	Inoculated using a sterile spoon and placed in sterile plates	4°C, 14 d Covered with immobilized nisin film or mixed with nisin solution	Reduction of 0.91 and 1.86 log CFU/cm² on samples covered with film (500 or 1000 IU/mL, resp.) After 14 days: N solution (500 or 1000 IU/g) mixed with ground beef reduced to 2.2 and 2.81 log CFU/g, respectively; N (500 or 1000 IU/g) in AAB reduced to 1.77 and 1.93 log CFU/g, respectively	[28]
Roast beef	Chitosan (CH, high or low molecular weight) coatings dissolved in lactic acid or acetic acid	Chitosan, lactic, or acetic acid (0.5 and 1%; w/v)	Listeria monocytogenes (10^6 CFU/g)	1 mL culture onto 5 g cubed meat, air dried 10 min Then dipped in chitosan for 30 s, dried for 1 h and placed into sterile Whirl-Pack bags	4°C, 28 d 5 g cubed roast beef samples placed into sterile bags	Reduction of 1–3 log CFU/g for low molecular weight chitosan in acetic and lactic acids, respectively, after 28 d	[32]
Frankfurters	Cellulose (produced by G. xylinus) films	Nisin (N), 625 and 2500 IU/mL	L. monocytogenes Scott A serotype 4b (10^6 CFU/mL)	Dipping in 0.85% saline sln. containing L. monocytogenes for 2 s	4°C, 14 d Samples wrapped in a single layer of film and vacuum-sealed for 2.5 s	Films containing 625 IU/mL N not significantly reduced L. monocytogenes populations. Films with 2500 IU/mL N decreased 2 log CFU/g compared to the control	[33]
Pork loins	Gelidium corneum–gelatin (GCG) films	Grapefruit seed extract (GFSE, 0.08% w/v) or green tea extract (GTE, 2.80% w/v)	E. coli O157:H7 (NCTC12079) and L. monocytogenes (KCTC 3710) (10^3 CFU/g each one)	Spread with a sterile glass rod and allowed to drain for 10 min	4°C, 10 d Samples were packed in direct contact to films and stored in sterile polystyrene trays	Samples packed with the GCG film containing GFSE or GTE decreased population of E. coli O157:H7 and L. monocytogenes in 1 and 2 log CFU/g, respectively, compared to the control	[34]

Product	Coating material	Antimicrobial compound	Target microorganism	Inoculation technique	Conditions	Results	Reference
Ham	Cellulose acetate films	Pediocin (ALTA 2351) (25% and 50%, w/v)	L. innocua (10^6 CFU/mL) and Salmonella sp. (10^5 CFU/mL)	Immersion in a 0.1% w/v peptone solution of L. innocua or Salmonella sp. for 10 min	12°C, 15 d Samples sterilized under UV (5 min each side) Films intercalated with ham slices packed in plastic bags and vacuum sealed	The 50% pediocin-film reduced L. innocua 2 log relative to the control. The 25% and 50% pediocin-films had similar performance on Salmonella sp. inhibition, both presenting 0.5 log reduction relative to the control	[35]
Fresh ground beef patties	Soy protein films	Oregano (OR), thyme (TH), or OR-TH essential oils (5%)	E. coli O157:H7, Staphylococcus aureus, Pseudomonas aeruginosa, and Lactobacillus plantarum	No inoculation	4°C, 12 d Film applied to the upper and bottom surfaces of patties and vacuum packaged in plastic bags	Pseudomonas spp. in samples coated with TH and OR films decreased in 1.13 and 1.27 log CFU/g, respectively. Coliforms were reduced by 1.6, 1.9 and 2.0 log CFU/g with addition of OR, OR-TH, and TH, respectively	[36]
Salami	Sodium caseinate (SC) films and coatings	Chitosan (CH) (2%)	Mesophilic and psychrotrophic aerobic bacteria and yeast and mold	No inoculation	10°C, 5 d, 65% RH. Film added by immersion and as wrapper Immersed slices air dried at 30°C and 50% RH for 50 min All food faces were contacted with wrapping film	CH and SC/CH films applied as both, coatings and wrappers, exerted a strong bactericidal action on 3 microbial populations analyzed, with reductions of 2 to 4.5 log CFU/g	[37]
Ground beef patties	Zein films	Lysozyme (LY) (43 mg/g) and disodium Ethylene diamine tetra acetic acid (Na₂EDTA, 19 mg/g)	Mesophilic microorganisms (TVC) and coliforms (TCC)	No inoculation	4°C, 7 d Films at both sides of each piece Wrapped with stretch plastic film and with aluminum foil	After 5 and 7 d, TVC of patties with LY and Na₂EDTA films were significantly lower (0.75–1.9 log CFU/g) than control films. After 5 d, TCC of patties with LY and Na₂EDTA films were significantly lower than the control but after 7 d, no significant difference in TCC of patties was found	[38]
Pork meat hamburgers	High molecular weight chitosan (1% w/v), acetic acid (1% w/v), lactic acid (1% w/v) films	Sunflower oil (1%)	Mesophilic bacteria, coliforms	No inoculation	5°C, 8 d Surface of both sides of the hamburgers coated with the films placed in PET trays	Reduction of 0.5–1 log for mesophilic microorganisms, 1 log CFU/g for coliforms	[39]

Product	Coating material	Antimicrobial compound	Target microorganism	Inoculation technique	Conditions	Results	Reference
Cooked ham and bologna	High methoxyl pectin + Apple, carrot or hibiscus puree films	Carvacrol (CV) or cinnamaldehyde (CM) (0.5%, 1.5%, and 3.0%, w/v)	L. monocytogenes (101M; serotype 4b) 10^4 CFU/mL	Dispersed on the surface as droplets. Inoculated samples dried under the biohood (30 min), flipped over, and inoculated on the other side Inoculated samples were dried again (30 min) and then surface wrapped with one of the test films	4°C, 7 d Samples were kept frozen and thawed before use. Sample wrapped in 2 pieces of circular films and parts of the films were not directly in contact with the meat surface	Films containing 3% CV showed 3 log reductions on ham at day 7. Bologna, films with 3% CV reduced 2 log CFU/g at day 7. Reductions with 1.5% CV were 0.5–1, 1–1.5, and 1–2 logs at day 0, 3, and 7, respectively. Films containing 3% CM, only 0.5–1.5 and 0.5–1.0 log CFU/g reductions were seen at day 7 on ham and bologna, respectively. Limited reduction (0.2–0.3 log CFU/g) was observed with 1.5% CM films	[40]
Bacon	Red algae (RA) films	1% w/v, grapefruit seed extract (GFSE)	Escherichia coli O157:H7 (10^6 CFU/g) and L. monocytogenes (10^7 CFU/g)	Spread separately on the surface of bacon with a sterile glass rod and allowed to rest for 30 min	4°C, 15 d Packed by wrapping	E. coli O157:H7 decreased 0.45 log CFU/g and L. monocytogenes decreased by 0.76 log CFU/g respect to the controls	[41]
Pork sausages	Chitosan films	Green tea extract 20% (w/v) in the chitosan film-forming solution	Mesophilic bacteria, yeasts and molds, lactic acid bacteria (LAB), not inoculated	No inoculation	4°C, 20 d Sausages wrapped with films, packaged into a pouch of low density polyethylene coated with polyamide plastic bag and heat-sealed	On day 12, faster growth in control samples for total viable count and molds and yeast was found; no difference for LAB	[42]
Cooked cured ham	Polylactic acid (PLA) films	Lauric arginate (LAE) (0% to 2.6%, w/v)	Listeria monocytogenes and Salmonella enterica serovar typhimurium (10^2 CFU/mL).	Inoculum of both L. monocytogenes and S. typhimurium onto surface of the sliced ham	4°C, 7 d Slices sterilized with UV on each side prior to inoculation Inoculated samples wrapped with LAE-coated PLA film and stored in closed plates	LAE-coated PLA film (2.6%) showed a significantly greater antibacterial activity, with L. monocytogenes and S.typhimurium reduced to <2 log CFU/film after 24 h exposure and remaining at this low level for the next 6 d	[43]

Table 2: Use of antimicrobial films and coatings in poultry

Product	Coating material	Antimicrobial compound	Target microorganism	Inoculation technique	Conditions	Results	Ref.
Chicken breast (ready-to-eat cooked chicken cubes)	Zein coatings dissolved in propylene glycol (ZP) or ethanol (ZE)	Nisin (N) (1000 IU/g) and/or calcium propionate (CP) (1% w/v)	L. monocytogenes V7 (low and high inoculum 2.67 log CFU/g and 6.89 log CFU/g, respectively, at 4°C)	Cubes immersed in 24 h broth cultures for 30 s, allowed to drip free of excess inoculum, and dried. Frozen samples were irradiated (3.0 kGy) and kept frozen until used	4°C or 8°C, 24 d. Cubes boiled in water bath for 20 min. were inoculated, then dried, followed by dipping in edible ZP or ZE, with and without antimicrobials. Air dried samples (20 min) stored in sterile bags	L. monocytogenes was reduced by 4.5–5 log CFU/g relative to the control after high dose and 16 d at 4°C, with more significant effect when N was added to the films. Low inoculum dose and using ZPNCP film caused complete inhibition from 4 to 24 d, either at 4° or at 8°C	[18]
Turkey frankfurter	WPI coatings	Grape seed extract (GSE, 1.0–3.0% w/v), nisin (N, 6–18 kIU/g), malic acid (MA 1.0–3.0%; w/v), EDTA (1.6 mg/mL), and their combinations	L. monocytogenes, E. coli O157:H7, and Salmonella typhimurium (10^6 CFU/g)	Samples were defrosted and dipped into cultures of 10^6 CFU/mL of L. monocytogenes, E. coli O157:H7, or S. typhimurium for 1 min at room temperature. Inoculated samples were then air dried under laminar flow conditions	4°C, 28 d Samples were dipped in film-forming solutions (1 min) and dried (10 min, room temperature). Samples were then packed individually in sterile bags, and stored	L. monocytogenes decreased to 2.3 log/g (N, 6000 IU/g; GSE 0.5%; MA 1.0%). S. typhimurium decreased to 5 log CFU/g using any antimicrobial, whereas E. coli decreased to 4.6 log cycles using N, MA and EDTA. All reductions were relative to the control	[53]
Chicken breast	High methoxyl pectin 11400 with apple puree films	Carvacrol (C) or cinnamaldehyde (CM) at 0.5–3% (w/w).	Salmonella enterica serovar Enteritidis or E. coli O157:H7 (ATCC 35150) (10^7 CFU/g)	Inoculum was dispersed on the surface as droplets	23°C or 4°C, 3 d Samples were dipped in boiling water (40 s), plated and exposed in a bio-hood for drying (30 min). Sample was flipped over and inoculated in a similar way. Meat was wrapped using appropriate edible films	At 23 °C, films with 3% antimicrobials showed the highest reductions (4.3–6.8 log CFU/g) of both S. Enteritidis and E. coli O157:H7. At 4°C, C exhibited greater activity than CM. Relative to control samples, films with 0.5–3% C reduced S. Enteritidis by 1.6–3 log CFU/g, whereas 1–3% CM films reduced its population by 1.2–2.8 log CFU/g. Films with 0.5–3% C reduced 1–3 log E. coli whereas 1–3% CM films inhibited 0.2–1.2 log CFU. Treatments were at 4°C	[55]

Product	Coating material	Antimicrobial compound	Target microorganism	Inoculation technique	Conditions	Results	Ref.
Chicken breast	k-carrageenan (kCF) films	Ovotransferrin, OTf (25 mg), EDTA (5 mM), and potassium sorbate (PS, 10 mg/g of k-carrageenan)	E. coli	No inoculation	5°C, 7 d Samples were wrapped with k-carrageenan-based films and packed in plastic bags	Samples wrapped with kCF added with 5 mM EDTA alone or mixed with 25 mg OTf allowed 2.7 log CFU/g reduction of E. coli, compared to the control, at day 7. Addition of 25 mg of OTf or 10 mg PS slightly inhibited microbial growth	[60]
Turkey bologna	Gelatin films	Nisaplin based films (GNF) (0.025–0.5 %; w/v nisin) and Guardian CSI-50 based films (GGF) (0.5–4 %, w/v).	L. monocytogenes (10⁶ cfu/mL)	Inoculated by surface spreading. Samples were thawed at 4°C for 18 h and then inoculated and covered with antimicrobial film. Each sample was vacuum-sealed	4°C, 56 d Samples were irradiated at 4°C (2.4 mrad for 521 min) and stored at −70°C	Both 0.5% GNF and 1% GGF inhibited L. monocytogenes by 4 log CFU/cm² and 3 log CFU/cm², respectively, relative to the control, during storage at 4°C for 56 d. GGF inhibited L. monocytogenes by 2.17 log CFU/cm² at 7 d	[61]
Roasted turkey	Starch, chitosan, alginate, or pectin coatings	Sodium lactate (SL) and sodium diacetate (SD). OptiForm PD4 (OF4), NovaGARDCB1 (NG), Protect-M (PM), and Guardian NR100 (GN)	A cocktail of five strains of L. monocytogenes (PSU1 serotype 1/2a, F5069 serotype 4b, ATCC19115 (serotype 4b), PSU9 serotype 1/2b, and Scott A serotype 4b) 10³ UFC/mL.	Spreading on both sides of the turkey surface, 10³ CFU/cm². After inoculation, turkey samples were kept at 4°C for 20 min	4°C, 8 weeks Coatings on each side were dried in a laminar-flow hood for 20 min each side. All samples were inserted into nylon/polyethylene pouches and vacuum sealed	OF4 (2.5%) alone or mixed with PM (0.12%) in films made from alginate, chitosan or pectin were the most effective, reaching L. monocytogenes reduction by 3.5 log/cm² relative to the control after storage	[46]
Chicken breasts	3% solution of high methoxyl pectin added with golden delicious apple puree films	Carvacrol (C) and cinnamaldehyde (CM) (0.5–3.0 %; w/v)	Campylobacter jejuni (D28a, H2a and A24a), 10⁷ CFU/mL	Samples dipped in boiling water for 40 s and dried in a biohood for 1 h. Chicken was dip-inoculated for 5 min	23°C and 4°C, 72 h in anaerobiosis. Samples placed in sterile plates, dried in a 42°C, 10% CO₂ incubator for 1 h and then wrapped with apple films and stored	Films with ≥1.5% CM reduced populations of both strains to below detection at 23°C at 72 h. Films with 3% C reduced populations of A24a and H2a to below detection. Using 3% C, films reduced to 0.5 log CFU/g of both strains A24a and D28a and 0.9 logs for H2a at 4°C	[56]

Product	Coating material	Antimicrobial compound	Target microorganism	Inoculation technique	Conditions	Results	Ref.
Chicken breast fillets	Chitosan (CH) coating, deacetylation degree of 75–85%	Chitosan (1.5 % w/v) and/or oregano oil 0.25% v/w (OO)	Mesophilic microorganisms, Pseudomonas spp. and Brochothrix thermosphacta	No inoculation	12, 18 y 21 d, 4°C Samples were dipped into the chitosan solution (1.5 min) and drained. Sterile OO was added to the surface. Fillets were packed in plastic pouches, and stored in a modified atmosphere	Shelf life of chicken fillets can be extended using either OO and/or CH, by approximately 6–21 d	[49]
Chicken breasts fillets	Chitosan (CH) films	CH or CH+LAE (1, 5 or 10%,by weight)	Mesophiles, psychrophiles, yeast, moulds, Pseudomonas, coliforms, LAB, and hydrogen sulfide-producing bacteria	No inoculation	0, 2, 6 and 8 d, 4°C Slices wrapped with CH or CH-5% LAE films, then packed in polyethylene films	CH films reduced 0.47–2.96 log population of fillets, depending on time and microbial group studied. Incorporation of LAE (5%) increased antimicrobial activity to 1.78–5.81 log reduction, and maintaining the initially low microbial fillets load for 8 d	[50]
Sliced turkey deli meat	Chitosan (CH, 2–5% w/w) films and coatings added with 2% solution of either acetic, lactic or levulinic acids	Lauric arginate (LAE, 50–200 mL/mL) and nisin (NIS, 25 mg/mL) alone or in combination	L. innocua (6–7 log CFU/cm²)	Even spread over the meat surface (3 × 3 cm²) using sterile spreaders	48 h, 37°C. Films were placed on top of inoculated turkey; coatings applied by spreading. The product was vacuum packed and stored at 10°C for 24 h prior to microbiological analysis	High CH levels reduced 4.6 log CFU/cm². NIS addition (486 IU/cm²) reduced Listeria by 2 and 2.4 log CFU/cm² for 2% and 5% CH, respectively. Combination of CH, LAE and NIS had similar reductions as only CH with LAE, suggesting no additive or synergistic effect by NIS. Despite no statistical difference (P < 0.05), coatings showed more microbial reduction than films	[47]

Table 3: Use of antimicrobial films and coatings in fish and seafood

Product	Coating material	Antimicrobial compound	Target microorganism	Inoculation technique	Conditions	Results	Ref.
Atlantic cod (*Gadus morhua*) and herring (*Clupea harengus*)	Chitosan coatings	Chitosan (CH) with different molecular weights and viscosities (14, 57 or 360 mPa s)	Psychrotrophic microorganisms (PT) and total plate count (TPC)	No inoculation	12 d, 4°C Samples immersed in CH solution (5°C) 30 s and after 2 min a second immersion for 30 s. Then they were dried at 40°C for 2 h in a forced air oven and stored	Herring fillets treated with 57 and 360 cP CH showed lower PT population than CH treatments reduced to 10³ and 10² TPC of herring and cod samples, respectively, after 12 d	[51]
Smoked salmon	Whey protein isolate (WPI) coatings	Lactoperoxidase system (LPO) (0–0.5%, w/v)	*L. monocytogenes* (V7 serotype 4b, LCDC 81-861 serotype 4b, Scott A serotype 4b, 101M and 108M) (10²–10⁴ CFU/g).	Spotted directly onto the salmon (I + C) or on top of applied coating (C + I) and spread with a hockey stick	4°C and 10°C, 35 d. I + C samples were inoculated and then dried for 0.5 h and coated. C + I samples dried for 1 h and then inoculated	Samples coated by LPO-WPI showed <1.0 log CFU/g of *L. monocytogenes* at 4°C for 35 d (both treatments). *L. monocytogenes* was completely inhibited in C + I samples stored during 35 d at 10°C	[54]
Cold-smoked sardine (*Sardina pilchardus*)	Gelatin (G) films	Oregano extract (OE) (*Origanum vulgare* 1.5%, v/v) or rosemary (RM) (*Rosmarinus officinalis*, 20% w/v) or Chitosan (CH) (1.5% w/v), high pressure (300 MPa/20°C/15 min) (HP).	Total viable count (TVC), H₂S-reducing organisms, luminescent bacteria, and Enterobacteriaceae	No inoculation	5°C, 20 d Fish slices were placed between two layers of edible films and were stored in clean bags	Fish coated with OE-G and RM-G films reduced TVC by 1.99 and 1.34 log CFU/g respectively, on d 16. H₂S-reducing bacteria followed a similar pattern. OE and RM had no effect, but CH reduced to ≤10² CFU/g in all cases. Pressurized samples produced undetectable levels of all microorganisms for 20 d, except uncoated sample whose TVC was 10⁵ CFU/g at d 20	[58]
Cod (*Gadus morhua*)	Gelatin (G) or in combination with chitosan (CH) films	Clove essential oil (CO)	Total bacterial count (TVC), H₂S-producers organisms, luminescent organisms, *Pseudomonas*, Enterobacteriaceae (EB), and lactic acid bacteria (LAB)	No inoculation	2°C, 11 d Fillets were covered with the G-CH film containing CO, and vacuum-packed in plastic bags	TVC count was 6.1 log CFU/g. Luminescent bacteria reached 6 log CFU/g after 3 d, but later were undetected. H₂S producers were completely inhibited from d 3 onwards. LAB and EB increased during storage despite storage temperature	[48]

Product	Coating material	Antimicrobial compound	Target microorganism	Inoculation technique	Conditions	Results	Ref.
Cold smoked salmon (CSS) slices and fillets	Alginate (AL), κ-carrageenan, pectin, gelatin, or starch coatings	Sodium lactate (SL, 0–2.4% w/v) and sodium diacetate (SD, 0–0.25% w/v), OptiForm (OF, 2.5% w/v)	Mixture of *L. monocytogenes* strains: PSU1, PSU9, F5069, ATCC 19115, and Scott A. 10³ CFU/g	Surface-inoculated	4°C, 30 d Samples coated with AL incorporating SL/SD or OF dried 20 min and stored at 4°C in vacuum sealed bags	AL coatings with 2.4% SL/0.25% SD and OF reduced *L. monocytogenes* by 3.2 and 4 log CFU/g (slices) and 2.4 and 3 log CFU/g (fillets), respectively, relative to control sample	[62]
Bream (*Megalobrama amblycephala*)	Alginate (AL) coatings	Vitamin C (VC, 5% w/v) and tea polyphenols (TP, 0.3% w/v)	TVC	No inoculation	21 d, 4°C Bream was dipped in AL-antimicrobial solutions (1 min), air-dried (1 min) and immersed in 2% (w/v) CaCl₂ (1 min) to obtain gels. Samples were packed and stored	After 4 d of storage, The TVC of VC and TP decreased by 1.6 and 1.5 log CFU/g, respectively, on day 21	[63]
Sea bass slices	Gelatin extracted from the skin of unicorn leatherjacket (*Aluterus monoceros*) films	Lemongrass essential oil (LEO) 25% (w/w)	Mesophilic (TVC) and psychrophilic (PS) microorganisms, enterobacteria (EB), and H₂S-producing bacteria LAB	No inoculation	12 d, 4°C For each slice, films were placed on both sides. Subsequently, the samples were placed in polystyrene trays wrapped with extensible polypropylene film	TVC of unwrapped sample increased to 7.2 log CFU/g at d 4 reaching 7.9 log CFU/g at d 12. TVC of LEO-film wrapped samples was 5.6 log CFU/g at 12. PS count for control, G and LEO films was 6.0, 5.5 and 4.0 log CFU/g, respectively. LAB increased to 7.2, 6.7 and 5.9 log CFU/g at the end of storage. LEO-film showed the lowest EB counts (2.2 log CFU/g), as compared to control	[57]
Sea bass (*Dicentrarchus labrax*)	Chitosan (CH) films	CH with vacuum packaging	Total mesophilic aerobic bacteria (TVC) and psychrotrophic (PS) aerobic bacteria	No inoculation	4°C until end of shelf life. Fish fillets were covered using CH films, wrapped and vacuum packaged using polyethylene bags	The acceptable limit of 6 log CFU/g and 7 log CFU/g for PS and TVC bacteria, respectively, was reached after 25 d at 4°C. Control samples reached this limit after 5 d	[52]

Product	Coating material	Antimicrobial compound	Target microorganism	Inoculation technique	Conditions	Results	Ref.
Indian oil sardine (*Sardinella longiceps*)	Chitosan (CH) (1 and 2% w/v) coatings	Chitosan (1 and 2% w/v)	Mesophilic microorganisms (TVC)	No inoculation	11 d, 1-2°C. Fillets were dipped in 1 and 2% CH at 1-2°C for 10 min, drained for 5 min and placed in trays for 24 h, then sealed using HDPE	Eating quality was maintained for 8 and 10 d for 1 and 2% CH respectively, whereas untreated samples lasted 5 d. The limit of 10^7 CFU/g of TVC was exceeded after 7, 9 and 11 d for untreated, 1% and 2% CH treated samples, respectively	[64]
Salmon	Barley bran protein and gelatin (BBG) films	Grapefruit seed extract (GFSE) (0.5–1.2% w/v)	*E. coli* O157:H7 and *L. monocytogenes* (10^6 CFU/mL)	*E. coli* O157:H7 and *L. monocytogenes* were spread individually on sample surface and allowed to rest for 30 min	4°C, 15 d Samples wrapped using the BBG film. Samples packed in polyethylene terephthalate film were used as control	After 15 d, populations of *E. coli* and *L. monocytogenes* inoculated salmon with the BBG film containing GSE decreased by 0.53 and 0.50 log CFU/g, respectively, compared to the control	[59]
Cold-smoked salmon	Potato processing waste (PPW) films	Oregano essential oil (OO) 0.97% and 1.92% (185 and 289 mg oil/g film)	*L. monocytogenes* V7 (6.7–6.9 log CFU/g)	*L. monocytogenes* Overnight culture (100 μL) was spotted at 25–30 locations on salmon fillet, spread and dried in a biological hood (30 min)	4°C, 28 d Salmon samples were wrapped with edible films and were vacuum packed	Coated samples with PPW-OO, reduced *Listeria* population by 0.4–2.4 log CFU/g as compared to control samples, after storage period	[65]

Composition and Properties of Protein-Based Films and Coatings

Film-forming proteins are derived from animals (casein, whey protein concentrate and isolate, collagen, gelatin, and egg albumin) or plant sources (corn, soybean, wheat, cottonseed, peanut, and rice). Protein-based films adhere well to the meat hydrophilic surfaces and provide barrier for oxygen and carbon dioxide but do not resist water diffusion [27]. Plasticizers, such as polyethylene glycol or glycerol, are added to improve flexibility of the protein network, whereas water permeability can be overcome by adding hydrophobic materials such as beeswax or oils like oleic that can affect films properties such as crystallinity, hydrophobicity, surface charge, and molecular size, improving films characteristics and their application [6, 31, 34]. Despite their advantages, protein films may be susceptible to proteolytic enzymes present in meat products or allergenic protein fractions may cause adverse reactions to susceptible people [24].

Composition and Properties of Polysaccharides-Based Film and Coatings

Polysaccharide coatings are generally poor moisture barriers, but they have selective permeability to O_2 and CO_2 and resistance to fats and oils [25]. Polysaccharide films can be made of cellulose, starch (native and modified), pectins, seaweed extracts (alginates, carrageenan, and agar), gums (acacia, tragacanth, and guar), pullulan, and chitosan. These compounds impart hardness, crispness, compactness, viscosity, adhesiveness, and gel-forming ability to a variety of films [24, 44, 45]. Polysaccharide films and coatings can be used to extend the shelf life of muscle foods by preventing dehydration, oxidative rancidity, and surface browning. When applied to wrapped meat products and exposed to smoke and steam, the polysaccharide film actually dissolves and becomes integrated into the meat surface resulting in higher yields, improved structure and texture, and reduced moisture loss [27].

Materials recently used to obtain AEFC in meat and meat products, poultry, and fish and fish products are shown in column 2 of Tables 1, 2, and 3, respectively. Chitosan based AEFC were the most commonly reported in recent years and have been used to wrap pork meat hamburgers and sausages [39, 42] (Table 1), as films and coatings on roasted and sliced turkey [46, 47] (Table 2) and as films on cod fillets [48] (Table 3). Chitosan was used as both polymeric material and antimicrobial agent, for roast beef coating [32] (Table 1) and chicken breast fillets (wrapping and coating) [49, 50] (Table 2) and as coating of Atlantic cod and herring [51] and as films on sea bass fillets [52] (Table 3). WPI and cellulose (or its acetate salt), despite being less reported, are also materials used in AEFC for meat and meat products [29, 30, 33, 35] (Table 1): turkey frankfurters [53] (Table 2) and smoked salmon [54] (Table 3). Several reports mention pectin for production of AEFC to wrap cooked ham and bologna [40] (Table 1) and chicken breast [55, 56] (Table 2), whereas other reports show gelatin based antimicrobial films placed on top or between slices of fish products [48, 57–59] (Table 3).

COMMON ANTIMICROBIALS USED IN EFC

Incorporation of antimicrobial compounds into EFC as an alternative to their direct application onto the meat surface has the advantage of gradual release of the antimicrobial compound from the AEFC leading to a reduction of added antimicrobial and to reduced sensory changes. Antimicrobial compounds within AEFC are less exposed to interaction with meat surface components than those added directly to the surface and thus maintaining their activity [66–68].

Antimicrobial agents recently incorporated in AEFC for meat and meat products, poultry, and fish and fish products are shown in column 3 of Tables 1, 2, and 3, respectively. Target microorganisms aimed by recently developed AEFC as well as inoculation technique for meat and meat products, poultry, and fish and fish products are shown in columns 4 and 5 of Tables 1, 2, and 3, respectively.

The characteristics and mode of action of most common antimicrobials used to promote meat safety are described below.

Organic Acids

The antimicrobial effect of organic acids depends on concentration of undissociated form, which can penetrate the bacterial cell membrane. Inside the cell, their dissociation leads to interference with membrane transport and disruption of proton motive force [30]. Organic acids incorporated into EFC include lactate and acetate [46], propionate [18], and p-aminobenzoic acid [30]. WPI coatings added with malic acid, nisin, and grape seed extract applied on turkey frankfurters decreased to 2.3 log CFU/g of L. monocytogenes and 5 log CFU/g S. typhimurium after 28 d of storage at 4°C [53] (Table 1). Zein based AEFC using calcium propionate combined with nisin, reduced up to 5 log CFU/g of L. monocytogenes after 14 d at 4°C, when used to coat chicken breast [18]. Sodium lactate combined with other commercial antimicrobials reduced to 3.5 log/cm² of this pathogen when roasted turkey was stored at 4°C for 8 weeks [46] (Table 2). Thus, organic acids, especially when acting combined with other antimicrobial agents, have an important role in maintaining microbiological quality of meat and meat products.

Essential Oils and Plant Extracts

Essential oils are complex mixtures of volatile compounds obtained from plants, which mainly include terpenes, terpenoids, and aliphatic chemicals, all characterized by low molecular weight [69]. Oils containing phenols such as thymol, carvacrol, and eugenol exhibit the highest activity against all kind of microorganisms. Essential oils usually show higher antibacterial activity than mixtures of their major antimicrobial components, suggesting that minor components are critical for enhanced activity [69]. The antimicrobial mechanism is attributed to the disturbance of the cytoplasmic membrane disrupting the proton motive force; active transport and coagulation of cell contents may occur [70]. Direct incorporation of essential oils in the formulation of AEFC applied to meat products is expected to reduce bacterial population but may alter their sensory characteristics [68]. Microencapsulation of essential oils or their ingredients may be an alternative to protect them from interaction with environmental factors, avoiding their oxidation or volatilization while exerting their antimicrobial effect. Moreover, encapsulation increases the oil solubility in water, prevents its release at an undesired stage, and makes it easier to handle [71, 72]. Essential oils or their constituents that may be incorporated in AEFC on AOF include those extracted from lemongrass, oregano, pimento, thyme, or cinnamon [40, 57, 65]. Oregano essential oil has been the most commonly reported in recent years including a 1.5% extract (v/v), successfully used to reduce total viable count by 2 log CFU/g of cold smoked sardine covered with an AEFC after 20 d storage at 5°C [58], whereas at 1.9% it achieved L. monocytogenes population reduction by 2.4 log CFU/g after 28 d, at 4°C in wrapped cold smoked salmon [65] (Table 3). Oregano essential oil combined with thyme extract, was incorporated into a film placed on top and bottom of fresh ground beef patties reducing Pseudomonas spp. and coliforms populations [36], whereas mixed with pimento essential oil, the films covering beef muscle slices reduced to 1 log of E. coli O157:H7 after 7 d of storage at 4°C [31] (Table1). Grapefruit seed extract (GSE) incorporated into AEFC was found to inhibit E. coli O157:H7 and L. monocytogenes from pork loins [34], bacon [41], and salmon [59] (Tables 1 and 3). However, some commercial GSE is adulterated with synthetic preservatives such as benzalkonium and benzethonium chlorides, which are solely responsible for the antimicrobial activity of GSE. These compounds show toxicity and allergenicity to humans, and it is unlikely that they are formed during any extraction and/or processing of grapefruit seeds and pulp [73, 74].

Bacteriocins

Bacteriocins from lactic acid bacteria are peptides produced by bacteria that inhibit or kill other related and unrelated microorganisms [75]. These agents are generally heat-stable, apparently hypoallergenic and readily degraded by proteolytic enzymes in the human intestinal tract [68]. Class I bacteriocins, such as nisin, bind to plasma membranes via nonspecific electrostatic interactions and have a dual mode of action. The antibacterial activity results from pore formation in the bacterial plasma membrane, leading to dissipation of the transmembrane potential and vital solute gradients. The high efficiency of pore formation is the result of a second mechanism involving the cell wall precursor Lipid II which increases the affinity of nisin for the membrane, stabilizes a transmembrane orientation of nisin, and forms and integral part of the nisin pore. The pore structure involves a complex made up of four lipid II and 8 nisin molecules, which interferes with peptidoglycan biosynthesis [76, 77]. Other bacteriocins such as pediocin have been widely studied in food systems, but nisin remains the only one approved by European Union (EU) and the USA where it enjoys GRAS status [68, 78]. The effect of nisin incorporation into AEFC is the most studied, either to protect beef and turkey frankfurters, or turkey bologna against L. monocytogenes [33, 53, 61] (Tables 1 and 2); but pediocin has also been tested [35].

Proteins

Lysozyme is a naturally produced enzyme active against gram-positive bacteria, by hydrolyzing N-glycosidic bonds connecting N-acetyl muramic acid with the fourth carbon atom of N-acetyl glucosamine of the peptidoglycan molecule in the cell wall. This antimicrobial has been formulated in whey protein isolate (WPI) films and tested for its diffusivity and antimicrobial effect on salmon slices [79] and also tested in ground beef patties using zein films [38] (Table 1).

Chitosan

Chitosan is a linear polysaccharide composed of randomly distributed β-(1-4)-linked D-glucosamine and N-acetyl-D-glucosamine. Chitosan is believed to chelate certain ions from the lipopolysaccharide (LPS) layer of the outer membrane of bacteria or to exhibit electrostatic interactions among its NH_3^+ groups and the negative charges of microbial cell membrane. In both cases cell permeability increases releasing key cellular components of bacteria. The antimicrobial action of chitosan is influenced by type of chitosan, degree of polymerization, and environmental conditions. Chitosan coatings act as barrier against oxygen transfer leading to growth inhibition of aerobic bacteria [42]. In addition to the functionality of chitosan as polymeric material and antimicrobial agent (Section 3.3), it has been used as coating and wrapper in salami [37] and as film and coating combined with lauric arginate and nisin to reduce L. monocytogenes population in sliced turkey deli meat [47] (Tables 1 and 2) and also in seafood and fish [48, 52].

Lauric Arginate

Lauric arginate (LAE) is a food-grade cationic surfactant that is highly active against a wide range of food pathogens and spoilage microorganisms including bacteria, yeasts, and molds. It is obtained through the reaction of L-arginine, hydrochloric acid, ethanol, thionyl chloride, sodium hydroxide, lauryl chloride, and deionized water [80]. LAE affects cells viability by disturbing membrane potential and causing structural changes, although no disruption of cells is detected. In gram-negative cells, LAE alter both the cytoplasm membrane and the external membrane, while in gram-positive cells, alterations were observed in the cell membrane and in the cytoplasm. However, in both cases, cells remained intact and cell lysis is not observed [81]. LAE is nontoxic and is metabolized to naturally occurring amino acids, mainly arginine and ornithine, after consumption. Effectiveness of LAE, alone or in combination with other antimicrobials, has been tested against L. monocytogenes, S. enterica, and L. innocua in cooked ham and sliced turkey deli meat producing 2 log reductions in all cases [43, 47] (Tables 1 and 2).

Antimicrobial agents recently incorporated in AEFC for meat and meat products, poultry, and fish and fish products are shown in column 3 of Tables 1, 2, and 3, respectively, whereas application conditions and effect of AEFC are shown in columns 6 and 7 of the same tables, respectively.

MIGRATION OF ANTIMICROBIAL AGENTS FROM FILMS

Few reports have considered the migration extent of antimicrobial agents from edible films to the food surface. A study showed the effect of film thickness, solution pH, and temperature on nisin migration from an active WPI edible film to an aqueous solution. Results indicated that nisin is able to migrate from the film where diffusivity increased at lower pH and thickness, while it increased at higher temperatures [82]. Sorbic acid migration from an active cellulose film into pastry dough was evaluated for 40 days and it was not significantly affected by film thickness, achieving a migration of 0.07%, (w/v) [83]. Nisin release measured from low density polyethylene film was unpredictable but it was affected by temperature and pH [84]. Migration of lysozyme from WPI-glycerol films indicated that the diffusion coefficient decreased as the WPI-glycerol ratio increased or storage temperature decreased [79]. Chitosan-glycerol films incorporated with 1–10% (w/v) lauric arginate showed full release of the agent and followed a Fickian behavior in a few hours at 4° and 28°C. Films were active in liquid and solid media against bacteria, yeast and fungi achieving 1.8–5.8 log reductions [50]. These findings lead us to consider that antimicrobial agents incorporated into AEFC may prevent microbial contamination of food surfaces.

APPLICATION AND EFFECT OF AEFC ON MEAT PRODUCTS

Antimicrobial packaging can be a promising tool for protecting meat from pathogens contamination by preventing microbial growth by direct contact of the package with its surface. The gradual release of an antimicrobial substance from a packaging film to the food surface for extended period of time may be more advantageous than incorporating the antimicrobial into foods [85].

Studies using chitosan films incorporated in meat products demonstrated that lipid oxidation is reduced, suggesting that it may be due to the antioxidant activity of chitosan [52], as well as its low oxygen permeability characteristic [42]. Similar results have been obtained when other compounds were incorporated such as essential oils [57], grapefruit extracts [41, 59], and lysozyme [38]. In all cases, the oxidation rates decreased maintaining an acceptable quality in meat, poultry, or fish products. However, even when the coating may confer protection against lipid oxidation, other characteristics may have changed, leading to modified sensory attributes that made the food unacceptable for consumers. Application of films on meat surface in some cases could increase the stability of the red meat color [57], but if coatings act as gas barriers undesirable color changes may occur [38]. Sensory studies on fish indicated that not only bacterial number is critical for fish acceptance, but other factors such as bacterial types, autolytic activity, biochemical properties of fish, and storage conditions are significant [76]. In other studies, using chitosan film incorporated with oregano essential oil did not negatively influence the taste of chicken samples, extending the shelf-life of chicken fillets by 14 days, maintaining acceptable sensory characteristics [49]. Therefore, each particular application should be evaluated to establish the conditions leading to maintain meat safety without altering sensory characteristics.

Potential benefits of using AEFC for the meat industry are prevention of moisture loss, avoiding texture, flavor, and color changes, producing a significant economic impact by increasing saleable weight of products. Other advantages include reduction of dripping enhancing products presentation and reduced use of absorbent pads at the bottom of trays. Low oxygen permeability leads to decreased lipids oxidation and brown color-causing myoglobin oxidation, reduced load of spoilage and pathogenic microorganisms, and partial inactivation of deteriorative proteolytic enzymes at the surface of coated meat. Volatile flavor loss and foreign odors pick-up by meat, poultry, or seafood could be restricted by using edible films and coatings and incorporation of additives such as antimicrobial agents can be used for direct treatment of meat surface. There are, however, some factors that may represent disadvantages of using AEFC; there is wide diversity of meat products whose characteristics may vary making it difficult to standardize a single application procedure. Composition and properties of AEFC will provide different functionality and may affect scaling up of application methods for coatings.

Selection of the appropriate AEFC for a specific meat product will depend on its nature, characteristics, specific needs, costs, and benefits that this technology can offer to the manufacturers and the consumer. Thus, more research is needed to improve production and application processes of AEFC intended for the meat industry to be economically feasible and appropriate for each product.

CONCLUSIONS

The application and effects of AEFC of different nature have been investigated in several AOF. Effectiveness shown by each one depends on meat source, polymer used, film barrier properties, target microorganism, antimicrobial substance, and conditions of storage among others. EFC are a good alternative to improve the quality and safety of food and also to add value to food industry by-products. However, some challenges remain such as the need to improve and standardize coating procedures according to industry requirements aiming to reduce costs and increase shelf life to meet consumer demands without altering sensory characteristics of meat and meat products.

ACKNOWLEDGMENTS

The authors are grateful to PROMEP for a PhD grant to ISO and to CONACYT for financial support to project no. 166751.

REFERENCES

1.　P. D. Warriss, Meat Science: An Introductory Text, CAB International Publishers, New York, NY, USA, 2010.

2.　R. L. Scharff, "Economic burden from health losses due to foodborne illness in the united states," Journal of Food Protection, vol. 75, no. 1, pp. 123–131, 2012.

3.　CDC, 2012, http://www.cdc.gov/features/dsFoodborneOutbreaks/.

4.　J. C. Buzby and T. Roberts, "Economic costs and trade impacts of microbial foodborne illness," World Health Statistics Quarterly, vol. 50, no. 1-2, pp. 57–66, 1997. ·

5.　J. H. Han and A. Gennadios, "Edible films and coatings: a review," in Innovations in Food Packaging, J. H. Han, Ed., pp. 239–262, Elsevier Science, New York, NY, USA, 2005.

6.　Cagri, Z. Ustunol, and E. T. Ryser, "Antimicrobial edible films and coatings," Journal of Food Protection, vol. 67, no. 4, pp. 833–848, 2004. ·

7.　E. Pavlath and W. Orts, "Edible films and coatings: why, what, and how?" in Edible Films and Coatings for Food Applications, M. E. Em buscado and K. C. Huber, Eds., pp. 57–112, Springer, New York, NY, USA, 2009.

8.　E. A. Baldwin and R. Hagenmaier, "Introduction," in Edible Coatings and Films to Improve Food Quality, E. A. Baldwin, R. Hagenmaier, and J. Bai, Eds., pp. 1–12, CRC Press, Boca Raton, Fla, USA, Second edition, 2012.

9.　Z. Ustunol, "Edible films and coatings for meat and poultry," in Edible Films and Coatings for Food Applications, M. E. Embuscado and K. C. Huber, Eds., pp. 245–268, Springer, New York, NY, USA, 2009.

10.　D. Dave and A. E. Ghaly, "Meat spoilage mechanisms and preservation techniques: a critical review," The American Journal of Agricultural and Biological Sciences, vol. 6, no. 4, pp. 486–510, 2011.

11.　Codex Alimentarius, Code of hygienic practice for meat. Codex Alimentarius Commision/Recommended Code of Practice. 58-2005.

New Zealand, FAO/WHO, 2005.

12. USDA, United States Department of Agriculture, 2014,http://www.usda. gov/wps/portal/usda/usdahome.

13. FAO, 2012, http://www.fao.org/index_en.htm.

14. ICMSF, Microorganisms in Foods 6. Microbial Ecology of Food Commodities, 2nd edition, 2005.

15. J. Cerveny, J. D. Meyer, and P. A. Hall, "Microbiological spoilage of meat and poultry products compendium of the microbiological spoilage, of foods and beverages," in Food Microbiology and Food Safety, W. H. Sperber and M. P. Doyle, Eds., pp. 69–868, Springer, NY, NY, USA, 2009.

16. P. E. Simitzisand and S. G. Deligeorgis, "Lipid oxidation of meat and use of essential oils as antioxidants in meat products," 2010,http://www. scitopics.com/Lipid_Oxidation_of_Meat_and_Use_of_Essential_Oils_ as_Antioxidants_in_Meat_Products.html.

17. K. Kuwahara and K. Osako, "Effect of sodium gluconate on gel formation of Japanese common squid mantle muscle," Nippon Suisan Gakkaishi (Japanese Edition), vol. 69, no. 4, pp. 637–642, 2003. ·

18. M. E. Janes, S. Kooshesh, and M. G. Johnson, "Control of Listeria monocytogenes on the surface of refrigerated, ready-to-eat chicken coated with edible zein film coatings containing nisin and/or calcium propionate," Journal of Food Science, vol. 67, no. 7, pp. 2754–2757, 2002. ·

19. CDC, 2013, http://www.cdc.gov/features/dsFoodborneOutbreaks/.

20. EFSA (European Food Safety Authority), European Centre for Disease Prevention and Control, The European Union Summary Report on Trends and Sources of Zoonoses, Zoonotic Agents and Food-borne Outbreaks in 2010, http://www.efsa.europa.eu/efsajournal.

21. S. DeWaal, X. A. Tian, and F. Bhuiya, Outbreak Alert! 2008 Center for Science in the Public Interest (CSPI) Washington, 2013, http://www. cspinet.org/.

22. K. Dangaran, P. M. Tomasula, and P. Qi, "Structure and function of protein-based edible films and coatings," in Edible Films and Coatings for Food Applications, M. E. Embuscado and K. C. Huber, Eds., pp. 25– 56, Springer, New York, NY, USA, 2009.

23. T. H. McHugh and R. J. Avena-Bustillos, "Applications of edible films and coatings to processed foods," inEdible Coatings and Films to Improve Food Quality, E. A. Baldwin, R. Hagenmaier, and J. Bai, Eds.,

pp. 291–318, CRC Press, Boca Raton, Fla, USA, 2012.

24. Gennadios, M. A. Hanna, and L. B. Kurth, "Application of edible coatings on meats, poultry and seafoods: a review," LWT—Food Science and Technology, vol. 30, no. 4, pp. 337–350, 1997.

25. R. Soliva-Fortuny, M. A. Rojas-Graii, and O. Martin-Belloso, "Polysaccharide coatings," in Edible Coatings and Films To Improve Food Quality, E. Baldwin, R. Hagenmaier, and J. Bai, Eds., pp. 103–136, CRC Press, Boca Raton, Fla, USA, 2012.

26. F. Debeaufort and A. Voilley, "Lipid based edible films and coatings," in Edible Films and Coatings for Food Applications, M. E. Embuscado and K. C. Huber, Eds., pp. 135–168, Springer, New York, NY, USA, 2009.

27. N. Cutter, "Opportunities for bio-based packaging technologies to improve the quality and safety of fresh and further processed muscle foods," Meat Science, vol. 74, no. 1, pp. 131–142, 2006.

28. M. Millette, C. le Tien, W. Smoragiewicz, and M. Lacroix, "Inhibition of Staphylococcus aureus on beef by nisin-containing modified alginate films and beads," Food Control, vol. 18, no. 7, pp. 878–884, 2007.

29. Cagri, Z. Ustunol, and E. T. Ryser, "Inhibition of three pathogens on bologna and summer sausage using antimicrobial edible films," Journal of Food Science, vol. 67, no. 6, pp. 2317–2324, 2002. ·

30. Cagri, Z. Ustunol, W. Osburn, and E. T. Ryser, "Inhibition of Listeria monocytogenes on hot dogs using antimicrobial whey protein-based edible casings," Journal of Food Science, vol. 68, no. 1, pp. 291–299, 2003. ·

31. M. Oussalah, S. Caillet, S. Salmiéri, L. Saucier, and M. Lacroix, "Antimicrobial and antioxidant effects of milk protein-based film containing essential oils for the preservation of whole beef muscle," Journal of Agricultural and Food Chemistry, vol. 52, no. 18, pp. 5598–5605, 2004. ·

32. R. L. Beverlya, M. E. Janes, W. Prinyawiwatkula, and H. K. No, "Edible chitosan films on ready-to-eat roast beef for the control of Listeria monocytogenes," Food Microbiology, vol. 25, no. 3, pp. 534–537, 2008.

33. V. T. Nguyen, M. J. Gidley, and G. A. Dykes, "Potential of a nisin-containing bacterial cellulose film to inhibit Listeria monocytogenes on processed meats," Food Microbiology, vol. 25, no. 3, pp. 471–478, 2008.

34. Y. H. Hong, G. O. Lim, and K. B. Song, "Physical properties of Gelidium corneum-gelatin blend films containing grapefruit seed extract or green

tea extract and its application in the packaging of pork loins,"Journal of Food Science, vol. 74, no. 1, pp. 6–10, 2009.

35. P. Santiago-Silva, N. F. F. Soares, J. E. Nóbrega et al., "Antimicrobial efficiency of film incorporated with pediocin (ALTA 2351) on preservation of sliced ham," Food Control, vol. 20, no. 1, pp. 85–89, 2009.

36. Z. K. Emiro□lu, G. P. Yemiş, B. K. Coşkun;, and K. Cando□an, "Antimicrobial activity of soy edible films incorporated with thyme and oregano essential oils on fresh ground beef patties," Meat Science, vol. 86, no. 2, pp. 283–288, 2010.

37. M. D. R. Moreira, M. Pereda, N. E. Marcovich, and S. I. Roura, "Antimicrobial effectiveness of bioactive packaging materials from edible chitosan and casein polymers: assessment on carrot, cheese, and salami,"Journal of Food Science, vol. 76, no. 1, pp. M54–M63, 2011.

38. U. Ünalan, F. Korel, and A. Yemenicio□lu, "Active packaging of ground beef patties by edible zein films incorporated with partially purified lysozyme and Na2EDTA," International Journal of Food Science and Technology, vol. 46, no. 6, pp. 1289–1295, 2011.

39. M. Vargas, A. Albors, and A. Chiralt, "Application of chitosan-sunflower oil edible films to pork meat hamburgers," in Proceedings of the 11th International Congress on Engineering and Food (ICEF ‹11), vol. 1, pp. 39–43, Procedia Food Science, 2011.

40. S. Ravishankar, D. Jaroni, L. Zhu, C. Olsen, T. McHugh, and M. Friedman, "Inactivation of Listeria monocytogenes on ham and bologna using pectin-based apple, carrot, and hibiscus edible films containing carvacrol and cinnamaldehyde," Journal of Food Science, vol. 77, no. 7, pp. 377–382, 2012.

41. Y. J. Shin, H. Y. Song, Y. B. Seo, and K. B. Song, "Preparation of red algae film containing grapefruit seed extract and application for the packaging of cheese and bacon," Food Science and Biotechnology, vol. 21, no. 1, pp. 225–231, 2012.

42. U. Siripatrawan and S. Noipha, "Active film from chitosan incorporating green tea extract for shelf life extension of pork sausages," Food Hydrocolloids, vol. 27, no. 1, pp. 102–108, 2012.

43. P. Theinsathid, W. Visessanguan, J. Kruenate, Y. Kingcha, and S. Keeratipibul, "Antimicrobial activity of lauric arginate-coated polylactic acid films against Listeria monocytogenes and Salmonella typhimuriumon cooked sliced ham," Journal of Food Science, vol. 77, no. 2, pp. 142–149, 2012.

44. M. B. Nieto, "Structure and function of polysaccharide gum-based edible

films and coatings," in Edible Films and Coatings for Food Applications, M. E. Embuscado and K. C. Huber, Eds., pp. 57–112, Springer, New York, NY, USA, 2009.

45. Eroglu, M. Torun, C. Dincer, and A. Topuz, "Influence of pullulan-based edible coating on some quality properties of strawberry during cold storage," Packaging Technology and Science, 2014. ·

46. Z. Jiang, H. Neetoo, and H. Chen, "Efficacy of freezing, frozen storage and edible antimicrobial coatings used in combination for control of Listeria monocytogenes on roasted turkey stored at chiller temperatures," Food Microbiology, vol. 28, no. 7, pp. 1394–1401, 2011.

47. M. Guo, T. Z. Jin, L. Wang, O. J. Scullen, and C. H. Sommers, "Antimicrobial films and coatings for inactivation of Listeria innocua on ready-to-eat deli turkey meat," Food Control, vol. 40, pp. 64–70, 2014.

48. J. Gómez-Estaca, A. López de Lacey, M. E. López-Caballero, M. C. Gómez-Guillén, and P. Montero, "Biodegradable gelatin-chitosan films incorporated with essential oils as antimicrobial agents for fish preservation," Food Microbiology, vol. 27, no. 7, pp. 889–896, 2010.

49. S. Petrou, M. Tsiraki, V. Giatrakou, and I. N. Savvaidis, "Chitosan dipping or oregano oil treatments, singly or combined on modified atmosphere packaged chicken breast meat," International Journal of Food Microbiology, vol. 156, no. 3, pp. 264–271, 2012.

50. L. Higueras, G. López-Carballo, P. Hernández-Muñoz, R. Gavara, and M. Rollini, "Development of a novel antimicrobial film based on chitosan with LAE (ethyl-Nα-dodecanoyl-L-arginate) and its application to fresh chicken," International Journal of Food Microbiology, vol. 165, no. 3, pp. 339–345, 2013.

51. Y. J. Jeon, J. Y. V. A. Kamil, and F. Shahidi, "Chitosan as an edible invisible film for quality preservation of herring and Atlantic cod," Journal of Agricultural and Food Chemistry, vol. 50, no. 18, pp. 5167–5178, 2002.

52. Günlü and E. Koyun, "Effects of vacuum packaging and wrapping with chitosan-based edible film on the extension of the shelf life of sea bass (Dicentrarchus labrax) fillets in cold storage (4°C)," Food and Bioprocess Technology, vol. 6, no. 7, pp. 1713–1719, 2013.

53. V. P. Gadang, N. S. Hettiarachchy, M. G. Johnson, and C. Owens, "Evaluation of antibacterial activity of whey protein isolate coating incorporated with nisin, grape seed extract, malic acid, and EDTA on a turkey frankfurter system," Journal of Food Science, vol. 73, no. 8, pp. 389–394, 2008.

54. S. Min, L. J. Harris, and J. M. Krochta, "Listeria monocytogenes inhibition

by whey protein films and coatings incorporating the lactoperoxidase system," Journal of Food Science, vol. 70, no. 7, pp. 317–324, 2005. ·

55. S. Ravishankar, L. Zhu, C. W. Olsen, T. H. McHugh, and M. Friedman, "Edible apple film wraps containing plant antimicrobials inactivate foodborne pathogens on meat and poultry products," Journal of Food Science, vol. 74, no. 8, pp. 440–445, 2009.

56. R. M. Mild, L. A. Joens, M. Friedman et al., "Antimicrobial edible apple films inactivate antibiotic resistant and susceptible Campylobacter jejuni strains on chicken breast," Journal of Food Science, vol. 76, no. 3, pp. 163–168, 2011.

57. M. Ahmad, S. Benjakul, P. Sumpavapol, and N. P. Nirmal, "Quality changes of sea bass slices wrapped with gelatin film incorporated with lemongrass essential oil," International Journal of Food Microbiology, vol. 155, no. 3, pp. 171–178, 2012.

58. J. Gómez-Estaca, P. Montero, B. Giménez, and M. C. Gómez-Guillén, "Effect of functional edible films and high pressure processing on microbial and oxidative spoilage in cold-smoked sardine (Sardina pilchardus)," Food Chemistry, vol. 105, no. 2, pp. 511–520, 2007.

59. H. Y. Song, Y. J. Shin, and K. B. Song, "Preparation of a barley bran protein-gelatin composite film containing grapefruit seed extract and its application in salmon packaging," Journal of Food Engineering, vol. 113, no. 4, pp. 1736–1743, 2012.

60. K. H. Seol, D. G. Lim, A. Jang, C. Jo, and M. Lee, "Antimicrobial effect of κ-carrageenan-based edible film containing ovotransferrin in fresh chicken breast stored at 5°C," Meat Science, vol. 83, no. 3, pp. 479–483, 2009.

61. J. Min, I. Y. Han, and P. L. Dawson, "Antimicrobial gelatin films reduce Listeria monocytogenes on turkey bologna," Poultry Science, vol. 89, no. 6, pp. 1307–1314, 2010.

62. H. Neetoo, M. Ye, and H. Chen, "Bioactive alginate coatings to control Listeria monocytogenes on cold-smoked salmon slices and fillets," International Journal of Food Microbiology, vol. 136, no. 3, pp. 326–331, 2010.

63. Y. Song, L. Liu, H. Shen, J. You, and Y. Luo, "Effect of sodium alginate-based edible coating containing different anti-oxidants on quality and shelf life of refrigerated bream (Megalobrama amblycephala)," Food Control, vol. 22, no. 3-4, pp. 608–615, 2011.

64. O. Mohan, C. N. Ravishankar, K. V. Lalitha, and T. K. Srinivasa Gopal, "Effect of chitosan edible coating on the quality of double filleted

Indian oil sardine (Sardinella longiceps) during chilled storage,"Food Hydrocolloids, vol. 26, no. 1, pp. 167–174, 2012.

65.   N. Tammineni, G. Ünlü, and S. C. Min, "Development of antimicrobial potato peel waste-based edible films with oregano essential oil to inhibit Listeria monocytogenes on cold-smoked salmon," International Journal of Food Science and Technology, vol. 48, no. 1, pp. 1–4, 2013.

66.   P. Appendini and J. H. Hotchkiss, "Review of antimicrobial food packaging," Innovative Food Science and Emerging Technologies, vol. 3, no. 2, pp. 113–126, 2002.

67.   S. Quintavalla and L. Vicini, "Antimicrobial food packaging in meat industry," Meat Science, vol. 62, no. 3, pp. 373–380, 2002.

68.   V. COMA, "Bioactive packaging technologies for extended shelf life of meat-based products," Meat Science, vol. 78, no. 1-2, pp. 90–103, 2008.

69.   H. N. Bassolé and H. R. Juliani, "Essential oils in combination and their antimicrobial properties,"Molecules, vol. 17, no. 4, pp. 3989–4006, 2012.

70.   S. Burt, "Essential oils: their antibacterial properties and potential applications in foods—a review,"International Journal of Food Microbiology, vol. 94, no. 3, pp. 223–253, 2004.

71.   Arana-Sánchez, M. Estarrón-Espinosa, E. N. Obledo-Vázquez, E. Padilla-Camberos, R. Silva-Vázquez, and E. Lugo-Cervantes, "Antimicrobial and antioxidant activities of Mexican oregano essential oils (Lippia graveolens H. B. K.) with different composition when microencapsulated inβ-cyclodextrin,"Letters in Applied Microbiology, vol. 50, no. 6, pp. 585–590, 2010.

72.   C. Liolios, O. Gortzi, S. Lalas, J. Tsaknis, and I. Chinou, "Liposomal incorporation of carvacrol and thymol isolated from the essential oil of Origanum dictamnus L. and in vitro antimicrobial activity," Food Chemistry, vol. 112, no. 1, pp. 77–83, 2009.

73.   G. R. Takeoka, L. T. Dao, R. Y. Wong, and L. A. Harden, "Identification of benzalkonium chloride in commercial grapefruit seed extracts," Journal of Agricultural and Food Chemistry, vol. 53, no. 19, pp. 7630–7636, 2005.

74.   G. Takeoka, L. Dao, R. Y. Wong, R. Lundin, and N. Mahoney, "Identification of benzethonium chloride in commercial grapefruit seed extracts," Journal of Agricultural and Food Chemistry, vol. 49, no. 7, pp. 3316–3320, 2001.

75.   M. Balciunas, F. A. Castillo Martinez, S. D. Todorov, B. D. G. D. M. Franco, A. Converti, and R. P. D. S. Oliveira, "Novel biotechnological

applications of bacteriocins: A review," Food Control, vol. 32, no. 1, pp. 134–142, 2013.

76. C. Chatterjee, M. Paul, L. Xie, and W. A. van der Donk, "Biosynthesis and mode of action of lantibiotics,"Chemical Reviews, vol. 105, no. 2, pp. 633–683, 2005.

77. Breukink, "A lesson in efficient killing from two-component lantibiotics," Molecular Microbiology, vol. 61, no. 2, pp. 271–273, 2006.

78. FDA US Food and Drug Administration. Nisin preparation: affirmation of GRAS status as direct human ingredient, 21 Code of Federal Regulations Part 184, Federal Register, 53, 1988.

79. S. Min, T. R. Rumsey, and J. M. Krochta, "Diffusion of the antimicrobial lysozyme from a whey protein coating on smoked salmon," Journal of Food Engineering, vol. 84, no. 1, pp. 39–47, 2008.

80. N. Terjung, M. Loeffler, M. Gibis, H. Salminen, J. Hinrichs, and J. Weiss, "Impact of lauric arginate application form on its antimicrobial activity in meat emulsions," Food Biophysics, vol. 9, pp. 88–98, 2014.

81. Rodríguez, J. Seguer, X. Rocabayera, and A. Manresa, "Cellular effects of monohydrochloride of L-arginine, Nα- lauroyl ethylester (LAE) on exposure to Salmonella typhimurium and Staphylococcus aureus," Journal of Applied Microbiology, vol. 96, no. 5, pp. 903–912, 2004.

82. Rossi-Márquez, J. H. Han, B. García-Almendárez, E. Castaño-Tostado, and C. Regalado-González, "Effect of temperature, pH and film thickness on nisin release from antimicrobial whey protein isolate edible films," Journal of the Science of Food and Agriculture, vol. 89, no. 14, pp. 2492–2497, 2009.

83. M. F. A. Silveira, N. F. F. Soares, R. M. Geraldine, N. J. Andrade, and M. P. J. Gonçalves, "Antimicrobial efficiency and sorbic acid migration from active films into pastry dough," Packaging Technology and Science, vol. 20, no. 4, pp. 287–292, 2007.

84. Mauriello, E. de Luca, A. La Storia, F. Villani, and D. Ercolini, "Antimicrobial activity of a nisin-activated plastic film for food packaging," Letters in Applied Microbiology, vol. 41, no. 6, pp. 464–469, 2005.

85. M. Ye, H. Neetoo, and H. Chen, "Control of Listeria monocytogenes on ham steaks by antimicrobials incorporated into chitosan-coated plastic films," Food Microbiology, vol. 25, no. 2, pp. 260–268, 2008.

Chapter 12

ANTIMICROBIAL ACTIVITY OF BACTERIOCIN-PRODUCING LACTIC ACID BACTERIA ISOLATED FROM CHEESES AND YOGURTS

En Yang,[1,3] Lihua Fan,[2] Yueming Jiang,[1] Craig Doucette,[2] and Sherry Fillmore[2]

[1]South China Botanical Garden, Chinese Academy of Sciences, Guangzhou 510650, China

[2]Agriculture and Agri-Food Canada, Atlantic Food and Horticulture Research Centre, 32 Main Street, Kentville, NS B4N 1J5, Canada

[3] Kunming University of Science and Technology, Yunnan, China

ABSTRACT

The biopreservation of foods using bacteriocinogenic lactic acid bacteria (LAB) isolated directly from foods is an innovative approach. The objectives of this study were to isolate and identify bacteriocinogenic LAB from various cheeses and yogurts and evaluate their antimicrobial effects on selected spoilage and pathogenic microorganisms *in vitro* as well as on a food commodity.

LAB were isolated using MRS and M17 media. The agar diffusion bioassay was used to screen for bacteriocin or bacteriocin-like substances (BLS) producing LAB using *Lactobacillus sakei* and *Listeria innocua* as indicator organisms. Out of 138 LAB isolates, 28 were found to inhibit these bacteria and were identified as strains of *Enterococcus faecium, Streptococcus thermophilus, Lactobacillus casei* and *Lactobacillus sakei* subsp. *sakei* using 16S rRNA gene sequencing. Eight isolates were tested for antimicrobial activity at 5°C and 20°C against *L. innocua, Escherichia coli, Bacillus cereus, Pseudomonas fluorescens, Erwinia carotovora,* and *Leuconostoc mesenteroides subsp. mesenteroides*using the agar diffusion bioassay, and also against *Penicillium expansum, Botrytis cinerea* and *Monilinia frucitcola* using the microdilution plate method. The effect of selected LAB strains on *L. innocua* inoculated onto fresh-cut onions was also investigated.

Twenty percent of our isolates produced BLS inhibiting the growth of *L. innocua* and/or *Lact. sakei*. Organic acids and/or H_2O_2 produced by LAB and not the BLS had strong antimicrobial effects on all microorganisms tested with the exception of *E. coli*. *Ent. faecium*, *Strep. thermophilus* and *Lact. casei* effectively inhibited the growth of natural microflora and *L. innocua* inoculated onto fresh-cut onions. Bacteriocinogenic LAB present in cheeses and yogurts may have potential to be used as biopreservatives in foods.

INTRODUCTION

Lactic acid bacteria (LAB) are generally recognized as safe (GRAS microorganisms) and play an important role in food and feed fermentation and preservation either as the natural microflora or as starter cultures added under controlled conditions. The preservative effect exerted by LAB is mainly due to the production of organic acids (such as lactic acid) which result in lowered pHs ([Daeschel 1989]). LAB also produce antimicrobial compounds including hydrogen peroxide, CO_2, diacetyl, acetaldehyde, D-isomers of amino acids, reuterin and bacteriocins (Cintas et al. [2001]).

Bacteriocins are ribosomally synthesized antimicrobial peptides that are active against other bacteria, either of the same species (narrow spectrum), or across genera (broad spectrum) (Bowdish et al. [2005]; Cotter et al. [2005]). Bacteriocins may be produced by both gram negative and gram positive bacteria (Savadogo et al. [2006]). In recent years, bacteriocin producing LAB have attracted significant attention because of their GRAS status and potential use as safe additives for food preservation (Diop et al. [2007]). Nisin, produced by *Lactococcus lactis*, is the most thoroughly studied bacteriocin to date and has been applied as an additive to certain foods worldwide (Delves-Broughton et al. [1996]). Substantial work has been done on the effectiveness of nisin on various spoilage and pathogenic microorganisms such as *L. monocytogenes* and its application in different food products ([Staszewski and Jagus 2008]; Freitas et al. [2008]; Schillinger et al. [2001]). Other bacteriocins such as pediocin, may also have potential applications in foods, though they are not currently approved as antimicrobial food additives (Naghmouchi et al. [2007]).

Fresh fruits and vegetables harbor various microorganisms, some of which are psychrotrophic. *L. monocytogenes* is one of the pathogenic bacteria capable of growing at refrigeration temperatures. Moreover, it is also tolerate to acidic pH and salt concentrations up to 10% ([Tasara and Stephan 2006]; Vescovo et al. [2006]). Therefore, it is important to seek biopreservatives that control both spoilage and pathogenic microorganisms including *L. monocytogenes*. Although several studies have indicated the presence of LAB species with antagonistic activity for improving the quality and safety of meat and dairy products (Gagnaire et al. [2009]; [Stiles and Holzapfel 1997]), few reports have involved fresh produce (Trias et al. [2008a, b]). Since the isolation and screening of microorganisms from natural sources has always been the most powerful means for obtaining useful and genetically stable strains for industrially important products (Ibourahema et al. [2008]), in the prsent study, we isolated and identified bacteriocinogenic LAB from cheeses and yogurts, then further evaluated their antimicrobial effects in *vitro* and on fresh-cut produce inoculated with *L. innocua*, a surrogate bacteria for *L. monocytogenes*.

MATERIALS AND METHODS

Isolation and Screening of LAB

LAB were isolated from 7 commercial cheeses: [Tre Stelle® bocconcini cheese (TSB); Tre Stelle® fromage romano cheese (TSR); Saputo® feta cheese (SF); Agropur® signature rondoux pure goat cheese (ASR); Agropur® signature OKA cheese (OKA); Arla® fontina cheese (Arla) and Jarlsberg® firm ripened cheese (JFR)]; plus 3 commercially available yogurts: (Danone, Activia®; Astro® BioBest yogurt; and Bioghurt® Liberte yogurt), and one in- house produced yogurt.

A 25 g sample of cheese was weighed into filtered stomacher bags (Fisher Scientific, Nepean, ON, Canada) and mixed with 225 ml of sterile 0.1% (w/v) peptone water (Fisher Scientific). Samples were blended at 280 rpm for 3 min (400C stomacher circulator, Seward, England). For yogurt samples, 1 ml of sample was added to 99 ml of sterile 0.1% peptone water. All samples were serially diluted and 50uL of each dilution was spiral plated onto de Man, Rogosa and Sharp (MRS) agar, (Oxoid, Basingstoke, UK) and M17 agar (Oxoid). MRS plates were incubated at 37°C under both aerobic and anaerobic conditions for 48 h and M17 plates at 44°C under anaerobic condition for 48 h. All gram positive, catalase negative (3% v/v H_2O_2) isolates were purified and observed under a light microscope. All isolates were coded and stored in MRS or M17 broth containing equal amounts of 30% sterile glycerol at - 80°C.

Antimicrobial Activity and Enzymatic Testing of Bacteriocin Producing LAB

The agar diffusion bioassay described by Herreros et al. ([2005]) was used to screen for bacteriocin producing LAB among the 138 isolates.*L. innocua* (ATCC

33090™) and *Lact. sakei* (ATCC 15521™) were used as indicator bacteria. *L. innocua* was incubated overnight in Brain Heart Infusion broth (BHI, Fisher Scientific, ON, Canada) at 37°C and *Lact. sakei* was cultured anaerobically in MRS broth at 37°C.

One ml of each indicator organism (5×10^5 cfu ml^{-1}) was inoculated into 15 ml of semisolid BHI or MRS agar (BHI or MRS broth plus 0.7% bacteriological agar) maintained at 50°C and then poured into a petri dish. After solidification, three wells (5 mm diameter) were cut and 35 µl of cell-free supernatant (CFS) from each LAB isolate and appropriately adjusted was added to each well. CFS were prepared as follows: one ml of frozen LAB isolate was cultured overnight in 20 ml MRS or M17 broth then 1 ml culture was sub-cultured overnight in 20 ml MRS broth. Cells were removed by centrifuging at 14,000 g for 5 min (Sorvall RC6 PLUS, Thermo-electron Corporation, Asheville, NC, USA). The supernatant was filtered through a sterile 0.22 µm syringe filter (Chromatographic Specialties Inc., ON, Canada) and 35 µl of the unadjusted aliquot of CFS was added to the first well. The remaining CFS was adjusted to pH 6.0 with 1 mol l^{-1} NaOH in order to rule out possible inhibition effects due to organic acids. 35 µl of the pH adjusted CFS was filtered and added to the second well. The neutralized CFS was then treated with 1 mg ml^{-1} of catalase (Sigma-Aldrich Corporation, USA) at 25°C for 30 min to eliminate the possible inhibitory action of H_2O_2 and filtered, then was placed in the third well. If inhibitions zones were found in the third well, the isolates were considered to be able to produce BLS.

The BHI or MRS plates were incubated at 37°C aerobically for 24 h or at 37°C anaerobically for 24 h, respectively. Inhibition zones were measured using an electronic caliper with digital display (Mastercraft^{MD}, Miami, FL, USA). Screenings for bacteriocin producing LAB were repeated twice for each isolate. To confirm production of a proteinaceous compound, CFS displaying antimicrobial potential after acid neutralization and H_2O_2 elimination were treated with 1 mg ml^{-1} of proteolytic enzymes, including proteinase K (33 U mg^{-1}), α-chymotripsin (66 U mg^{-1}), and trypsin (105 U mg^{-1}) (Sigma- Aldrich Corporation, USA) at 37°C for 2 h (Bonadè et al. [2001], Herreros et al. [2005]). Antimicrobial activity of treated CFS was determined by the agar diffusion bioassay as described above.

Identification of BLS Producing LAB using 16S rRNA gene Sequencing

Near full-length 16S rRNA gene sequencing was used to identify unknown bacteriocin producing LAB strains based on the method of Abnous et al.

([2009]). Briefly, genomic DNA was extracted from isolates using the UltraClean™ Microbial DNA Isolation kit (MO Bio laboratories, Inc., Carlsbas, CA, USA). Universal primers F44 (5'RGTTYGATYMTGGCTCAG-3') and R1543 (5'-GNNTACCTTKTTACG ACTT-3') (Abnous et al., [2009]) were used for the amplification of the 16S rRNA gene by PCR. PCR reactions were carried out using a Biometra thermal cycler (Montreal Biotech Inc., Kirkland, QC, Canada) with the following cycle parameters: an initial denaturation at 94°C for 2 min, followed by 35 cycles of denaturation at 94°C for 30 s, annealing at 52°C for 30 s, and elongation at 72°C for 1 min. A final elongation step was performed at 72°C for 5 min. PCR amplicons were separated by agarose gel electrophoresis (0.8% w/v) and visualized by staining with ethidium bromide.

The PCR products were cloned into the psc-A-amp/kan vector using the StrataClone PCR Cloning kit (Stratagene, La Jolla, CA) and transformed into *E. coli* according to the manufacturer's instructions. Transformants were grown overnight in Luria-Bertani broth supplemented with $100 \mu g ml^{-1}$ ampicillin. Plasmids were extracted and purified from selected *E. coli* clones with the UltraClean ™ mini plasmid prep kit (MO Bio Laboratories) according to the manufacturer's recommendations and then sequenced using the Big Dye Terminator v3.1 cycle sequencing kit. A homology search of the sequences was conducted using the BLAST program at the NCBI database.

Thermal Stability of Bacteriocins Produced by LAB Isolates

Based on identified LAB species, isolation source and the size of inhibition zones, eight LAB isolates were chosen for thermal stability tests. The pH adjusted and H_2O_2 eliminated CFS described above were treated at 80 and 100°C for both 60 and 90 min, and at 121°C for 15 min. pH adjusted and H_2O_2 eliminated CFS without any heat treatments served as a controls. Residual antimicrobial activity of heat-treated CFS was determined by the agar diffusion bioassay compared to the control using *L. innocua* as the indicator bacteria.

Antibacterial Effects of LAB Isolates *in vitro*

The antibacterial effects of eight selected LAB isolates on six common food borne pathogens or spoilage organisms at 5 and 20°C were investigated using the agar diffusion bioassay described above. Targeted indicator organisms and their respective media used were as follows: *L. innocua* (BHI, Difco™, Spark, MD), *E. coli* K-12 (ATCC 10798™) (Tryptic Soy Broth, Difco™), *B. cereus* (ATCC 14579™) (Nutrient Broth (NB), Difco™), *Ps. fluorescens* (A7B) (NB), *Erw. carotovora* (ATCC 15713™) (NB) and *Leuc. mesenteroides* subsp.*mesenteroides* (ATCC 8293™) (MRS, Difco™). All strains were cultured aerobically for 24 h at their optimal growth temperatures: 26°C for *Erw. carotovora*, *Ps. fluorescens* and *Leuc. mesenteroides*; 30°C for *B. cereus*; and 37°C for *E. coli* and *L. innocua*. LAB isolates were sub-cultured and the CFS was prepared as previously described. Following inoculation, plates were incubated at 5 and 20°C for 7 d and 24 h, respectively. Inhibition zones were measured as before.

Antifungal effects of LAB Isolates *in vitro*

The microdilution method described by Lavermicocca et al. ([2003]) with some modifications was used to test the eight LAB isolates against *P. expansum* (Pex 03–10.1), *B. cinerea* (B94-b) and *M. fructicola* (Mof 03–25) at 5 and 20°C. Fungi were obtained from the culture collection at our Research Centre, AAFC. The stock cultures were stored as spore or mycelial suspensions in 15% glycerol (v/v) at −80°C. Conidia of *P. expansum* were collected from 4 d-old cultures grown on potato dextrose agar (PDA) at 25°C. Conidia of *B. cinerea* were collected from 12 d-old cultures grown under a 12 h light/dark cycle on Pseudomonas Agar F at 22°C. Conidia of *M. fructicola* were collected from 12 d-old cultures grown on modified V-8™ medium at 25°C. Using sterile distilled water, the density of spore suspensions were diluted to 2×10^5 spores ml^{-1} as determined using a haemocytometer (Hausser Scientific, PA, USA).

Microdilution tests were performed in sterile 96-well micro dilution plates (Costar 3370, Corning Incorporated, Corning, NY, USA). 200 μl of test solution consisting of 185 μl LAB CFS inoculated with 15 μl of conidial suspension was dispensed into the wells. Microdilution plates were incubated at either 20 or 5°C and the optical density (OD) at 580 nm was recorded at specific time intervals using a microtiter plate reader/ spectrophotometer (Spectra MAX 190, Molecular Devices, CA, USA). For the 20°C plates, OD values were measured at 0, 24, 40, 48 or 72 h, while at 5°C, they were measured at 0, 48, 72, 96 or 120 h. For each fungus, both positive controls containing 185 μl MRS broth and 15 μl conidial suspension and negative controls containing 185 μl LAB CFS and 15 μl dead conidial suspension were prepared and monitored. Experiments were repeated three times.

In vivo Testing of Selected LAB on Microflora Naturally Occurring on fresh-cut Onions

Ent. faecium, *Strep. thermophilus* and *Lact. casei* were chosen for *in vivo* tests. Fresh-cut yellow onions processed at a commercial facility were supplied by Nova Agri Inc. (Canning, NS). *Ent. faecium* and *Lact. casei* were incubated anaerobically in MRS broth at 37°C for 16 h. *Strep. thermophilus* was incubated anaerobically in M17 broth at 43°C for 16 h. LAB were sub-cultured twice and then centrifuged at 14,000 g for 5 min. Pellets of LAB were washed using sterile distilled water, centrifuged, re-suspended and inoculated onto the batches of diced onions using a calibrated TLC sprayer to give a final concentration of 5×10^5 cfu g^{-1}. Diced onions inoculated with sterile distilled water served as a control. Samples were left to dry for 10 min, then 100 g was transferred to food grade bags (Golden Eagle-VH-62 190) and sealed using an electric sealer (FoodSaver V2490, Tilia International Inc. US). The densities of naturally occurring (indigenous) LAB, *Pseudomonas* sp., yeasts and moulds, *Listeria* sp. and *Enterobacteriaceae* were enumerated following 0, 3, 6, 9 and 12 d of storage at 4°C. LAB were cultured anaerobically on MRS Agar at 37°C for 48 h; *Pseudomonas* sp. on Pseudomonas selective medium at 30°C for 48 h; yeasts and moulds on PDA supplemented with chloramphenicol (Difco™) at 25°C for 48 h; and *Enterobacteriaceae* on Violet Red Bile Agar (VRBG)(Difco™) at 37°C for 48 h. Colonies were counted using an aCOLyte automated colony counter (Synbiosis, Cambridge, England) and expressed as cfu g^{-1} of diced onions.

In vivo testing of Selected LAB on *L. innocua* Inoculated on Fresh-cut onions

L. innocua was sub-cultured into BHI broth and incubated at 37°C for 24 h. The selected LAB strains and inoculation procedures were the same as described above. The final density of *L. innocua* inoculated on fresh-cut onions was 5×10^3 cfu g^{-1}. After drying for 10 min, the LAB species were inoculated respectively onto the fresh-cut onions at a density of 5×10^5 cfu g^{-1}. A batch of fresh-cut onions inoculated with *L. innocua* alone was used as the control. Samples (100 g/ bag) were stored at 4°C and *Listeria* sp. and LAB were enumerated following 0, 3, 6, 9 and 12 d. *Listeria* sp. was cultured on *Listeria* selective medium at 35°C for 48 h and LAB on MRS agar at 37°C for 48 h anaerobically.

Statistical Analysis

The initial screening for bacteriocin producing LAB from 138 LAB isolates was repeated twice. For antimicrobial tests, a three repetition split-split plot design was used with the eight LAB isolates on the main plot versus the six bacteria (or three fungi) on the sub plot, which was then split into two temperatures (5 and 20°C). For the *in vivo* testing, non-inoculation or inoculation of *L. innocua* was designed on the main plot with the three LAB strains used to investigate the treatment effect on naturally occurring microflora and introduced *L. innocua* on the onions. The sub plot was represented by sample removal at days 0, 3, 6, 9 and 12. Data were analyzed using the ANOVA directive and standard errors of mean (SEM) option of GenStat® (12th Edition, VSN International Ltd, Hemel Hempstead UK, 2009). The results of a Principal Components Analysis (PCA) were discussed in terms of component scores.

Results

Isolation of Potential LAB and Screening for Bacteriocin Producing LAB

A total of 160 potential LAB strains were isolated from 11 different cheeses and yogurts. Eighty seven isolates were obtained from MRS agar and 73 from M17 medium. Of these, 138 were gram positive and catalase negative, and were cocci or rod in shape. Twenty percent of isolates showed antimicrobial activity against at least one indicator that is presumed to be attributable to BLS, which was determined after the neutralization of pH, and the elimination of H_2O_2 from the CFS (Table 1). All 28 LAB isolates had BLS inhibitory effect on *L. innocua*, while BLS produced by 20 isolates were effective against *Lact. sakei*.

Table 1: Inhibition zones (mm) of 28 LAB isolated from commercial cheeses and yogurts using *L. innocua* and *Lact. sakei* as indicator bacteria

Isolate	GenBank accession number	L. innocua			Lact. sakei		
		Control[a]	-Acid[b]	BLS[c]	Control	-Acid	BLS
Arla-1	JX275802	8.65±0.23[d]	7.10±0.15	5.82±0.09	5.12±0.01	5.97±0.01	5.74±0.01
Arla-10	JX275803	8.96±0.30	8.15±0.21	6.47±0.03	6.51±0.18	5.59±0.12	5.50±0.03
Arla-11	JX275804	8.83±0.14	7.71±0.09	6.41±0.06	5.76±0.07	6.90±0.05	6.31±0.10
Arla-12	JX275805	9.26±0.09	8.57±0.10	6.47±0.14	5.82±0.18	6.67±0.10	5.83±0.15
Arla-14	JX275806	9.33±0.05	8.40±0.08	5.87±0.04	5.98±0.17	7.22±0.32	7.41±0.14
Arla-16	JX275807	8.47±0.07	7.84±0.20	5.81±0.17	5.09±0.01	5.98±0.03	5.15±0.01
Arla-17	JX275808	8.52±0.10	7.82±0.12	5.57±0.38	--[e]	--	--
Arla-18	JX275809	9.28±0.08	8.84±0.08	6.09±0.10	6.94±0.08	8.49±0.07	7.95±0.10
ASR-1	JX275810	9.02±0.03	7.88±0.07	5.54±0.03	6.85±0.11	7.68±0.07	6.52±0.07
ASR-5	JX275811	10.02±0.10	8.99±0.18	6.13±0.06	7.16±0.04	6.57±0.03	5.88±0.32
ASR-6	JX275812	10.38±0.07	9.50±0.15	6.59±0.06	8.30±0.10	8.59±0.14	8.06±0.07
ASR-7	JX275813	8.89±0.12	7.21±0.15	5.97±0.01	--	--	--
JFR-1	JX275814	8.62±0.01	7.82±0.16	6.01±0.25	5.10±0.01	5.82±0.01	5.51±0.09
JFR-3	JX275815	8.93±0.27	7.42±0.19	6.34±0.07	--	--	--

JFR-4	JX275816	9.31±0.07	7.74±0.11	5.46±0.04	--	--	--
JFR-5	JX275817	8.22±0.13	7.18±0.05	5.93±0.14	--	--	--
TSB-8	JX275827	10.91±0.15	9.32±0.14	6.28±0.08	10.15±0.11	9.88±0.16	8.90±0.16
OKA-1	JX275818	9.13±0.12	8.18±0.01	6.23±0.01	--	--	--
OKA-3	JX275819	8.47±0.20	7.36±0.31	5.42±0.01	--	--	--
OKA-4	JX275820	8.33±0.03	7.59±0.01	5.96±0.06	--	--	--
OKA-8	JX275821	8.84±0.19	8.24±0.26	5.93±0.07	5.97±0.12	6.14±0.17	5.68±0.09
OKA-9	JX275822	8.60±0.10	7.69±0.13	6.16±0.03	5.30±0.01	6.10±0.07	6.21±0.28
OKA-14	JX275823	10.12±0.20	9.08±0.30	6.03±0.09	8.31±0.09	8.65±0.15	8.01±0.10
OKA-15	JX275824	9.89±0.16	9.60±0.08	6.55±0.08	8.75±0.05	9.15±0.20	8.76±0.13
OKA-19	JX275825	11.24±0.06	9.62±0.10	6.54±0.11	8.78±0.06	9.05±0.22	8.67±0.13
OKA-22	JX275826	9.39±0.04	8.52±0.07	5.92±0.14	7.94±0.15	8.22±0.12	8.43±0.25
Yog-3 L	JX275800	10.42±0.16	9.26±0.03	6.11±0.08	8.36±0.24	8.65±0.12	7.79±0.13
Yog-3S	JX275801	10.65±0.09	9.14±0.10	6.17±0.03	8.23±0.09	8.83±0.13	8.12±0.15

[a] CFS of LAB isolates without any treatment.

[b] CFS with pH neutralized to 6.0.

[c] CFS with pH neutralized to 6.0 and H_2O_2 eliminated.

[d] Means±standard deviation of two replicate experiments.

[e] No inhibition detected.

The BLS from all 28 LAB isolates lost their anti-listerial activity following treatment with proteinase K, α-chymotripsin and/or trypsin (Table 2). However, when *Lact. sakei* was used as an indicator, the BLS produced by the different LAB stains had varying activity following treatment with these enzymes. Nevertheless, the lost antimicrobial ability following treatment with proteolytic enzymes indicated the proteinaceous nature of the BLS.

Table 2: The effect of α-chymotripsin, protease and trypsin treatment on inhibitory activity of BLS produced by the 28 LAB isolates against *L. innocua* and *Lact. sakei*

Isolates	control[a]	*L. innocua*(indicator)			control	*Lact. sakei*(indicator)		
		Enzyme treatment				Enzyme treatment		
		α-chymotripsin	protease	trypsin		α-chymotripsin	protease	trypsin
Arla-1	+[b]	--[c]	--	--	+	--	--	--
Arla-10	+	--	--	--	+	--	--	--
Arla-11	+	--	--	--	+	--	--	--
Arla-12	+	--	--	--	+	--	--	--
Arla-14	+	--	--	--	+	--	--	--
Arla-16	+	--	--	--	+	--	--	--
Arla-17	+	--	--	--	n.d	n.d	n.d	n.d
Arla-18	+	--	--	--	+	--	--	--
ASR-1	+	--	--	--	+	--	--	--
ASR-5	+	--	--	--	+	--	--	--
ASR-6	+	--	--	--	+	--	+	--
ASR-7	+	--	--	--	n.d	n.d	n.d	n.d
JFR-1	+	--	--	--	+	--	--	--
JFR-3	+	--	--	--	n.d	n.d	n.d	n.d
JFR-4	+	--	--	--	n.d	n.d	n.d	n.d
JFR-5	+	--	--	--	n.d	n.d	n.d	n.d
TSB-8	+	--	--	--	+	--	+	+
OKA-1	+	--	--	--	n.d	n.d	n.d	n.d
OKA-3	+	--	--	--	n.d	n.d	n.d	n.d
OKA-4	+	--	--	--	n.d	n.d	n.d	n.d
OKA-8	+	--	--	--	+	+	--	+
OKA-9	+	--	--	--	+	--	--	--
OKA-14	+	--	--	--	+	--	--	+
OKA-15	+	--	--	--	+	--	+	+
OKA-19	+	--	--	--	+	--	+	+
OKA-22	+	--	--	--	+	--	--	--
Yog-3 L	+	--	--	--	+	--	--	+
Yog-3S	+	--	--	--	+	--	--	+

[a] Control=pH-neutralized and H_2O_2-eliminated CFS without addition of enzymes.

[b] Inhibition zones present.

[c] No inhibition zone.

[d] n.d. = not determined.

Identification of LAB Isolates using 16S rRNA gene Sequences

Near-full length sequencing of the 16S rRNA gene for the 28 bacteriocin producing LAB identified the isolates as follows: 24 strains of *Ent. faecium*, two strains of *Strep. thermophilus*, one strain of *Lact. casei* and one strain of *Lact. sakei* subsp. *sakei*.

Thermal Stability of BLS Following heat Treatments

The BLS produced by the eight selected LAB isolates were heat-treated at 80 and 100°C for 60 and 90 min, respectively. BLS were thermal stable at these heat conditions above as their inhibitory effects against *L. innocua* were retained (Table 3). However, BLS were sensitive to autoclaving at 121°C for 15 min displaying either smaller or no inhibition zones compared to the control. Isolates *Strep. thermophilus* (ASR-1) and *Lact. casei* (JFR-5) totally lost their anti-listerial activity subsequent to exposure to 121°C for 15 min.

Table 3: Thermal stability of BLS produced by the selected LAB following various heat treatments (*L. innocua* was used as an indictor bacterium)

Anti-listerial activity of BLS after heat treatment							
Isolates	CFS*	BLS**	80°C, 60 min	80°C, 90 min	100°C, 60 min	100°C, 90 min	121°C, 15 min
Arla-18	+++[a]	++[b]	++	++	++	++	++
ASR-1	+++	++	++	++	++	+	--[d]
ASR-6	+++	++	++	++	++	++	++
JFR-1	+++	++	++	++	++	++	+
JFR-5	++	+[c]	+	+	+	+	--
TSB-8	+++	++	++	++	++	++	+
OKA-14	+++	++	++	++	++	++	+
Yog-3S	+++	++	++	++	++	++	+

[a] Inhibitory zone >10 mm.

[b] Inhibitory zone = 7–10 mm.

[c] Inhibitory zone = 5–7 mm.

[d] No inhibitory zone.

* CFS: cell free supernatant from LAB culture.

** BLS: CFS following the pH neutralization and H_2O_2 elimination.

Action of LAB Isolates Against Six Bacteria *in vitro*

The CFS from the eight LAB isolates significantly inhibited the growth of all bacteria tested ($p < 0.05$) at 5 and 20°C with the exception of *E. coli* (Table 4). Neutralized CFS from isolates Arla-18, ASR-6, JFR-1, OKA-14, and Yog-3S continued to inhibit the growth of *Leuc. mesenterioides* at 5°C. After pH neutralization and H_2O_2 elimination, the CFS from the eight LAB isolates had inhibitory effects only on *L. innocua* but not the other test bacteria suggesting that organic acids and /or H_2O_2 produced by LAB had strong antimicrobial effects on bacteria tested.

Table 4: Effect of 8 selected LAB isolates on various bacterial species *in vitro*

Isolates	Ps. fluorescens (Control[a])		B. cereus (Control)		Erw. carotovora (Control)		Leuc. mesenteroides (Control)		Leuc. mesenteroides (−Acid[b])		L. innocua (Control)		L. innocua (−Acid)		L. innocua (BLS[c])	
	5°C	20°C	5°C	20°C	5°C	20°C	5°C	20°C	5°C	20°C	5°C	20°C	5°C	20°C	5°C	20°C
Arla-18	7.95	9.50	9.17	8.12	10.80	8.57	5.60	5.26	5.48	−[d]	16.75	15.58	14.96	14.76	9.61	11.15
ASR-1	8.85	9.87	8.53	8.13	10.44	7.53	5.47	5.18	−	−	15.31	12.46	14.33	11.46	8.97	8.71
ASR-6	8.29	10.23	9.50	8.17	10.45	8.12	5.91	5.35	5.68	−	16.87	14.86	14.95	13.80	9.24	11.16
JFR-1	7.54	9.60	9.52	8.11	10.80	8.93	5.71	5.36	5.57	−	17.27	14.42	15.84	13.23	9.64	11.38
JFR-5	9.42	10.42	10.09	9.02	11.24	9.42	−	−	−	−	15.30	12.65	14.18	11.71	7.92	9.29
OKA-14	8.09	9.26	9.43	8.24	11.23	8.65	5.78	5.30	5.49	−	16.84	14.27	15.42	13.55	8.54	10.41
TSB-8	8.61	9.81	8.84	7.36	9.45	8.80	5.45	5.21	−	−	18.77	13.62	17.34	13.10	8.60	8.49
Yog-3S	8.02	9.96	10.35	8.60	11.62	9.37	5.64	5.47	5.46	−	16.76	14.14	15.12	13.33	8.62	11.07
SEM	0.62	0.62	0.63	0.63	0.58	0.58	0.10	0.10	0.81		0.90	0.90	8.45	8.45	0.87	0.87

[a] CFS without any treatment.

[b] CFS with pH neutralized to 6.0.

[c] CFS with pH neutralized to 6.0 and H_2O_2 eliminated.

[d] No inhibition zone shown.

Inhibition zones (mm) were determined using an agar diffusion assay when target bacteria were used as indicator organisms.

In conducting a principal components analysis, we intended to figure out that the principal component had as high a variance as possible (accounted for as much of the variability in our data as possible). It was found that score 1 was a contrast between *L. innocua* (control), *L. innocua* (pH neutralized), *Leuc. mesenteroides* (pH neutralized), *Erw. carotovora* (control), *B. cereus* (control) vs *Ps. fluorescens* (control) (Figure 1). A high score 1 was dominated by the values (inhibitory zones) at 5°C except for *Strep. thermophilus* and *Lact. casei*. All the CFS produced by *Ent. faecium* (Arla-18, ASR-6, JFR-1, OKA-14, and Yog-3S) had a similar antibacterial activity towards all the bacteria tested at both temperatures. The effects of untreated CFS and neutralized CFS from the eight LAB isolates on *L. innocua* highly correlated at 5 and 20°C.

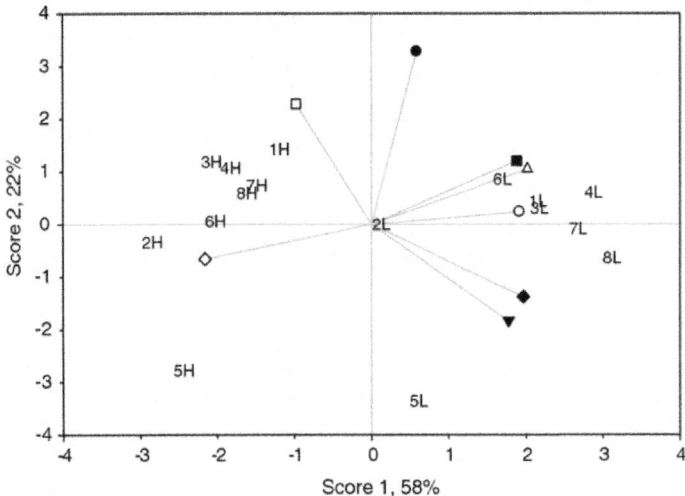

Figure 1: Principal component analysis (a biplot of antibacterial data of CFS produced by 8 selected bacteriocinogenic

LAB cultured at 5 and 20°C). Indicator organisms included: *Leuc. mesenteroides* CFS, untreated (●), *Leuc. mesenteroides* CFS, pH neutralized (○), *B. cereus* CFS, untreated (▼), *L. innocua* CFS, untreated (Δ), *L. innocua* CFS, pH neutralized (■), *L. innocua* CFS, pH neutralized and H_2O_2 eliminated (□), *Erw. carotovora* CFS, untreated (♦), and *Ps. fluorescens* CFS, untreated (◊). H-20°C; L-5°C; 1-Arla-18 (*Ent. faecium*); 2-ASR-1 (*Strep. thermophilus*); 3-ASR-6 (*Ent. faecium*); 4-JFR-1 (*Ent. faecium*); 5-JFR-5 (*Lact. casei*); 6-TSB-8 (*Strep. thermophilus*); 7-OKA-14 (*Ent. faecium*); 8-Yog-3S (*Ent. faecium*).

Action of LAB Isolates on Three fungi *in vitro*

No significant difference among the eight LAB isolates was observed with respect to inhibition of spore germination of *M. fructicola, B. cinerea* and *P. expansum* at either 5 or 20°C ($p > 0.05$). Therefore, data for antifungal ability of untreated CFS, pH neutralized CFS, and pH neutralized and H_2O_2 eliminated CFS from the eight isolates were averaged, respectively, and compared to the positive controls. At 5 and 20°C, untreated CFS was the most effective at inhibiting fungal growth compared to their controls (Figure 2). Neutralized CFS also had an inhibitory effect on the growth of these fungi ($p < 0.001$). However, pH neutralized and H_2O_2 eliminated CFS showed no significant effect on fungal growth at 20°C (Figure 2a1, 2b1, and 2c1). Similar results were found at 5°C as the CFS, but not the BLS, produced by the eight LAB were effective in controlling the fungal growth (Figure 2a2, 2b2, and 2c2).

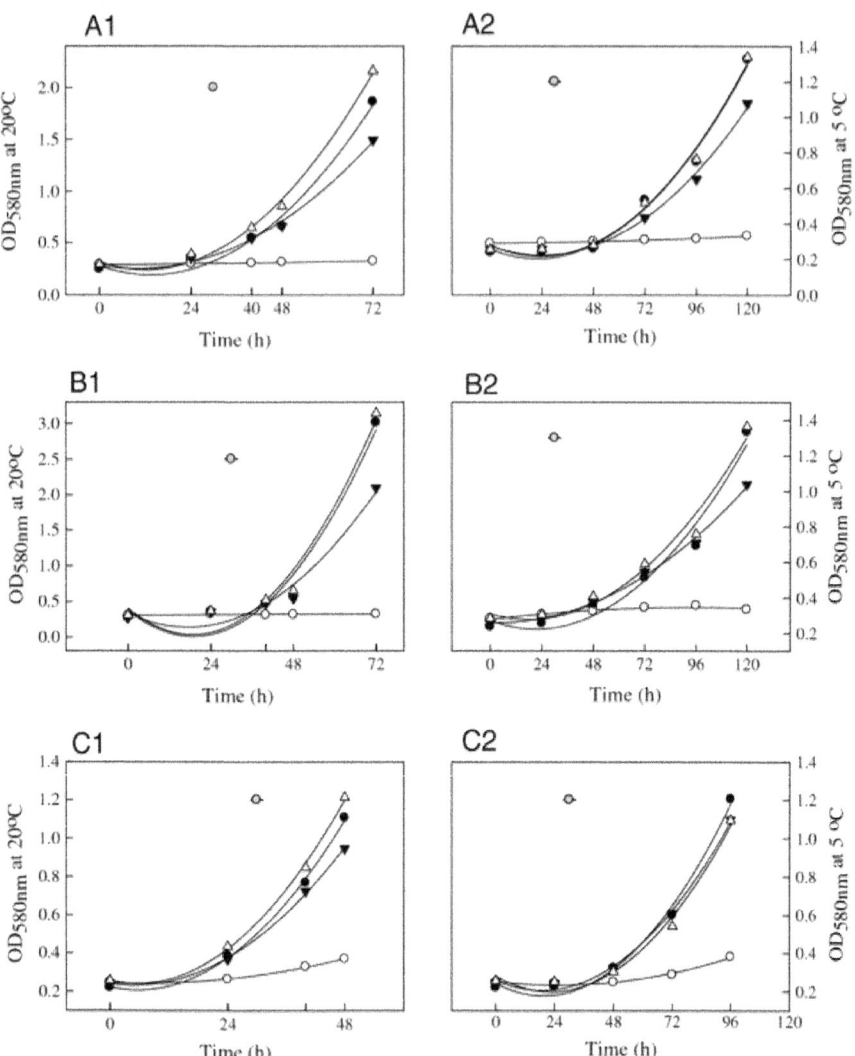

Figure 2: Influence of BLS-producing LAB on fungal growth (a1, a2 = *M. fructicola* at 20 or 5°C ; b1, b2 = *B. cinerea* at 20 or 5°C; c1, c2 = *P. expansum* at 20 or 5°C) after incubation at 20°C for 72h or 5°C for 120h.

Positive control = Fungal spores grown in MRS broth (●), fungal spores in untreated LAB CFS (○), fungal spores in pH neutralized CFS (▼), fungal spores in CFS after pH neutralization and H_2O_2 elimination (△). The vertical bar represents standard errors of mean.

Effect of selected LAB on Microflora naturally occurring on Fresh- cut Onions

Three LAB species inoculated on fresh-cut onions significantly inhibited *Pseudomonas* sp. ($p=0.02$) (Figure 3a) and lactose-positive (Lac$^+$)*Enterobacteriaceae* ($p=0.042$) (Figure 3b) during 12 d when stored at 5°C compared to non-LAB inoculated controls. On day 12, Lac$^+$*Enterobacteriaceae* increased to 4.5 log cfu g^{-1} on control samples while levels of 0.3, 2.2 and 1.1 log cfu g^{-1} were obtained in *Strep. thermophilus*, *Lact. casei* and *Ent. facium* treated samples, respectively. However, no significant differences were found in yeast and mould levels between LAB inoculated and control samples (Figure 3c). Moreover, no dramatic decrease in the LAB inoculated on the fresh-cut onions was observed over the 12 d storage period (Figure 3d) and no naturally occurring *Listeria* sp. was detected in this study.

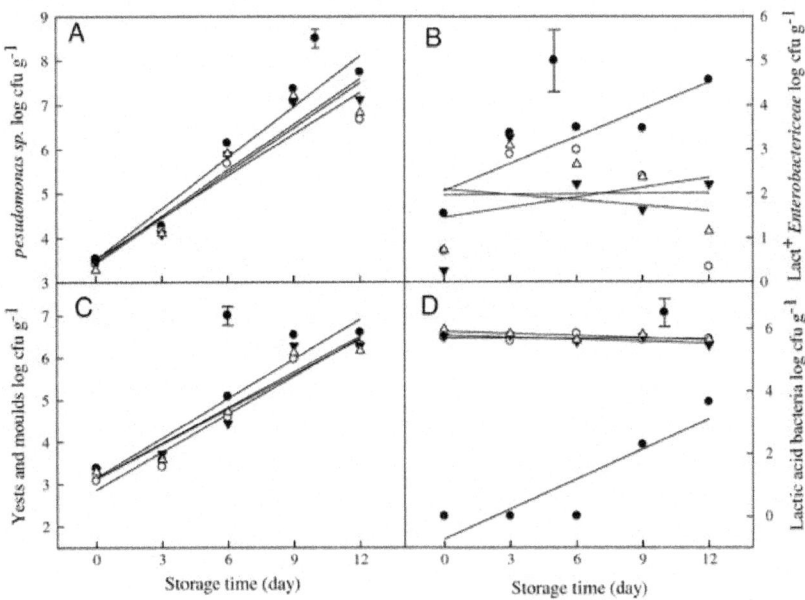

Figure 3: Influence of BLS-producing LAB on the growth of naturally occurring microorganisms.

Pseudomonas sp. (**a**), lactose-positive*Enterobacteriaceae* (**b**), yeasts and moulds (**c**), and LAB viability (**d**) on fresh-cut onion during 12 d storage at 5°C. Fresh- cut onions without LAB inoculation (●), inoculated with *Strep. thermophilus* (○), inoculated with *Lact. casei* (▼), inoculated with *Ent. facium* (Δ). Fresh-cut onions were inoculated with LAB to a density of 5×10^5 cfu g^{-1}. The vertical bar represents standard errors of mean.

Listeria challenge test on fresh-cut onions

The initial density of *L. innocua* subsequent to their introduction onto fresh-cut onions was 3.2 log cfu g⁻¹ on day 0. During the following 12 d at 5°C, the growth of *L. innocua* was significantly inhibited by the LAB ($p = 0.042$, Figure 4a) as its levels were reduced by 1.0, 1.6, and 1.6 log cfu g⁻¹ in samples treated with *Strep. thermophilus*, *Lact. casei* and *Ent. faecium*, respectively. In contrast, levels of LAB inoculated on the fresh-cut onions remained at the initial inoculation level of 5×10^5 cfu g⁻¹ over the 12 d storage at 5°C (Figure 4b).

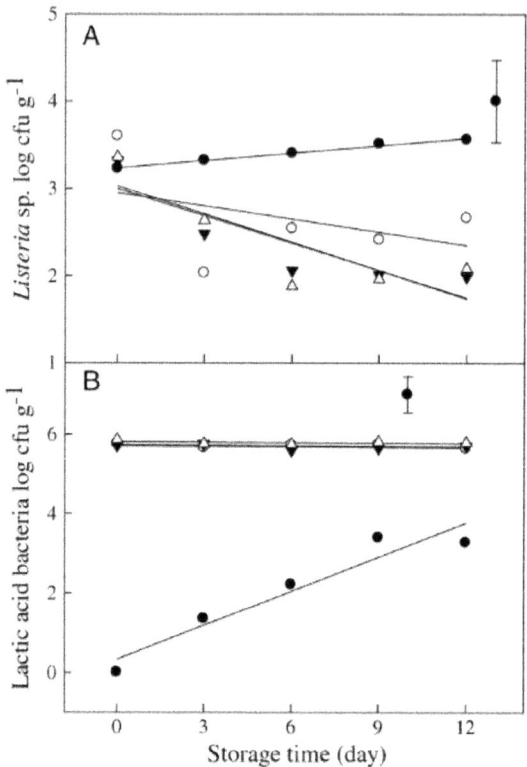

Figure 4: The viability of *L. innocua* (a) and LAB (b) inoculated onto fresh-cut onions during 12 d storage at 5°

Control (in Figure 4a)=fresh-cut onions inoculated with *L. innocua* at 5×10^3 cfu g⁻¹ (●); control (in Figure 4b)=LAB naturally occurring on fresh-cut onions (●). Samples inoculated with *Strep. thermophilus* at 5×10^5 cfu g⁻¹ after introduction of *L. innocua* (○), samples inoculated with *Lact. casei* at 5×10^5 cfu g⁻¹ after introduction of *L. innocua* (▼), samples inoculated with *Ent. faecium* at 5×10^5 cfu g⁻¹ after introduction of *L. innocua* (Δ). The vertical bar represents standard errors of mean.

DISCUSSION

In the present work we isolated, identified and characterized bacteriocinogenic LAB indigenous to cheese and yogurt and explored their potential as biopreservatives. Two types of media were chosen for the isolation of LAB. First, MRS was used as a medium for LAB which gave a general scope of the flora present in the samples ([Reuter 1985]). M17 agar was also used as a selective medium for the isolation of*Streptococci* sp. Twenty percent of our isolates produced BLS effective against *L. innocua* and 16S rRNA gene sequencing determined that these isolates belonged to four LAB species. ([Sharpe 2009]) reported detecting 8.7% bacteriocinogenic strains among 92 LAB isolated from fresh-cut vegetable products, whereas Sezer and Güven ([2009]) screened 12,700 LAB isolates from milk and meat products and found only 35 exhibited bacteriocin production. Therefore, the choice of food source and media are important for the successfully isolation of bacteriocinogenic LAB.

Using 16S rRNA gene sequencing to identify the 28 bacteriocinogenic LAB isolates, we found that the genus *Enterococcus* was predominant representing 85.7% of the isolates. Another 7.1% were *Streptococcus* sp. whereas the other 7.1% were *Lactobacillus* sp. Furthermore, all the*Enterococcus* sp. were identified as *Ent. faecium*, which was also found to be the most frequently isolated species in cured and semi-cured cheese (Cogan et al. [1997]). López-Díaz et al. ([2000]) found that 40.4% of nearly 500 strains isolated from Valdeón cheese were*Enterococcus* sp.

Bacteriocins can be broken down by some proteolytic enzymes leading to a loss in their antimicobial activity. In our study, we used *L. innocua*or *Lact. sakei* as indicators, and the BLS produced by different LAB stains had various inhibitory effects following treatment with proteolytic enzymes (Table 2). Similar behavior was observed by Khalil et al. ([2009]) with a *Bacillus megaterium*19 strain isolated from a mixture of fermented vegetable wastes. They found that pepsin and trypsin treatment inhibited the bacteriocin activity against *Staphylococcus aureus*more than *Salmonella typhimurium*. Cherif et al. ([2001]) used pepsin, papain, trypsin, chymotrypsin, proteinase K, lysozyme, catalase, DNase and RNase to treat thuricin 7, a bacteriocin produced by *Bacillus thuringiensis* BNG 1.7. They found that the inhibitory activity was only susceptible to proteinase K.

The thermal stability at 80 and 100°C (up to 90 min) (Table 3) of BLS produced by our bacteriocinogenic LAB isolates may constitute an advantage for potential use as biopreservatives in combination with thermal processing in order to preserve food products. However, it should be noted that the antimicrobial effect of these BLS on *L. innocua* and/or *Lact. sakei* was markedly decreased or completely lost after treatment at 121°C for 15 min. [Todorov and Dicks (2009)] reported that bacteriocin ST44AM remained stable at 25, 30, 45, 60 and 100°C for 120 min. However, the activity of this bacteriocin against *L. ivanovii* subsp. *ivanovii* ATCC 19119 was reduced from 3.3×10^6 AU ml^{-1} to 4.1×10^5 AU ml^{-1} after exposure at 121°C for 20 min. Similar results were reported for a bacteriocin produced by *Lactobacillus* CA44 (Joshi et al. [2006]) and also thuricin 7 from *B. thuringiensis* BMG1.7 (Cherif et al. [2001]).

In the present study, eight BLS producing LAB isolates were tested for their antimicrobial effects on three gram-negative, and three gram-positive bacteria, as well as three common spoilage fungi. The results showed that untreated CFS inhibited all test bacteria and fungi except for *E. coli*. However, after pH neutralization and H_2O_2 elimination, the CFS inhibited only *L. innocua*. Similar results were reported by [Sharpe (2009)] as they found that the BLS produced by *Lact. lactis* and *Ent. faecium* were able to control *Lact. sakei* and *L. innocua*. However, the strong antimicrobial effects associated with *Lact. lactis* and *Ent. faecium* in our study appeared to be a direct result of the organic acids and the H_2O_2 present in the CFS rather than the BLS. In other research, the antimicrobial activity of 12 enterococci strains was confirmed by Hajikhani et al. ([2007]), showing that *Ps. aeruginosa* and *Proteus vulgaris* were sensitive to compounds produced by enterococci but *E. coli* and *Yersinia enterocolitica* were not affected. Trias et al. ([2008a]) treated apple wounds and lettuce cuts with the LAB strains resulting in reduced counts of *Salmonella typhimurium* and *E. coli* by 1 to 2 log cfu g^{-1}, whereas the growth of *L. monocytogenes* was completely inhibited. Cheikhyoussef et al. ([2010]) investigated bifidin I from *Bifidobacterium infantis* BCRC 14602, and reported an increase in bacteriocin activity from 2. 6×10^2 AU mg^{-1} for neutralized CFS to 3.7×10^5 AU mg^{-1} for the purified bacteriocin. Simova et al. ([2009]) achieved a 10^5 -fold increase in bacteriocin activity after a single peak was assessed upon C2/C18 reversed-phase liquid chromatography purification. The relative low concentration of bacteriocin in the CFS likely contributed to the BLS not being able to inhibit all bacteria and fungi examined in this study. Further investigation is needed to establish the method of direct use of LAB as protective cultures or that of purified bacteriocins on foods.

It has been hypothesized that organic acids act on the cytoplasmic membrane by neutralizing its electrochemical potential and increasing its permeability, thus leading to bacteriostasis and eventual death of susceptible bacteria (Dalié et al. [2010]). The same hypothesis could also explain the susceptibility of some fungal cultures to organic acids (Batish et al. [1997]). However, [Elsanhoty (2008)] suggested that the antifungal effect of LAB could not simply be the result of low pH but is most probably due to the formation and secretion of pH dependent antifungal metabolites. Magnusson and Schnürer ([2001]) found that the antifungal metabolite produced by the *Lact. coryniformis* subsp. *coryniformis* Si3 strain was a small highly heat stable peptide and its activity was stable at pHs between 3.0 and 4.5, but rapidly decreased between 4.5 and 6.0. No inhibitory activity was detected above pH 6.0.

In the present study, selected LAB strains were also inoculated onto fresh-cut onions to investigate *in vivo* application of these isolates. These isolates significantly inhibited the growth of *Pseudomonas* sp. and Lac⁺ *Enterobacteriaceae* during storage at 5°C (Figure 3a, 3b). Similar results were reported by [Sharpe (2009)]; when *L. lactis* and *Ent. faecium* were inoculated onto fresh-cut salads, the growth of *Pseudomonas* sp., yeasts and total coliforms were remarkably reduced. Vescovo et al. ([1995, 1996]) reported that the inoculation of ready-to-use vegetables with selected strains of LAB effectively controlled the growth of undesirable bacteria. In our challenge tests, *L. innocua* inoculated onto fresh-cut onions was reduced by 1 to 1.6 log cfu g^{-1} after 12 d storage at 5°C due to the presence of selected LAB strains. Although the inoculated LAB loads did not increase during the storage period, they still significantly inhibited the growth of *L. innocua*. Further investigations into the mechanisms of inhibition and determination of the optimal growth conditions for LAB to produce BLS are necessary. Moreover, more studies are required to envisage the effectiveness of LAB and BLS on other food products.

ACKNOWLEDGMENTS

Authors would like to thank Drs. Greg Bezanson and Tim Ells at AAFC for their critical review and helpful suggestions in preparing the manuscript, thank Judy Kwan and Steve Brooks at Health Canada for doing the gene sequencing, and thank the MOE-AAFC Ph.D training program for supporting Ph.D students.

COMPETING INTEREST

The authors declare that they have no competing interests.Contribution no. 2379 of the Atlantic Food and Horticulture Research Centre, Agriculture and Agri-Food Canada (AAFC).

REFERENCES

1.   Abnous K, Brooks SPJ, Kwan J, Matias F, Green-Johnson J, Selinger LB, Thomas M, Kalmokoff M: Diets enriched in oat bran or wheat bran temporally and differentially alter the composition of the fecal community of rats. *J Nutr* 2009, 139:2024–2031.

2.   Batish VK, Roy U, Lal R, Grover S: Antifungal attributes of lactic acid bacteria – A review. *Critical Reviews in Biotechnology* 1997,17:2009–2225.

3.   Bonadè A, Murelli F, Vescovo M, Scolari G: Partial characterization of a bacteriocin produced by Lactobacillus helveticus. *Lett Appl Microbiol* 2001, 33:153–158.

4.   Bowdish DM, Davidson DJ, Hancock RE: A re-evaluation of the role of host defence peptides in mammalian immunity. *Curr Protein Pept Sci* 2005, 6:35–51.

5.   Cheikhyoussef A, Cheikhyoussef N, Chen H, Zhao J, Tang J, Zhang H, Chen W: Bifidin I – A new bacteriocin produced by Bifidobacterium infantis BCRC 14602: Purification and partial amino acid sequence. *Food Control* 2010, 21:746–753.

6.   Cherif A, Ouzari H, Daffonchio D, Cherif H, Slama KB, Hassen A, Jaoua S, Boudabous A: Thuricin 7: a novel bacteriocin produced by Bacillus thuringiensis BMG1.7, a new strain isolated from soil. *Lett Appl Microbiol* 2001, 32:243–247.

7.   Cintas LM, Casaus MP, Herranz C, Nes IF, Hernández PE: Review: Bacteriocins of Lactic Acid Bacteria. *Food Sci Tech Int* 2001,7:281–305.

8.   Cogan TM, Barbosa M, Beuvier E, Bianchi-Salvadori B, Cocconcelli PS, Fernandes I, Gomez J, Gomez R, Kalantzopoulos G, Ledda A, Medina M, Rea MC, Rodriguez E: Characterization of the lactic acid bacteria in artisanal dairy products. *J. Dairy Res* 1997, 64:409–421.

9.   Cotter PD, Hill C, Ross RP: Bacteriocins: developing innate immunity for food. *Nat Rev Microbiol* 2005, 3:777–778.

10.   Daeschel MA: Antimicrobial substances from lactic acid bacteria for use as food preservatives. *Food Technol* 1989, 43:164–167.

11.   Dalié DKD, Deschamps AM, Richard-Forget F: Lactic acid bacteria – Potential for control of mould growth and mycotoxins: A review. *Food Control* 2010, 21:370–380.

12.   Delves-Broughton J, Blackburn P, Evans RJ, Hugenholtz J: Applications of the bacteriocin, nisin. *Anton Leeuw Int J G* 1996, 69:193–202.

13.   Diop MB, Dibois-Dauphin R, Tine E, Jacqueline AN, Thonart P: Bacteriocin producers from traditional food products. *Biotechnol Agron Soc Environ* 2007, 11:275–281.

14.   Elsanhoty RM: Screening of some lactobacillus strains for their antifungal activities against aflatoxin producing aspergilli in vitro and maize. *J Food Agric Environ* 2008, 6:35–40.

15. Freitas WC, Souza EL, Sousa CP, Travassos AER: Anti- staphylococcal effectiveness of nisaplin in refrigerated pizza doughs. *Braz Arch Biol Technol* 2008, 51:95–599.

16. Gagnaire V, Jardin J, Jan G, Lortal S: Invited review: Proteomics of milk and bacteria used in fermented dairy products: From qualitative to quantitative advances. *J Dairy Sci* 2009, 92:811–825.

17. Hajikhani R, Beyatli Y, Aslim B: Antimicrobial activity of enterococci strains isolated from white cheese. *Int J Dairy Technol* 2007,60:105–108.

18. Herreros MA, Sandoval H, González L, Castro JM, Fresno JM, Tornadijo ME: Antimicrobial activity and antibiotic resistance of lactic acid bacteria isolated from Armada cheese (a Spanish goats' milk cheese). *Food Microbiol* 2005, 22:455–459.

19. Ibourahema C, Dauphin RD, Jacqueline D, Thonart P: Characterization of lactic acid bacteria isolated from poultry farms in Senegal.*African J Biotechnol* 2008, 7:2006–2012.

20. Joshi VK, Sharmaand S, Rana NS: Production, purification, stability and efficacy of bacteriocin from isolates of natural lactic acid fermentation of vegetables. *Food Technol Biotechnol* 2006, 44:435–439.

21. Khalil R, Elbahloul Y, Djadouni F, Omar S: Isolation and partial characterization of a bacteriocin produced by a newly isolated Bacillus megaterium 19 strain. *Pak J Nutr* 2009, 8:242–250.

22. Lavermicocca P, Valerio F, Visconti A: Antifungal activity of phenyllactic acid against molds isolated from bakery products. *Appl Environ Microbiol* 2003, 69:634–640.

23. López-Díaz TM, Alonso C, Román C, García-López ML, Moreno B: Lactic acid bacteria isolated from a hand-made blue cheese. *Food Microbiol* 2000, 17:23–32.

24. Magnusson J, Schnürer J: Lactobacillus coryniformis subsp. coryniformis strain Si3 produces a broad-spectrum proteinaceous antifungal compound. *Appl Environ Microbiol* 2001, 67:1–5.

25. Naghmouchi K, Kheadr E, Lacroix C, Fliss I: Class I /Class IIa bacteriocin cross-resistance phenomenon in Listeria monocytogenes. *Food Microbiol* 2007, 24:718–727.

26. Reuter G: Elective and selective media for lactic acid bacteria. *Int J Food Microbiol* 1985, 2:55–68.

27. Savadogo A, Ouattara CAT, Basssole IHN, Traoer SA: Bacteriocins and lactic acid bacteria- a minireview. *Afr J Biotechnol* 2006,5:678–683.

28. Schillinger U, Becker B, Vignolo G, Holzapfel WH: Efficacy of nisin

in combination with protective cultures against Listeria monocytogenes Scott A in tofu. *Int J Food Microbiol* 2001, 71:159–168.

29. Sezer G, Güven A: Investigation of bacteriocin production capability of lactic acid bacteria isolated from foods. *Lafkas Univ Vet Fak Derg* 2009, 15:45–50.

30. Sharpe VD: *Bioproservation of fresh-cut salads using bacteriocinogenic lactic acid bacteria isolate from commercial produce*. Dalhousie University, Halifax, Nova Scotia, Canada, Master's Thesis; 2009.

31. Simova ED, Beshkova DB, Dimitrov ZP: Characterization and antimicrobial spectrum of bacteriocins produced by lactic acid bacteria isolated from traditional Bulgarian dairy products. *J Appl Microbiol* 2009, 106:692–701.

32. Stiles M, Holzapfel W: Lactic acid bacteria of foods and their current taxonomy. *Int J Food Microbiol* 1997, 36:1–29.

33. Staszewski M, Jagus RJ: Natural antimicrobials: Effect of MicrogardTM and nisin against Listeria innocua in liquid cheese whey.*Int Dairy J* 2008, 18:255–259.

34. Tasara T, Stephan R: Review cold stress tolerance of Listeria monocytogenes: A review of molecular adaptive mechanisms and food safety implications. *J Food Protection* 2006, 69:1473–1484.

35. Todorov SD, Dicks LMT: Bacteriocin production by Pediococcus pentosaceusisolated from marula (Scerocarya birrea). *Int J Food Microbiol* 2009, 132:117–126.

36. Trias R, Bañeras L, Badosa E, Montesinos E: Bioprotection of Golden Delicious apples and Iceberg lettuce against foodborne bacterial pathogens by lactic acid bacteria. *Int J Food Microbiol* 2008, 123:50–60.

37. Trias R, Bañeras L, Badosa E, Montesinos E: Lactic acid bacteria from fresh fruit and vegetables as biocontrol agents of phytopathogenic bacteria and fungi. *Int Microbiol* 2008, 11:231–236.

38. Vescovo M, Orsi C, Scolari G, Torriani S: Inhibitory effect of selected lactic acid bacteria on microflora associated with ready-to-use vegetables. *Letters in Applied Microbiol* 1995, 21:121–125.

39. Vescovo M, Torriani S, Orsi C, Macchiarolo F, Scolari G: Application of antimicrobial producing lactic acid bacteria to control pathogens in ready-to-use vegetables. *J Applied Bacteriol* 1996, 81:113–119.

40. Vescovo M, Scolari G, Zacconi C: Inhibition of Listeria innocua growth by antimicrobial-producing lactic acid cultures in vacuum-packed cold-smoked salmon. *Food Microbiol* 2006, 23:689–693.

Chapter 13

SELECTION OF LACTOBACILLUS SPECIES FROM INTESTINAL MICROBIOTA OF FISH FOR THEIR POTENTIAL USE AS BIOPRESERVATIVES

Mahdi Ghanbari[1], Mansoureh Jami[2] and Masoud Rezaei[3]

[1]University of Zabol, Faculty of Natural Resources, Department of Fishery; Zabol, Iran

[2]BOKU -University of Natural Resources and Life Sciences, Department of Food Sciences and Technology, Institute of Food Sciences; Vienna,, Austria

[3]Tarbiat Modares University, Faculty of Marine Science, Department of Seafood Science and Technology, Noor, Mazandaran,, Iran

INTRODUCTION

Despite recent advances in seafood production, seafood safety is still an important public health issue. It is clear that indigenous bacteria present in marine environment as well as resulting from post contamination during processing are responsible for many cases of illnesses [1-3]. In the last years, traditional processes applied to seafood like salting, smoking and canning have decreased in favor of mild technologies involving lower salt content, lower heating temperature and vacuum (VP) or modified atmosphere packing (MAP, 3-5]. Most of these treatments are usually not sufficient to destroy microorganisms and in some cases psychrotolerant pathogenic such as *Listeria monocytogenes*or spoilage causing bacteria can develop during prolonged shelf-life of these products [2,5,6]. As several of these products are eaten raw, it is therefore essential that adequate precautious and preservation technologies are applied to maintain their safety and quality. Among alternative preservation technologies, particular attention has been paid to biopreservation to extend the shelf-life and to enhance the hygienic quality of perishable food products such as seafood, thereby minimizing the impact on nutritional and organoleptic properties [1,7,8]. In this context, lactic acid bacteria (LAB) possess a major potential in biopreservation strategies, since they are safe to consume, and during storage they naturally dominate the microbiota of many foods [7-11]. Lactic acid bacteria are gram-positive, non-sporulating and catalase negative rods or cocci that ferment various carbohydrates mainly to lactate and acetate [12]. Accordingly, they are commonly associated with nutritious environments like foods, decaying material and the mucosal surfaces of the gastrointestinal and urogenital tract [12-14], where they enhance the host protection against pathogens [13]. Their antagonistic and inhibitory properties are due to the competition for nutrients and the production of one or more antimicrobially active metabolites such as organic acids (lactic and acetic acid), hydrogen peroxide, and antimicrobial peptides like bacteriocins [8-11,15-17]. Bacteriocins are ribosomally synthesized peptides that exert their antimicrobial activity against either strains of the same species as the bacteriocin producer (narrow range), or to more distantly related species (broad range) [7,15,18]. An important reason for research on LAB based bacteriocins is due to their activity at nanomolar concentrations against number of bacterial pathogens [1,3,5,6,19,20]. Some bacteriocins even exhibit their activities against multidrug-resistant nosocomial pathogens such as methicillin-resistant *Staphylococcus aureus* (MRSA) and vancomycin-resistant enterococci [VRE, 17, 21]. Thus they also may have some big potential in medical and veterinary applications. Fermented food and plant material have been a well-known source for bacteriocin-producing LAB, but isolates from the intestinal of animals and humans has become an increasingly important source for such strains due to an increased awareness of their importance as probiotics. In fish the presence of LAB is meanwhile well documented and the bio-protective potential of some strains and/or their bacteriocin has been highlighted in the last years [4-6,16,18,22-26]. Kvasnikov et al. [12] described the presence of lactic acid bacteria, including *Lactobacillus* in the intestines of various fish species at larval, fry and fingerling stages inhabiting ponds in Ukraine. They give information on the changes in their composition as a function of the season of the year and life-stage of the fish. However, it was discussed that some human activities like artificial feeding in ponds would have had an effect on the bacterial composition and load in some fish, like carp (*Cyprinus carpio*) which showed the highest content of lactic acid bacteria in the intestines. Cai et al. [27] described the lactic acid bacteria in *Cyprinus carpio* collected from the Thajin river in Thailand. They reported the presence of *Enterococcus* spp. and the dominance of *Lactococcus garviae*, an emerging zoonotic pathogen, in *Cyprinus carpio*. Bucio Galindo et al. [23] studied the distribution of lactobacilli in the intestinal content of river fish and reported that various species of lactobacilli were present in relatively high numbers in the intestines of edible freshwater fish from the river, especially in warm season but in low numbers in cold season. There are no reports on the presence of *Lactobacillus* in the intestines of sturgeon fish inhabiting Caspian Sea, whereas other groups of bacteria have been studied in more details. In comparison with other food products of dairy or meat origin, only few bacteriocinogenic LAB strains have been recovered from seafood. The present study focuses on the characterization of antimicrobial compounds produced by the lactobacilli isolates, in addition, their ability to inhibit the growth of relevant food borne pathogens as well as of spoilage bacteria and last but not least, of contaminants in aquaculture.

MATERIALS AND METHODS

Fish Intestine Samples

Two species of Persian sturgeon (*Acipenser persicus*) and Beluga (*Huso huso*) were collected from the south coast of Caspian Sea in Iran. Twenty two individuals of these fish in adult stage were selected. The weight and length of the fish were measured before dissection. The fish were sacrificed by physical destruction of the brain, and the number of incidental organisms was reduced by washing the fish skin with 70% ethanol. Then, the ventral surface was opened with sterile scissors. After dissecting the fish, 1 g of the intestinal tract content of each fish was removed under aseptic condition and placed into previously weighed flasks containing storage medium.

Media and Culture Condition

Intestinal content was homogenized in a storage medium using a vortex mixer. One milliliter was transferred to reduced neutralized bacterial peptone (NBP, Oxoid L34, Hampshire, England) 0.5 g/L, NaCl 8 g/L, cysteine.HCl 0.5 g/L, pH adjusted to 6.7 [29]. Afterwards serial dilutions were spread on plates of selective media and incubated at the following conditions. Columbia blood agar (CAB, Oxoid CM 331) was used as a selective medium to make an estimation of the cultivable total anaerobic counts [29]. All the inoculated plates were incubated anaerobically at 30°C for 48 h. The following two media were used to isolate lactic acid bacteria (LAB). MRS (MRS, Merck, Darmstadt, Germany) with 1.5% agar (M641, HiMedia, Mumbai, India) and pH adjusted to 4.2 (MRS 4.2) and incubated anaerobically at 30°C for 96 h was used as a selective medium for lactic acid bacteria. MRS is an inhibitory medium for *Carnobacterium*. Anaerobic MRS with Vancomycin and Bromocresol green (LAMVAB), incubated at 30°C for 96 h was used as an elective and selective medium for*Lactobacillus* spp. [30]. Anaerobic incubation of the three media was made in an anaerobic Gas-Pack system (LE002, HiMedia, Mumbai, India) with a mixture of 80% N_2, 10% H_2 and 10% CO_2. Colonies were selected either randomly, or in case of less than 10 colonies per each plate, all the samples were counted according to the method described by Thapa et al. [31]. Purity of the isolates was checked again by streaking them onto fresh agar plates of the isolation media, followed by microscopic examinations. Identified strains of lactobacilli were kept in MRS broth with 15% (v/v) glycerol at -20°C.

Characterization Procedures for Lactic Acid Bacteria

Eighty four strains were randomly selected for identification procedures based on the phenotypical characteristics. Cell morphology and motility of all isolates were observed using a phase contrast microscope (CH3-BH-PC, Olympus, Japan). Isolates were gram-stained and tested for catalase production test. Preliminary identification and grouping was based on the cell morphology and phenotypic properties such as CO_2 production from glucose, hydrolysis of arginine, growth at different temperatures (10, 15 and 45°C), and at different pH (3.9 and 9.6). As well as the ability to grow in different concentrations of NaCl (6.5% (w/v), 10% (w/v) and 18% (w/v)) in MRS broth was checked as well. The configuration of lactic acid produced from glucose was determined enzymatically using d-lactate and l-lactate dehydrogenase test kits (Roche Diagnostic, France). The presence of diaminopimelic acid (DAP) in the cell walls of LAB was determined using thin-chromatography on cellulose plates. Fermentation of carbohydrates was determined using API 50 CHL (API 50 CH is a standardized system, associating 50 biochemical tests for the study of carbohydrate metabolism in microorganisms. API 50 CH is used in conjunction with API 50 CHL Medium for the identification of*Lactobacillus* and related genera) strips according to the manufacturer's instructions (Biomerieux, Marcy l' Etoile, France). The APILAB PLUS database identification software (bioMe´rieux, France) was used to interpret the results. Identification was undertaken according to the method described by Kandler and Weiss [12] and Hammes and Vogel [32].

Statistical Analysis

Statistical analysis using Student's t-test (SPSS, Version 11.0) was performed to find significant difference on lactobacilli count between LAMVAB and MRS 4.2. Pearson's correlation coefficient was used to investigate the correlation of lactobacilli count between LAMVAB and MRS 4.2 (SPSS Inc., Version 11.0, Chicago, USA). A significance level of $p < 0.05$ was used.

Screening of *Lactobacillus* Strains for Their Inhibitory Potential

In a first test series, the ability of each of the *Lactobacillus* isolates to exert an antibacterial effect against *Listeria monocytogenes* ATCC 19115 and *Salmonella* Typhimurium PTCC 1186 were examined by using three methods: the spot-on-lawn method, standardized agar disk diffusion method and the well diffusion method as described by Schillinger and Lucke [33], Benkerroum et al. [34] and Tagg & Mc Given [35]. Throughout, cell-free supernatants (CFS) of strains were obtained by centrifugation at 10,000 ×g for 20 min and then adjusted to pH 6.5 by applying NaOH (to exclude the effect of organic acid) before sterilization by filter (0.2 μm, Sigma, UK). Based on the screening tests, the inhibitory spectrum of potential bacteriocin-producing isolates was assessed against 42 indicator strains using a standardized agar disk diffusion test. The strains were kept frozen in 20% (v/v) glycerol at -20°C. For this purpose, an aliquot of 20 ml CFS was applied on disks (6 mm) and set on agar plates previously inoculated with each individual indicator strain suspension, which corresponded to a 10^5 CFU/ml. Plates were incubated 24 h at optimum temperatures of the test organism. Antimicrobial activity was detected as a translucent halo in the bacterial lawn surrounding the disks.

Characterization of the Inhibitory Effect

In order to determine the biological nature of the antimicrobial activity of bacteria, CFS (pH 6.0) of 24-h lactobacilli cultures of two selected isolates (*Lactobacillus casei* AP8 & *Lactobacillus plantarum*H5) incubated at 30°C, were tested for their sensitivity to the proteolytic enzymes. One ml of CFS was treated for 2 h with 1 mg ml^{-1} final concentration of the following enzymes: papain, trypsin, proteinase K, pronase E and α-amylase (Sigma, London). To clarify whether the antimicrobial activity detected derives from the production of hydrogen peroxide, 2600 IU/ml of catalase (Sigma, London) were added to 1 ml portions of extracellular extracts of LAB exhibiting antimicrobial activity and incubated for 24 h at ambient temperature. Chemicals were added to the CFS and the samples incubated for 5 h before being tested for antimicrobial activity. To determine the sensitivity of potential bacteriocin activities to the temperatures, samples of CFS were incubated under defined conditions. The effect of pH on bacteriocin activity was determined by adjusting the pH of the CFS (cell free supernatant (pH 6.5) of 24-h lactobacilli cultures incubated at 30°C) with diluted appropriate volumes of HCl and NaOH (Table 3). After incubating for 2 h, the pH of the samples was readjusted to 6.5 followed by sterilization (0.2 μm, Sigma, UK). In all cases, the remaining bacteriocin activity was assessed exemplarily by using strain *L. monocytogenes* ATCC 19115 as the indicator bacterium and by applying the agar disk diffusion plate bioassay. Untreated cell-free supernatants were used as controls and experiments were performed in duplicate.

Growth Dynamics and Antimicrobial Compounds Production

The time course of inhibitory substance production was performed by inoculating 10 mL of an overnight culture of selected *Lactobacillus* isolates into 100 mL of MRS broth followed by incubation at 30°C. Cells were subsequently removed by centrifugation at 10,000 ×g for 20 min. At appropriate intervals, changes in pH and optical density (600 nm) of the cultures were measured to monitor bacterial growth using a spectrophotometer (Hitachi U 1100, Tokyo, Japan). Antibacterial activity was evaluated every hour by using serial twofold dilutions of each culture used as a neutralized cell-free supernatant (CFS) tested against *L. monocytogenes* ATCC 19115 based on the agar disk diffusion plate bioassay. In a separate experiment, the inhibitory effect of CFSs of lactobacilli strains on target cells in liquid medium was also examined against *L. monocytogenes* ATCC 19115 as indicator strain. For this purpose, 20 mL of each filter-sterilized bacteriocin-containing cell-free supernatant were added to a 100 mL culture of the indicator organism at early exponential phase (4 h old). These experiments were also repeated with stationary-phase cells. The optical density at 600 nm and viable cell count were determined every hour during an observation period of 20 h. Indicator cells without CFSs were used as control.

Adsorption of Bacteriocin to Producer Cells

Bacteriocin-producing cells were cultured for 18 h at 30 °C. The pH of the cultures was adjusted to 6.0 with 1 M NaOH to allow maximal adsorption of the bacteriocin to the producer cells, according to the method described by Yang et al. [36]. The cells were then harvested (10,000 ×g 20 min, 4 °C) and washed with sterile 0.1 M phosphate buffer (pH 6.5). The pellet was re-suspended in 10 ml of 100 mM NaCl (pH 2.0) and stirred slowly for 1 h at 4 °C. The suspension was then centrifuged (10,000 ×g 20 min, 4 °C), the CFS was neutralized to pH 7.0 with sterile 1 M NaOH followed by testing the bacteriocin activity as described above.

Partial Purification and Characterization of The Bacteriocin

Bacteriocin producer strains were grown in MRS broth, and incubated without agitation for 18 h at 30°C. The cells were harvested (10,000 ×g, 20 min, 4 °C) and the bacteriocin precipitated from the CFS with 60% saturated ammonium sulphate [45]. The precipitate in the pellet and floating on the surface were collected and re-suspended in one-tenth volume 25 mM ammonium acetate buffer (pH 6.5). The sample was stored at -20 °C for one week and activity tests were performed as described above. For the determination of the molecular size of the bacteriocins, precipitated peptides re-suspended in 25 mM ammonium acetate buffer (pH 6.5) were separated by Tricine-SDS-PAGE, according to Schägger and Von Jagow [38]. Low molecular weight markers, ranging from 2.5 to 45 kDa (Pharmacia, Sweden) were used. One half of the gel containing the molecular marker was fixed for 20 min in 5% (v/v) formaldehyde, then rinsed with water and stained with Coomassie Brilliant Blue R250 (Bio-Rad) overnight. The other half of the gel (not stained and extensively pre-washed with sterile distilled water) was overlaid with a culture of 10^6 cfu/ml *L. monocytogenes* ATCC 19115 embedded in BHI agar. The position of the active bacteriocin was visualized by an inhibition zone around the active protein band [39].

RESULTS

Isolation of Lactobacilli

Intestinal content of 22 fish were analysed for the presence of lactobacilli. To determine the most appropriate medium for isolating lactobacilli from fish intestines, two media (MRS agar, LAMVAB) were used. LAMVAB was highly selective to quantify lactobacilli, as 99% of 143 randomly picked colonies and purified isolates were identified as *Lactobacillus* spp. and confirmed according to [12] (Table 1). Counts of intestinal lactobacilli for Persian sturgeon and beluga were detected at the range of approximately $10^{5.3}$ to $10^{6.4}$ cfu/g, respectively. The physiological and biochemical characterization of *Lactobacillus* isolates and the presumptive *Lactobacillus* species found in two fish species are shown in Table 2. From 84 isolates, 2 metabolic groups of *Lactobacillus* were recovered: facultative and obligate heterofermentatives. *L. sakei* and *L. plantarum* were the most often found isolates (Table 2). MRS 4.2 was suitable to quantify lactobacilli. As 30 randomly picked colonies on the highest dilution were identified as lactobacilli and coccoid forms were not found. Means of counts of 90 samples were not statistically different to LAMVAB counts in the Student's t-test (P=0.29) and were correlated with LAMVAB counts (r = 0.85; P<0.001). The correlation of counts on MRS 4.2 with those on LAMVAB and the absence of coccoids suggests that lactobacilli were the most important acidophilic lactic acid bacteria in the samples analysed. Facultative anaerobic flora recovered in CAB medium provided the highest counts in the samples analysed (Table 1).

Table 1: Average bacterial counts of intestinal bacteria (Log cfu/g of intestinal content) for Persian sturgeon and beluga in different media

Fish species	No.	CAB (cfu/g)	LAMVAB (cfu/g)	MRS 4.2 (cfu/g)
Acipenser persicus	12	7.84	5.32	4.85
Huso huso	10	8.21	6.45	5.64

Screening of *Lactobacilli* Strains for Antimicrobial Activity and Bacteriocin Production

Eighty four lactobacilli strains previously isolated from two species of Sturgeon fish identified and their cell free supernatant extracts were assayed for antimicrobial activity and bacteriocin production against *Listeria monocytogenes* ATCC 19115 and *Salmonella* Typhimurium PTCC 1186 by using spot-on-lawn method, standardized agar disk diffusion method and well diffusion method. In each instance, diameters of inhibition were quantified. Fifteen strains (18%) exhibited inhibitory activity against both indicator organisms. Consequently, all candidate isolates (Inhibition zone> 8mm) subjected to different tests such as growth at different temperatures, pH, salt content, antibiotic resistance, etc. Based on the result of aforementioned tests, two strains *Lactobacillus* casei AP8 and *Lactobacillus plantarum* H5, isolated from Persian sturgeon and beluga respectively, were chosen as active strains and were subjected to further examinations.

Table 2: Biochemical characteristics of *Lactobacillus* species isolated from the intestines of Persian sturgeon and beluga

Presumptive Lactobacillus species	L. sakei	L. plantarum	L. coryneformis	L. alimentarius	L. brevis	L. casei	L. oris
No. of isolates	30	18	12	10	7	5	2
Diaminopimelic acid	ND	+	ND	ND	ND	ND	ND
CO₂ from glucose	-	-	-	-	+	-	+
NH₃ from arginine	-	-	-	-	+	-	+
10°C	+	+	+	+	+	+	+
15°C	+	+	+	+	+	+	+
45°C	-	-	-	2	-	-	-
Glycerol	-	+	-	1	-	+	-
L-Arabinose	+	+	-	2	2	-	+
Ribose	+	-	-	+	+	+	+
D-Xylose	26	-	-	-	-	-	+
Galactose	29	-	-	-	-	+	-
Rhamnose	-	-	+	-	2	+	-
Inositol	-	+	-	-	-	+	+
Mannitol	-	+	5	-	+	+	-
Sorbitol	-	+	-	-	-	+	-
1-Methyl-D-mannoside	-	+	-	-	-	-	+
1-Methyl-D-glucoside	-	+	-	7	+	-	+
N-Acetyl glucosamine	28	+	+	+	+	+	+

Amygdaline	10	+	-	+	-	+	+
Arbutine	1	+	-	+	-	+	+
Esculine	+	+	+	+	1	+	+
Salicin	+	+	-	+	-	+	+
Cellobiose	27	+	-	+	-	+	+
Maltose	19	+	-	+	+	+	+
Lactose	26	+	-	+	-	+	+
Melibiose	+	+	+	2	+	-	+
Sucrose	+	+	+	8	+	+	+
Trehalose	+	+	-	+	-	+	+
Melezitose	-	+	-	-	+	+	+
D-Raffinose	29	-	-	2	+	-	-
Starch	-	-	-	-	-	+	-
Xylitol	-	+	3	-	-	-	+
2-Gentiobiose	+	+	-	+	-	+	+
D-Turanose	-	-	-	-	+	-	-
D-Tagatose	1	-	-	+	-	+	-
D-Arabitol	-	+	5	-	-	-	+
Gluconate	+	-	-	+	+	+	+
2-keto-gluconate	-	-	1	2	-	-	+
5-keto-gluconate	-	-	-	1	-	+	+
Lactic acid configuration	DL	DL	DL	DL	DL	DL	D

Table 3: *Lactobacillus* species isolated from the intestines of sturgeon fish

Presumptive Lactobacillus species	L. sakei	L. plantarum	L. coryneformis	L. alimentarius	L. brevis	L. casei	L. oris
Acipenser persicus	**	**	*	**	-	**	*
Huso huso	**	*	-	*	**	*	*

* = Presence of lactobacilli. ** = High number of lactobacilli presence

Inhibitory Spectrum Of Bacteriocin

As the results in screening test showed that greater inhibition was observed by agar disk diffusion tests of cell-free supernatant extracts, so this method was selected as the best technique for examining the antibacterial activity of *L. casei* AP8 and *L. plantarum* H5 CFSs against forty two Gram-positive and Gram-negative bacteria. The CFS preparations from both strains showed a broad inhibitory spectrum against a wide range of LAB of different species and some food-borne pathogens and spoilage bacteria including *Listeria innocua, L. monocytogenes, Staphylococcus aureus, Aeromonas hydrophila,Aeromonas salmonicida, Bacillus cereus, Bacillus pumilus, Bacillus subtilis, Brochotrix thermosphacta*, Gram-negative *E. coli, Salmonella* and *Pseudomonas, Clostridium perfringens* and *Vibrioparahaemolyticus* (Table 4). Result showed that the Gram-positive bacteria tested were more sensitive to the bacteriocin produced by the isolates than Gram-negative bacteria. The largest spectrum of inhibition was showed by *L. casei* AP8 bacteriocin, which inhibited 33 out of 42 indicator strains.

Characterization of Inhibitory Effect

Table 5 and table 6 depict the stability of inhibitory substances at different physic-chemical conditions. To determine the biological nature of the antimicrobial activity of bacteria, CFSs were tested for their sensitivity to the proteolytic enzymes. Antimicrobial activities exhibited by *L. casei* AP8 and *L. plantarum* H5 were sensitive to proteolytic enzymes since proteolytic, but not lipolytic or glycolytic enzymes, completely inactivated the antimicrobial effect of both cell-free supernatants, confirming the proteinaceous nature of the inhibitors (Table 3). The effect of several chemicals on the antimicrobial activity was also evaluated. Interestingly, the cell-free extracts remained active after treatment with chemicals such as catalase, SDS, Triton X-100, Tween 20, Tween 80 and EDTA after 5 h of exposure (Table 2). Enhancing the antimicrobial activity in case of *L. casei* AP8 bacteriocin was observed after treating by EDTA and SDS against *L. monocytogenes* ATCC 19115. The stability study of

Table 4: Antimicrobial activity of potential bacteriocin producing strain *L. casei* AP8 *and L. plantarum*H5 as examined with selected bacterial indicator strains

Indicator organism	Medium*	Temp. [°C]	Bac AP8	Bac H5
Gram Negative Group				
Aeromonas hydrophilus MJ 1120	BHI	37	++	0
Aeromonas hydrophilus MJ 1240	BHI	37	+++	+
Aeromonas salmonicida CC 1546	BHI	37	+	+
Aeromonas salmonicida RT 7895	BHI	37	++	+
Brochothrix thermosphacta RF 35	BHI	37	++	+
Escherichia coli ATCC 25922	BHI	37	++	0
Escherichia coli PTCC 1325	BHI	37	++	++
Photobacterium damselae ssp. Piscida	BHI	37	0	0
Pseudomonas aeruginosa PTCC 1310	BHI	37	++	+
Pseudomonas fluorescens HFC 1236	BHI	37	++	0
Salmonella enteritidis ATCC 13076	BHI	37	++	++
Salmonella spp SM 162	BHI	37	+++	++
Vibrio anguillarum MI12	BHI	37	++	+
Vibrio parahaemolyticus MI 23	BHI	37	+++	0
Vibrio parahaemolyticus MI 56	BHI	37	+++	+
Gram Positive Group				
Bacillus cereus ATCC 9634	BHI	37	+++	+++
Bacillus coagulans	BHI	37	+++	++
Bacillus licheniformis PTCC 1331	BHI	37	++	0
Bacillus subtilis ATCC 9372	BHI	37	+++	+
Clostridium perfringens ATCC 3624	RCM	37	++	+
Clostridium sporogenes PTCC 1265	RCM	37	++	+
Lactobacillus acidophilus ATCC 4356	MRS	30	++	+
Lactobacillus alimentarius AP 10	MRS	30	+	++
Lactobacillus brevis H56	MRS	30	++	++
Lactobacillus brevis AP 83	MRS	30	++	++
Lactobacillus casei PTCC 1608	MRS	30	0	++
Lactobacillus casei RN 78	MRS	30	0	0
Lactobacillus casei LB 10	MRS	30	0	+
Lactobacillus casei LB 46	MRS	30	0	+
Lactobacillus plantarum PTCC 1050	MRS	30	0	0
Lactobacillus plantarum AP 76	MRS	30	+	0
Lactobacillus plantarum H12	MRS	30	+	0

Lactobacillus sakei AP 43	MRS	30	0	+
Lactobacillus sakei	MRS	30	0	0
Lactococcous sp	MRS	30	+	0
Lactobacillus curvatus	MRS	30	0	+
Listeria innocua AN 15	BHI	37	++	++
Listeria monocytogenes ATCC 7644	BHI	37	+++	+++
Listeria monocytogenes PTCC 1163	BHI	37	++	++
Listeria monocytogenes PTCC1297	BHI	37	++	++
Staphylococcus aureus ATCC 25923	BHI	37	+++	+
Staphylococcus aureus PTCC 1112	BHI	37	+++	+

inhibitory compounds of *L. casei* AP8 and *L. plantarum* H5 in different conditions indicated the high resistance of these agents. The antimicrobial compounds were able to resist most of these factors to which it was exposed even during prolong incubation period (Table 6). Cell free extracts prepared from both the isolates are found to be thermo-stable. When *L. casei* AP8 bacteriocin was heated at 40-100° C for 30 min, it retained inhibitory activity against *L. monocytogenes* ATCC 19115. However, a loss in activity in the ranges of 35% was observed when heated at 120°C for 15 min (Table 6). The Antilisterial activity of *L. plantarum* H5 bacteriocin was resistant to heat treatments of 40-100°C for 30 min and remained constant after heating at 121°C for 15 min. Both investigated bacteriocins were most stable at 4°C and - 20°C and able to retain their antilisterial activity for 30 days without any decrease. *L. casei* AP8 bacteriocin was active in a wide range of pH, as full activity was retained at pH values between 3 and 10. *L. plantarum* H5 bacteriocin remained stable after incubation for 2 h at pH values between 2.0 - 12.0.

Table 5: Effect of enzymes and chemicals on the antimicrobial activity of two selected strains *L. casei*AP8 and *L. plantarum* H5. For details see text

Treatment	Concentration	%Residual antimicrobial activity	
		L. casei AP8	L. plantarum H5
Enzymes			
Trypsin	1 mg/ml^{-1}	0	0
Papain	1 mg/ml^{-1}	100	100
Proteinase K	1 mg/ml^{-1}	0	0
Pronase E	1 mg/ml^{-1}	0	0
α- amylase	1 mg/ml^{-1}	100	100
Catalase	1mg/ml^{-1}	100	100
Organic solvents			
Butanol	10% [v/v]	100	100
Ethanol	10% [v/v]	100	100
Methanol	10% [v/v]	92	100
Ethyl ether	10% [v/v]	100	100
EDTA	5 mmol l^{-1}	100	83
Sodium deoxycholate	1mg ml^{-1}	100	100
Sulphobetaine 14	1mg ml^{-1}	92	100
SDS	1% [w/v]	100	100
Tween 20	1% [v/v]	100	100
Tween 80	1% [v/v]	92	100

Table 6: of cold storage, different temperatures and pH on inhibitory activity against *Listeria monocytogenes* ATCC 19115. For details see text.

Treatment (Storage, Temperature and pH stability)	Residual antimicrobial activity	
	L. casei AP8	L. plantarum H5
4 ºC, -20ºC/ 30 d	+	+
40-100 ºC/30 min	+	+
121 ºC/10 min	+ [-35%]	+
121 ºC/15 min	-	+
pH= 2	-	+
pH= range 3-10	+	+
pH= 11	-	+
pH= 12	-	+

Growth and Bacteriocin Production

Figure 1 shows the growth and bacteriocin production curves of *L. casei* AP8 and *L. plantarum H5*cultured at 30°C. For *Lactobacillus* casei AP 8 cell growth reached the stationary phase at 12 h of cultivation. Kinetics of bacteriocin production showed that its synthesis and/or secretion started at 4 h growth in the exponential phase of growth and maximum activity was observed at the early stationary phase of growth (1800 AU ml⁻¹) and had stabled for 6 h before the bacteriocin activity decreased (Figure 1). The pH values decreased from 6.5 to 3.7 at the end of incubation. For *L. plantarum* H5, bacteriocin activity was detectable in the culture supernatant after 5hr when an absorbance of 0.55 at 600 nm of the culture broth. Production of bacteriocin increased throughout logarithmic growth. In the stationary phase, *L. plantarum* H5 showed maximum bacteriocin activity (3400 AU/mL) and stabilized for 2 hr. But since then, bacteriocin activity declined gradually and stabilized at 1600 AU/ml during the following 4 h. In the stationary phase, extracellular pH was maintained, however, bactericidal activity decreased, excluding a possibility of lactic acid as a bactericidal mechanism.

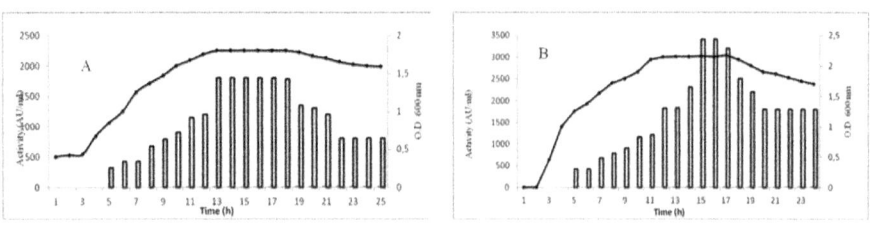

Figure 1: Antimicrobial activity [bars] against *L. monocytogenes* ATCC19115 of *L. casei* AP8 [A] and *L. plantarum* H5 [B] observed during growth in MRS medium [●] and expressed in AU/ml. Results are represent the mean of three independent experiments.

To investigate the reduction of viable cells of target organism in presence of inhibitory substances, twenty mL of each filter-sterilized bacteriocin-containing cell-free supernatant were added to 100 mL of *L. monocytogenes* ATCC 19115 (4 h old at 30°C). The optical density at 600 nm and viable cell count were determined every hour during 24 h. In the control samples inoculated with indicator strain the viable cell count reached to 10^{11}CFU/ml after 24 h incubation at 37°C. The inhibition kinetics using the bacteriocin AP8 (Figure 2) indicated a bactericidal mode of action against *L. monocytogenes*. Addition of the bacteriocin *L. casei* AP8 to early logarithmic-phase cells of indicator strain resulted in grows inhibition after 1h, followed by complete growth inhibition (slow decline) for the remaining time (20 h). In the case of *L. plantarum* H5 bacteriocin the inhibition kinetics showed a bacteriostatic mode of inhibition against indicator strain. Addition of bacteriocin H5 to culture of *L. monocytogenes*showed a growth inhibition after 1 h followed by slow growth. Experiment with stationary-phase cells did not showed any inhibition. No increase in the activity of bacteriocin AP8 and H5 were observed after treatment of the producer cells with 100 mmol/l NaCl at low pH, suggesting that these bacteriocins do not adhere to the surfaces of the producer cells.

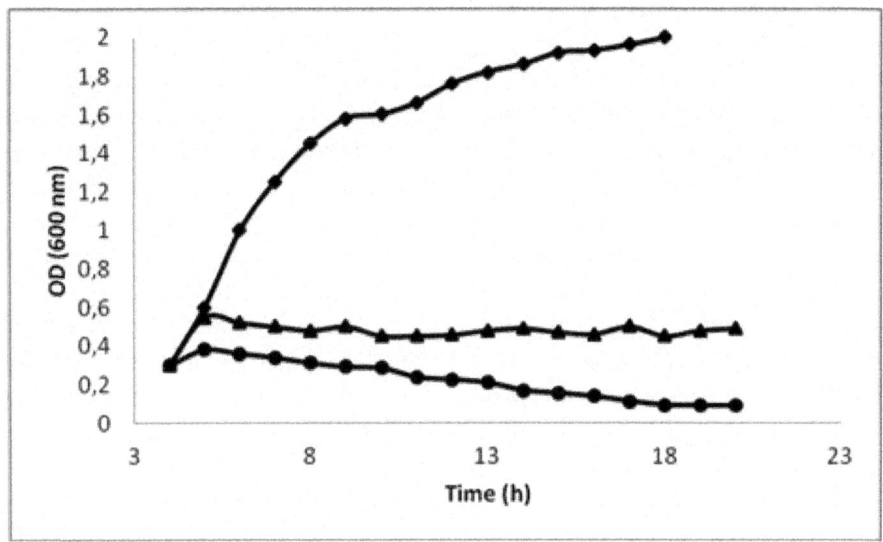

Figure 2: Antimicrobial effect of the CFS of *L. casei* AP8 [▲] and *L. plantarum* H5 [●] on the growth of *L. monocytogenes* ATCC 19115 at 30°C. Growth of *L. monocyto-genes* ATCC 19115 without added bacteriocins [control, ♦].

Partial Purification and Molecular Size of Bacteriocins Ap8 And H

Ammonium sulfate precipitation method with 60% saturated ammonium sulphate is used for partial purification of both bacteriocins. Results showed an increase (10-15%) in the inhibitory activity of both bacteriocins against *L. monocytogenes* ATCC 19115 after precipitation. The SDS-PAGE analysis of the partially purified samples showed peptide bands for bacteriocins AP8 and H5 in size of approximately 5 and 3 kDa respectively (Figure 3).

DISCUSSION

In this study, we isolated, quantified and characterized *Lactobacillus* species from two species of sturgeon fish inhabiting Caspian Sea to make a bank collection of strain for further research. These fishes are highly valuable species for fisheries and aquaculture in Iran. Presumptive lactobacilli species found in this study were relatively similar to the species described by Bucio Galindo et al. [28]. These authors reported *L. alimentarius, L. coryneformis, L. casei, L. sakei, L. pentosus, L. plantarum, L. brevis and L. oris,* as lactobacilli presented in the intestinal content of studied fish. However, the fish species analysed in that study were different from the two species in this study which were collected from a lake environment. The biochemical characteristics used for identification of *Lactobacillus* may suggest some ideas in relation to the occurrence of the strains in nature. Most of *Lactobacillus* examined in this study (80%) had the capacity to ferment lactose and galactose. Generally, most lactobacilli are able to ferment lactose, by uptake of this disaccharide by a specific permease and splitting it by S-galactosidase for further phosphorylation of galactose and glucose [12]. Because, lactose is only present in milk and milk derivates, it is possible that these strains have evolved from environments related with mammals, as was suggested for other lactose positive *Lactobacillus* [40]. Lactose may be present or was present in the environment as a waste; resulting from livestock production, and disposal effluents from dairy factories. Another component, often fermented by the strains was the amino-sugar N-acetyl-glucosamine, a compound present in peptidoglycans, in blood, chitin and as one of the main constituents of mucus in the gastrointestinal tract [41]. The carbohydrate portion constitutes above 40% of the weight of the mucus [42] or higher values [41].

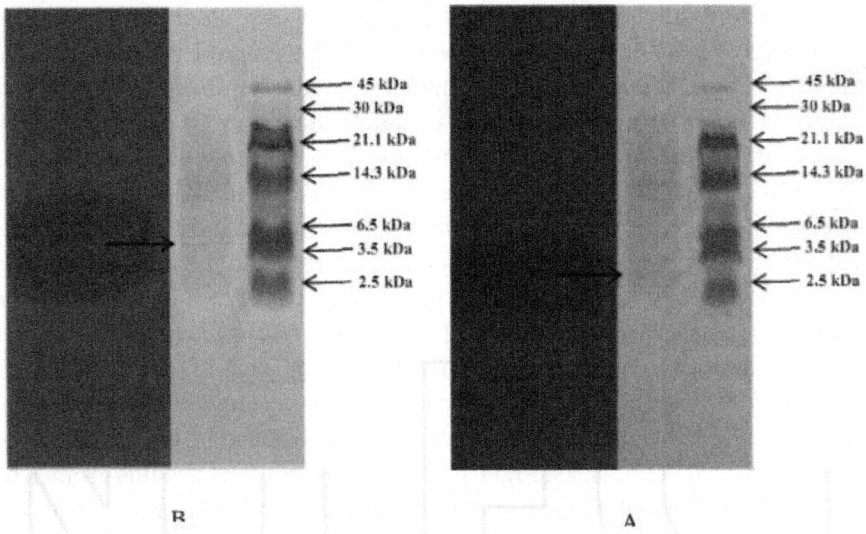

B   A

Figure 3: Tricine-SDS-PAGE gel of partially purified bacteriocins [precipitated by 60% saturated ammonium sulphate] *L. casei* AP8 [A] and *L. plantarum* H5 [B] along with the standard MW markers. The gel was overlaid with *L. monocytogenes* ATCC 19115 [approx. 10^6 CFU/ml], embedded in BHI agar, after incubation at 30 °C for 24.

It could be shown that two strains, *Lactobacillus casei* AP8 and *Lactobacillus plantarum* H5 isolated from intestinal bacterial flora of beluga (*Huso huso*) and Persian sturgeon (*Acipenser persicus*) were able to produce antibacterial substances. According to the findings it was likely that the antibacterial effect was due to the formation of bacteriocin. Results from enzyme inactivation studies demonstrated that antimicrobial activity of isolates AP 8 and H5 was lost or unstable after treatment with all the proteolytic enzymes, confirming the protein status of metabolites and indicating the presence of bacteriocins. Furthermore, treatment with lipolytic or glycolytic enzymes did not affect the activity of antimicrobial compound produced by strain, suggesting that produced bacteriocins do not belong to the controversial group IV of the bacteriocins, which contain carbohydrates or lipids in the active molecule structure [45-47]. It is important to note that, their activities were not due to hydrogen peroxide or acidity, as antimicrobial activity was not lost after treatment with catalase. Both of the presumptive were considered to be heat stable. Although heat stability of antibacterial substances produced by*Lactobacillus* spp. has been well established [39,48,49,50-53] heat stability of *L. casei* AP8 121°C for 10 min is novel. The result of pH stability were not coherent with previous report that had indicated the tolerance of bacteriocins to acidic pH rather than alkaline [36,54]. The loss of antimicrobial activity of AP8 bacteriocin at pH > 10 might be ascribed to proteolytic degradation, protein aggregation or instability of proteins at this extreme pH [39,48,55]. *L. casei* AP8 bacteriocin showed an increase in the inhibitory activity after treatments with SDS and EDTA, may be due to the ability of these compounds to break down the proteinaceous complex from its large form into smaller more active unite [21]. Similar to Lactocin RN78 and Plantaricin LC74, both bacteriocin *L. casei* AP8 and*L. plantarum* H5 were found to be stable after treatment with organic solvents like butanol, ethanol and methanol confirming their proteinaceous and soluble nature [18,21,56]. Pronounced inhibitory potential against various species of Gram-positive bacteria were shown, including pathogenic and spoilage microorganisms such as *A. hydrophila, A. salmonicida, C. perfringens, B. cereus* and *L. monocytogenes*. Observed effects were consistent with reports about bacteriocins produced by other strains of LAB [1,3,17,19,20,25,41,49,55,57 59,60]. Although bacteriocins from LAB usually are ineffective against Gram-negative bacteria and rather relate to a narrow antimicrobial spectrum [9,51,53], both presumptive bacteriocins AP8 and H5 showed broad antimicrobial activity against several genera of Gram-positive and Gram-negative bacteria. Even representatives of *Pseudomonas, Salmonella, E. coli, A. hydrophila, A. salmonicida and V. anguillarum* could be inhibited. Moreover a high level of inhibitory activity against *Listeria monocytogenes* was observed. Earlier studied have shown that several marine bacteria may produce inhibitory substances against bacterial pathogens in aquaculture systems [1,16,19]. Hence the use of such bacteria releasing antimicrobial substances in now gaining importance in fish farming as a natural alternative to administration of antibiotics [1,61-63]. In kinetic studies, both crude bacteriocins were continuously produced during logarithmic phase followed by optimal production during stationary growth phase, suggesting that these peptides may be secondary metabolites. Similar results were reported for some bacteriocins produced by some LAB isolates [5,64,65] and is contrary for other *Lactobacillus* species bacteriocins [1,16,25,55,60,66]. Bacteriocin H5 showed a decrease in activity towards the end of stationary growth may be due to proteolytic degration, protein aggregation, and feedback regulation as has been observed for Lactacin ST13BR, Lactacin B, Helveticin J and Enterocin1146 [53,55,67]. *L. casei* AP8 crude bacteriocin demonstrated a bactericidal mode of action, as the immediate decrease in the optical density of *L. monocytogenes* was observed in mix culture. In the case of H5 bacteriocin a bacteriostatic mode of action was observed. Crud H5 bacteriocin showed a growth inhibition, followed by decrease activity for remained time, suggesting that indicator organism became resistant to the bacteriocin or bacteriocin was destroyed by proteolytic enzymes [55]. Treating of bacteriocins AP 8 and H5 with NaCl at low pH did not result in increased levels of antilisterial activity, suggesting no adsorption of bacteriocins to their producer cells in agreement with result reported before for *Lactobacillus* strains bacteriocins [55,64,66].

More accurate techniques could be used to determine the molecular mass of molecules, yet the SDS-PAGE technique provides valuable information about the presence of the peptides [3]. In recent years, a large number of new bacteriocins produced by *L. plantarum* have been identified and characterized and the molecular masses of all the bacteriocins produced have been reported in the range of 3-10 kDa [5,39,55]. However, to our knowledge, there is no bacteriocin produced by any *L. casei* strain with a molecular mass of 5 kDa with similar characteristics to strain investigated in this study. Thus, it is possible that this bacteriocin may be a novel bacteriocin produced by *L. casei*. The physiochemical properties of bacteriocins from *L. casei* AP8 and *L. plantarum* H5 were similar to those of other bacteriocins of lactobacilli belonging to the group IIa lactic acid bacteria with respect to molecular weight, heat and pH stability and also sensivity to proteolytic enzymes [9,45,51]. Characteristics unifying all members of class IIa bacteriocins are 1) below 10kDa [1] their potent activity against*Listeria* spp., 2) their resistance to elevated temperatures and extreme pHs, and 3) their cystibiotic feature attributed to the presence of at least one disulfide bridge, which is crucial for antibacterial activity [15,45,51,55]. Class IIa bacteriocins were formerly considered as "narrow"-spectrum antibiotics, with antimicrobial activity directed against related strains. Recently, some class IIa bacteriocins, such as bacteriocin OR-7, enterocin E50-52, and enterocin E760, have been shown to be active against both Gram-negative and Gram-positive bacteria, including *Campylobacter jejuni,Yersinia* spp., *Salmonella* spp., *Escherichia coli* O157:H7, *Shigella dysenteriae, Staphylococcus aureus*, and *Listeria* spp. [15, 54, 45, 51, 3].

CONCLUSION

Bacteriocins AP8 and H5 showed a wide spectrum of antibacterial activity against seafood borne pathogens like *Listeria, Clostridium, Bacillus* spp, *S. aureus* and even Gram-negative pathogens like*Pseudomonas, Salmonella* and *E. coli*. Some of these foodborne pathogens can produce toxins resulting in human illness. In addition to the broad inhibition spectrum, theirs technological properties and especially cold, heat and storage stability, indicate that bacteriocins AP8 and H5 have potential for application not only as biopreservative agents to control pathogens in food products that are pasteurized and cook-chilled but also as bioprotect compounds at aquaculture. Accordingly *L. casei*may be of great interest as probiotics strains because of their ability to adhere to intestinal epithelial cells and being of human origin. Several authors have reported the production of bacteriocins by L. casei and L. plantarum strains from plant, dairy or meat origin. However, very few bacteriocins from L. plantarum have been reported to be isolated from fish and also based on our knowledge this is the first report of a L. casei bacteriocin isolated from fish.

ACKNOWLEDGEMENT

The authors thank the personnel of the Shahid Rajaei Aquaculture Center for their assistance. We are also grateful to Professor Mehdi Soltani from Faculty of Veterinary Medicine, University of Tehran, Dr Ali Shahreki from Medicine Science University of Zahedan, Iran, Professor Masoud Rezaei from Tarbiat Modares University, Iran, Professor Wolfgang Kneifel and Professor Konrad Domig from University of natural Resources and Life Sciences, Vienna, Austria for their helpful opinions. Some part of this research was funded by Tarbiat Modares University, Tehran, Iran

REFERENCES

1.   K. Indira, S. Jayalakshmi, A. Gopalakrishnan, M. Srinivasan, 2011 Biopreservative potential of marine Lactobacillus spp. African Journal of Biotechnology. 5[16]: 22872296

2.   N. Okafor, B. C. Nzeako, 1985 Microbial flora of fresh and smoked fish from Nigerian fresh waters. Food Microbiology. 2 7175 .

3.   A. L. Pinto, M. Fernandes, C. Pinto, H. Albano, F. Castilho, P. Teixeira, P. A. Gibbs, 2009 Characterization of anti-Listeria bacteriocins isolated from shellfish: Potential antimicrobials to control non-fermented seafood. International Journal of Food Microbiology. 129 5058

4.   Ghanbari, 2007 2007. The Use of Gram positive bacilli isolated from intestinal flora of fish to control growth of Listeria monocytogenes in cold smoked Roach. M.Sc Thesis. Trabiat Modares University. Iran.

5.   S. D. Todorov, P. Hob, M. Vaz-Velho, L. M. T. Dicks, 2010 Characterization of bacteriocins produced by two strains of Lactobacillus plantarum isolated from Beloura and Chouriço, traditional pork products from Portugal. Meat Science. 84 334343

6.   F. Leroi, 2010 Occurrence and role of lactic acid bacteria in seafood products. Food Microbiology. 27 112 .

7.   A. Glvez, H. abriouel, R. L. Lopez, N. B. Omar, 2007 Bacteriocin-based strategies for food biopreservation. International Journal of Food Microbiology. 120 5170 .

8.   W. H. Holzapfel, R. Geisen, U. Schillinger, 1995 Biological preservation

of foods with reference to protective cultures, bacteriocins and food grade enzymes. International Journal of Food Microbiology. 24 343362 .

9.   J. Cleveland, T. J., I. F. Montville, Nes, M. L. Chikindas, 2001 Bacteriocins: safe natural antimicrobials for food preservation. International Journal of Food Microbiology. 71 120 .

10.   O. Kandler, N. Weiss, 1986 In: Bergey's Manual of Systematic Bacteriology 2 Baltimore: Williams and Wilkins. 1209-1234.

11.   P. A. Maragkoudakis, G. Zoumpopoulou, C. Miaris, G. Kalantzopoulos, B. Pot, E. Tsakalidou, 2006 Probiotic potential of Lactobacillus strains isolated from dairy products. International Dairy Journal. 16 189199 .

12.   O. Kandler, N. Weiss, 1986 Genus Lactobacillus beijerinck 1901, 212AL. In: Sneath, PHA; Mair, NS; Sharpe, ME and Holt, JG [Eds.], Bergey's manual of systematic bacteriology. [Eds.], 2 Baltimore: Williams and Wilkins. 12091234 .

13.   R. Havenaar, Brink. B. Ten, in't. Huis, J. H. J. Veld, 1992 Selection of strains for probiotic use. In: Fuller, R [Ed.], Probiotics: the scientific basis. [1st Edn.], London, Chapman and Hall. 209224 .

14.   P. Walstra, T. J. Geurts, A. Noomen, A. Jellema, M. A. J. S. van Boekel, 1999 Dairy technology, principles of milk properties and processes. Eds., New York, Marcel Dekker, Inc., 727

15.   Y. Belguesmia, K. Naghmouchi, N. E. Chihib, D. Drider, 2011 Class IIa bacteriocins: current knowledge and perspectives, 171195 . In D. Drider and S. Rebuffat [ed.], Prokaryotic Antimicrobial Peptides: From Genes to Applications, DOI 10.1007/978-1-4419-7692-5_10, © Springer Science+Business Media, LLC 2011.

16.   C. A. Campos, O. Rodríguez, P. Calo-Mata, M. Prado, J. Barros-Velásquez, 2006 Preliminary characterization of bacteriocins from Lactococcus lactis, Enterococcus faecium and Enterococcus mundtii strains isolated from turbot [Psetta maxima]. Food Research International 39 356364 .

17.   B. Karska-Wysocki, M. Bazo, W. Smoragiewicz, 2010 Antibacterial activity of Lactobacillus acidophilus and Lactobacillus casei against methicillin-resistant Staphylococcus aureus [MRSA]. Microbiology Research 165 674686 .

18.   M. Ghanbari, M. Rezaei, M. Soltani, G. R. Shahosseini, 2009 Production of bacteriocin by a novel Bacillus sp. strain RF 140, an intestinal bacterium of Caspian Frisian Roach [Rutilus frisii kutum]. Iranian Journal of Veterinary Research 10 [3]: 267272 .

19.   O. B. Chahad, M. El Bour, P. Calo-Mata, A. Boudabous, J. Barros-

Velazquez, 2011 Discovery of novel biopreservation agents with inhibitory effects on growth of food-borne pathogens and their application to seafood products. Research in Microbiology. doi:10.1016/j.resmic.2011.08.005

20. J. Delves-Broughton, P. Blackburn, R. J. Evans, J. Hugenholts, 1996 Application of the bacteriocin, nisin. Antonie van Leeuwenhock. 69 193202 .

21. N. Mojgani, C. Amirinia, 2007 Kineitics of growth and bacteriocin production in L. casei RN 78 isolated from a dairy sample in IR Iran. International Journal of Dairy Science. 2[1]: 112

22. A. Bucio, R. Hartemink, J. W. Schrama, J. Verreth, F. M. Rombouts, 2006 Presence of lactobacilli in the intestinal content of freshwater fish from a river and from a farm with a recirculation system. Food Microbiology. 23 476482 .

23. M. Ghanbari, M. Rezaei, R. M. Nazari, M. Jami, 2009 Isolation and characterization of Lactobacillus species from intestinal contents of beluga [Huso huso] and Persian sturgeon [Acipenser persicus]. Iranian Journal of Veterinary Research 10 [2]: 152157 .

24. S. Itoi, T. Abe, S. Washio, E. Ikuno, Kanomata, H. Sugita, 2008 Isolation of halotolerant Lactococcus lactis subps. lactis from intestinal tract of coastal fish. International Journal of Food Microbiology. 121: 116121.

25. W. Noonpakdee, P. Jumriangrit, K. Wittayakom, J. Zendo, J. Nakayama, K. sonomoto, S. panyim, 2009 Two-peptide bacteriocin from Lactobacillus plantarum PMU 33 strainisolated from som-fak, a Thai low salt fermented fish product. Asia-Pacific Journal of Molecular Biology and Biotechnology. 17 1925

26. E. Ringø, F. J. Gatesoup, 1998 Lactic acid bacteria in fish: a review. Aquaculture 160 177203 .

27. Y. M. Cai, P. Suyanandana, P. Saman, Y. Benno, 1999 Classification and characterization of lactic acid bacteria isolated from the intestines of common carp and freshwater prawns. The Journal of General and Applied Microbiology 45 177184 .

28. Galindo. A. Bucio, R. Hartemink, J. W. Schrama, J. A. J. Verreth, F. M. Rombouts, 2006 Presence of lactobacilli in the intestinal content of freshwater fish from a river and from a farm with a recirculation system. Food Microbiology 23 476482 .

29. R. Hartemink, F. M. Rombouts, 1999 Comparison of media for the detection of bifidobacteria, lactobacilli and total anaerobes from faecal samples. Journal of Microbiology Methods 36 181192 .

30. R. Hartemink, V. R. Domenech, F. M. Rombouts, 1997 LAMVAB- a new selective medium for the isolation of lactobacilli from faeces. Journal of Microbiology Methods 29 7784 .

31. N. Thapa, J. Pal, J. P. Tamang, 2006 Phenotypic identification and technological properties of lactic acid bacteria isolated from traditionally processed fish products of the Eastern Himalayas. International Journal of Food Microbiology 107 3338 .

32. Hammes, WP and Vogel, RF 1995 The genus Lactobacillus. In: Wood, BJB and Holzapfel, WH [Eds.], The lactic acid bacteria, the genera of lactic acid bacteria. [Eds.], 2 London, Blackie Academic and Professional. 1954 .

33. U. Schillinger, F. Lucke, 1989 Antibactrial activity of Lactobacillus sake isolated from meat. Applied and Environmental Microbiology 55 19011906 .

34. N. Benkerroum, Y. Ghouati, W. E. Sandine, Tantaout-Elaraki, 1993 Methods to demonstrate the bactericidal activity of bacteriocins. Letters in Applied Microbiology 17 7881 .

35. J. R. Tagg, A. R. Mcgiven, 1971 Assay system for bacteriocins. Applied and Environmental Microbiology 21 943943 .

36. R. Yang, M. C. Johnson, B. Ray, 1992 Novel method to extract large amounts of bacteriocin from lactic acid bacteria. Applied and Environmental Microbiology 58 33553359 .

37. J. E. Sambrook, F. Eritsch, J. Maniatis.198, Cloning. A. Molecular, Manual. Laboratory, 2 ed. Cold Spring harbour Laboratory Press, Cold Spring Harbour, NY.

38. H. Schägger, G. Von, Jagow, 1987 Tricine-sodium dodecyl sulphate-polyacrylamide gel electrophoresis for the separation of protein in the range from 1 to 100 kDa. Analytical Biochemistry 166 368379 .

39. S. D. Todorov, H. Nyati, M. Meincken, L. M. T. Dicks, 2007 Partial characterization of bacteriocin AMA-K, produced by Lactobacillus plantarum AMA-K isolated from naturally fermented milk from Zimbabwe. Food Control 18 656664 .

40. 4. , E. I. Garvie, 1984 Taxonomy and identification of dairy bacteria. In: Davies, FL and Law, BA [Eds.], Advances in the microbiology and biochemistry of cheese and fermented milk, London: Elsevier Applied Science Publishers. 3565 .

41. S. J. Hicks, G. Theodoropoulos, S. D. Carrington, A. P. Corfield, 2000 The role of mucins in host-parasite interactions. Part I- Protozoan parasites.

Parasitol. Today. 16 476481 .

42. A. M. Stephen, 1985 Effect of food on the intestinal microflora. In: Food and the Gut [Hunter, J.O. and Alun Jones, V., Eds.]. Baillière Tindall. Sussex, England. 5777

43. A. C. Ouwehand, P. V. Kirjavainen, Grönlund, E. Isolauri, S. Salminen, 1999 Adhesion of probiotic micro-organisms to intestinal mucus. International Dairy Journal 9 623630 .

44. R. Fuller, 1989 Probiotics in man and animals. Journal of Applied Bacteriology 66 365378 .

45. P. D. Cotter, C. Hill, R. P. Ross, R. P. , 2005 Bacteriocins: Developing innate immunity for food. Nature Reviews Microbiology 3 777788 .

46. C. B. Lewus, S. Sun, T. J. Montville, 1992 Production of an amylase-sensitive bacteriocin by an atypical Leuconostoc paramesenteroides strain. Applied and Environmental Microbiology 58 143149 .

47. K. Keppler, R. Geisen, H. Holzapfelw, 1994 An amylase sensitive bacteriocin of Leuconostoc carnosum. Food Microbiology. 11 3945 .

48. S. Bhattacharya, A. Das, 2010 Study of Physical and Cultural Parameters on the Bacteriocins Produced by Lactic Acid Bacteria Isolated from Traditional Indian Fermented Foods. American Journal of Food Technology 5 111120 .

49. H. J. Chung, A. E. Yousef, 2010 Synergistic effect of high pressure processing and Lactobacillus casei antimicrobial activity against pressure resistant Listeria monocytogenes. New Biotechnology 27 [4]: 403408

50. D. Hernandez, E. Cardell, V. Zarate, 2005 Antimicrobial activity of lactic acid bacteria isolated from Tenerife cheese: initial characterization of plantaricin TF711, a bacteriocin-like substance produced by Lactobacillus plantarum TF711. Journal of Applied Microbiology 99 7784 .

51. D. Drider, G. Fimland, Y. Héchard, L. M. Mc Mullen, H. Prévost, 2006 The continuing story of Class IIa bacteriocins. Microbiology and Molecular Biology Reviews 70 [2]: 564582 .

52. S. T. Ogunbanwo, A. I. Sanni, A. A. Onilude, 2003 Characterization of bacteriocin produced by Lactobacillus plantarum F1 and Lactobacillus brevis OG1. African Journal of Biotechnology 2[8]: 219227 .

53. E. Parente, A. Ricciardi, 1999 Production, recovery and purification of bacteriocins from lactic acid bacteria. Applied Microbiology and Biotechnology 52 628638 .

54. A. K. Bhunia, M. C. Johnson, B. Ray, N. Kalchayanand, 1991 Mode of action of pediocin AcH from Pediococcus acidilactis H on sensitive

bacterial strains. Journal of Applied Bacteriology 70 2533 .

55.   S. D. Todorov, L. M. T. Dicks, 2004 Partial characterization of bacteriocins produced by four lactic acid bacteria isolated from regional South African barely beer. Annals of Microbiology 54[4]: 403413 .

56.   N. Rekhif, A. Atrih, M. Michel, G. Lefebvre, 1995 Activity of plantaricin SA6, a bacteriocin produced by Lactobacillus plantarum SA6 isolated from fermented sausages. Journal of Applied Bacteriology 78 34958 .

57.   S. A. Cuozzo, F. Sesma, J. M. Palacios, A. P. de Ruiz, Holgado, R. R. Raya, 2000 Identification and nucleotide sequence of genes involved in the synthesis of lactocin 705, a two-peptide bacteriocin from Lactobacillus casei CRL 705. FEMS Microbiology Letters 185 157161

58.   M. Rammelsberg, E. Miiller, F. Radler, 1990 Caseicin 80: purification and characterization of a new bacteriocin from Lactobacillus casei. Archives of Microbiology 154 249252

59.   C. A. Van Reenen, L. M. T. Dicks, M. L. Chikindas, 1998 Isolation, purification and partial characterization of plantaricin 423, a bacteriocin produced by Lactobacillus plantarum. Journal of Applied Microbiology 84 11311137 .

60.   G. Vignolo, S. Fadda, M. N. De Kairuz, A. A. P. de Ruiz, Holgado, G. Oliver, 1996 Control of Listeria monocytogenes in ground beef by Lactocin 705, a bacteriocin produced by Lactobacillus casei CRL 705. International Journal of Food Microbiology. 27: 397402.

61.   H. K. Venkat, N. P. Shau, K. J. Jain, 2004 Effect on feeding Lactobacillus-based probiotics on the gut microflora, growth and survival of postlarvae of Macrobrachium rosenbergii [de Man]. Aquaculture Research 35 501507 .

62.   L. Verschuere, G. Rombaut, P. Sorgeloos, W. Verstraete, 2000 Probiotic bacteria as biological control agents in aquaculture. Microbiology and Molecular Biology Reviews. 64 655671 .

63.   K. K. Vijayan, I. S. B. Singh, N. S. Jayaprakash, S. V. Alavandi, S. S. Pai, R. Preetha, J. J. S. Rajan, T. S. Santiago, 2006 A brackishwater isolate of Pseudomonas PS-102, a potential antagonistic bacterium against pathogenic vibrios in penaeid and non-penaeid rearing systems. Aquaculture 251 192200 .

64.   S. Todorov, B. Onno, O. Sorokine, J. M. Chobert, I. Ivanova, X. Dousset, 1999 Detection and characterization of a novel antibacterial substance produced by Lactobacillus plantarum ST 31 isolated from sourdough. International Journal of Food Microbiology. 48 167177

65. E. Toméa, V. L. Pereiraa, C. I. Lopesa, P. A. Gibbsc, P. C. Teixeiraa, 2008 In vitro tests of suitability of bacteriocin-producing lactic acid bacteria, as potential biopreservation cultures in vacuum-packaged cold-smoked salmon. Food Control 19[5]: 535543

66. V. Karthikeyan, S. W. Santosh, 2009 Isolation and partial characterization of bacteriocin produced from Lactobacillus plantarum. African Journal of Microbiology Research 3 [5]: 233239

67. H. Daba, S. Pandian, J. F. Gosselin, R. E. Simard, J. Huang, C. Lacroix, 1991 Detection and activity of bacteriocin produced by Leuconostoc mesenteriodes. Applied and Environmental Microbiology 57 34503455 .

68. M. E. Stiles, W. H. Holzaphel, 1997 Lactic acid bacteria of foods and their current taxonomy. International Journal of Food Microbiology. 36 129 .

69. H. J. Chung, A. E. Yousef, 2009 Screening of Lactobacilli Derived from Fermented Foods and Partial Characterization of Lactobacillus casei OSY-LB6A for Its Antibacterial Activity against Foodborne Pathogens. Journal of Food Science and Nutrition 14 162167

CITATION

CHAPTER 1

Mahdi Ghanbari and Mansooreh Jami (2013). Lactic Acid Bacteria and Their Bacteriocins: A Promising Approach to Seafood Biopreservation, Lactic Acid Bacteria - R & D for Food, Health and Livestock Purposes, Dr. J. Marcelino Kongo (Ed.), ISBN: 978-953-51-0955-6, InTech, DOI: 10.5772/50705.

CHAPTER 2

Rodríguez-Rubio L, Martínez B, Donovan DM, García P, Rodríguez A (2013) Potential of the Virion-Associated Peptidoglycan Hydrolase HydH5 and Its Derivative Fusion Proteins in Milk Biopreservation. PLoS ONE 8(1): e54828. doi:10.1371/journal.pone.0054828.

CHAPTER 3

Lipsy Chopra Gurdeep Singh Kautilya Kumar Jena and Debendra K. Sahoo, Sonorensin: A new bacteriocin with potential of an anti-biofilm agent and a food biopreservative, doi: 10.1038/srep13412.

CHAPTER 4

Djadouni Fatima, Kihal Mebrouk and Heddadji Miloud, Biopreservation of tomato paste and sauce with Leuconostoc spp. Metabolites, DOI: 10.5897/AJFS2015.1302.

CHAPTER 5

M. Ciani1 and f. Fatichenti, Killer Toxin of Kluyveromyces phaffii DBVPG 6076 as a Biopreservative Agent To Control Apiculate Wine Yeasts, DOI: 10.1128/AEM.67.7.3058–3063.2001.

CHAPTER 6

Antonio Gálvez, Hikmate Abriouel, Antonio Cobo and Rubén Pérez Pulido (2011). Natural Antimicrobials for Biopreservation of Sprouts, Soybean - Biochemistry, Chemistry and Physiology, Prof. Tzi-Bun Ng (Ed.), ISBN: 978-953-307-219-7, InTech, DOI: 10.5772/15746.

CHAPTER 7

Kamal Rai Aneja, Romika Dhiman, Neeraj Kumar Aggarwal, and Ashish Aneja, "Emerging Preservation Techniques for Controlling Spoilage and Pathogenic Microorganisms in Fruit Juices," International Journal of Microbiology, vol. 2014, Article ID 758942, 14 pages, 2014. doi:10.1155/2014/758942.

CHAPTER 8

Swarnadyuti Nath, S. Chowdhury, Prof. K.C. Dora and S. Sarkar, Role Of Biopreservation In Improving Food Safety And Storage, ISSN : 2248-9622, Vol. 4, Issue 1(Version 3), January 2014, pp.26-32.

CHAPTER 9

Backialakshmi S, Meenakshi RN, Saranya A, Jebil MS, Krishna AR, et al. (2015) Biopreservation of Fresh Orange Juice Using Antilisterial Bacteriocins101 and Antilisterial Bacteriocin103 Purified from Leuconostoc mesenteroides. J Food Process Technol 6:479. doi:10.4172/2157-7110.1000479.

CHAPTER 10

Tana Hintz, Karl K. Matthews, and Rong Di, "The Use of Plant Antimicrobial Compounds for Food Preservation," BioMed Research International, vol. 2015, Article ID 246264, 12 pages, 2015. doi:10.1155/2015/246264.

CHAPTER 11

Irais Sánchez-Ortega, Blanca E. García-Almendárez, Eva María Santos-López, Aldo Amaro-Reyes, J. Eleazar Barboza-Corona, and Carlos Regalado, "Antimicrobial Edible Films and Coatings for Meat and Meat Products Preservation," The Scientific World Journal, vol. 2014, Article ID 248935, 18 pages, 2014. doi:10.1155/2014/248935.

CHAPTER 12

En Yang, Lihua Fan, Yueming Jiang, Craig Doucette and Sherry Fillmore, Antimicrobial activity of bacteriocin-producing lactic acid bacteria isolated from cheeses and yogurts, DOI: 10.1186/2191-0855-2-48

CHAPTER 13

Mahdi Ghanbari, Masoud Rezaei and Mansoureh Jami (2013). Selection of Lactobacillus Species from Intestinal Microbiota of Fish for Their Potential Use as Biopreservatives, Lactic Acid Bacteria - R & D for Food, Health and Livestock Purposes, Dr. J. Marcelino Kongo (Ed.), ISBN: 978-953-51-0955-6, InTech, DOI: 10.5772/50166.

INDEX